Mechanochemical Synthesis of Composite Materials

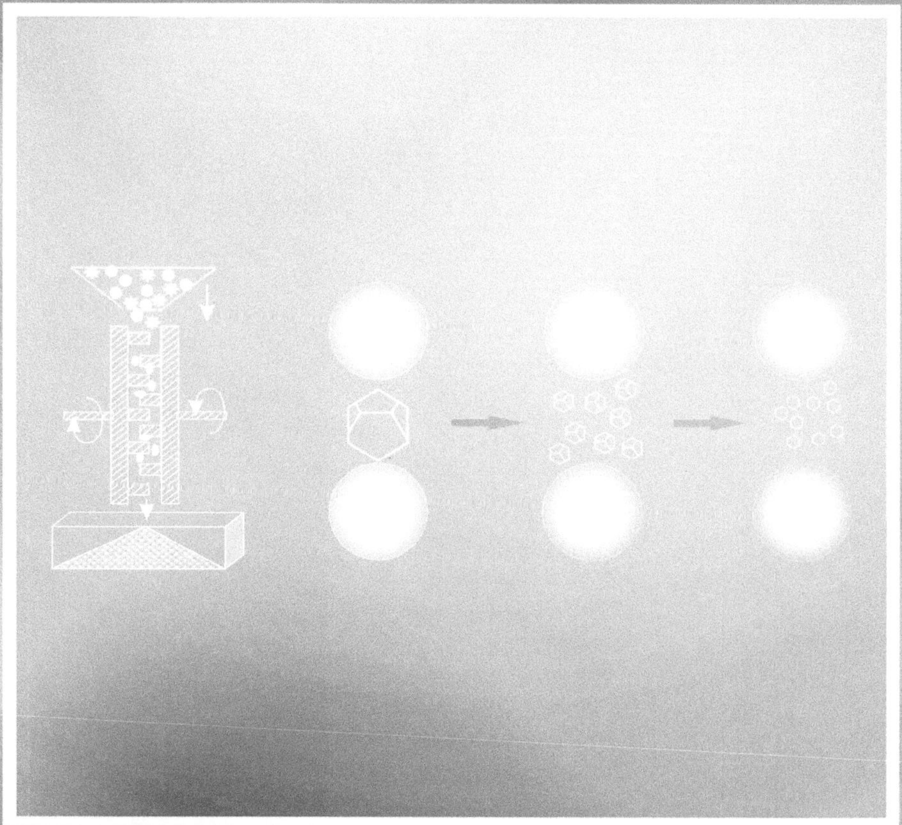

Mechanochemical Synthesis of Composite Materials

Zulkhair A. Mansurov
Nina N. Mofa
Tlek A. Ketegenov
Bakhtiyar S. Sadykov

JENNY STANFORD
PUBLISHING

Published by

Jenny Stanford Publishing Pte. Ltd.
101 Thomson Road
#06-01, United Square
Singapore 307591

Email: editorial@jennystanford.com
Web: www.jennystanford.com

British Library Cataloguing-in-Publication Data
A catalogue record for this book is available from the British Library.

Mechanochemical Synthesis of Composite Materials

ISBN 978-981-4800-88-4 (Hardcover)
ISBN 978-1-003-12081-0 (eBook)

Dedication

Prof. Nina N. Mofa (04.10.1946–16.08.2021)

Our sorrow today is as intense as our respect and love for Nina N. Mofa. Prof. Mofa lived a bright and worthy life and was a true intellectual, a person of brilliant intellect and great erudition. Since 1988, she worked at the Institute of Combustion Problems in various positions as a senior researcher, chief researcher, and head of the Laboratory of Mechanochemical Processes and Combustion Problems. She actively participated in the scientific and administrative life at the Institute. Her main scientific interest was in the study of the processes of creating nanostructured microcomposite systems by thermal and mechanochemical treatment. She authored more than 300 publications, including scientific articles and abstracts, and 4 monographs and held 28 certificates of authorship and patents of Kazakhstan and Russia. In 2014, she was awarded the badge of the Ministry of Industry and New Technologies of the Republic of Kazakhstan.

The directorate and the staff of the Institute of Combustion Problems express their deep condolences on the departure of Prof. Mofa, the outstanding scientist and unique organizer of science. May the departed soul rest in peace and be a driving force for all of us.

Contents

Introduction xiii

1. Basic Concepts and Representations of Mechanochemistry: Methods of Mechanical Action and Physicochemical Changes of the Substance 1

1.1 Historical Stages in the Development of Mechanochemistry as a Technological Process 1

1.2 Methods of Action and the Main Physicochemical Processes in the Mechanochemistry of Inorganic and Organic Substances 7

2. Mechanochemical Treatment, Activation, Synthesis, and Modification of the Surface of Particles of Inorganic Materials 15

2.1 Theoretical Basis of Crushing and Activation of Solids: Energy Intensity of Grinding Machines and Stored Energy of Materials Processed in Them 15

2.2 Physicochemical Processes of Substance Transformation during Mechanochemical Treatment 21

2.2.1 Deformation, Destruction, and Activation are the Main Stages in the Transformation of Matter under Mechanical Action 21

2.2.2 Models of Mechanochemical Processes: The Localization of Deformation and the Physicochemical Processes Accompanying It 25

2.2.3 Mechanisms of Initiation of Mechanochemical Reactions 33

2.2.4 Structural Rearrangement and Modification of the Surface of Dispersible Particles 41

2.3 Kinetics and Thermodynamics of Mechanochemical Treatment of Inorganic Materials 45
2.4 Features of Mechanochemistry of Organic Compounds and Systems of Inorganic and Organic Materials 52

3. Principles of Mechanochemical Activation in Technological Processes 63

3.1 The Influence of the Form of Grinding Bodies on the Parameters of Mechanochemical Treatment 63
3.2 Influence of the Coated Layer Thickness on the Parameters and Kinetics of Mechanical Activation 72
3.3 The Phenomenon of Abrasive-Reactive Wear in Mechanochemical Processes 81
3.4 The Effectiveness of the Implementation of the Abrasive-Reactive Wear in the Mechanochemical Processing of Mineral Raw Materials 90

4. Structural Changes of Silicon Dioxide under Thermal and Mechanical Impact 101

4.1 The Variety of Structural Forms and Features of the Surface Layers of Silicon Dioxide 101
4.2 Mechanochemistry of Quartz, Features of Structural Changes during Dispersion of Quartz—Surface Radicals and Their Transformation 106
4.3 Modification and Radical Polymerization of the Surface of Quartz Particles during Mechanochemical Processing 111

5. Activation and Modification of Quartz in Mechanical Reactors: Synthesis of Nanocomposition Quartz Particles Capsulated in Carbon-Containing Shells 119

5.1 Changes in the State, Structure, and Properties of Activated Quartz 119

5.2 Structure, State, and Properties of Quartz Powder after Mechanical Treatment with Modifiers 130
5.3 Morphological and Structural Features of Activated and Modified Quartz 140
5.3.1 Electron Microscopy of Modified Quartz 140
5.3.2 IR Spectroscopy of Activated and Modified Quartz 147
5.3.3 EPR, Mössbauer Spectroscopy and X-Ray Phase Analysis of Quartz Modified by Mechanochemical Processing 156
5.4 Features of Ferromagnetism of Quartz Powder Induced as a Result of Mechanochemical Processing 175

6. Theoretical Preconditions for Creation of Mechanochemical Synthesis of Composite Nanostructured Systems Based on Quartz 193

6.1 The Main Processes in Mechanochemistry of the Quartz Particle Modification 193
6.2 The Piezoelectric Effect of Quartz Is Part of the Process of Modifying the Surface of Particles with Organic Compounds 206
6.3 Simulation of the Process 214

7. Mechanochemical Synthesis of Disperse Composition Systems of Different Purpose 227

7.1 Composition Systems Quartz Core–Polymer Shell 227
7.2 Composite Systems Inorganic Core–Polymer Shell, Obtained by Mechanochemical Treatment of Calcite and Wollastonite 235

8. Mechanochemistry under the Conditions of Ultrasonic Treatment of Powder Systems 251

8.1 Ultrasonic Treatment of the Material Is a Way to Change the Structure and State and Obtaining Nanostructured Systems 251

8.2 Changes in the Structure, Properties, and Modification of Quartz and Calcite under the Influence of Ultrasound 256
8.3 Changes in the Structure, Properties, and Modification of Wollastonite under the Influence of Ultrasound 264
8.4 Simulation and Quantum-Chemical Calculations of the Formation of Surface Compounds in the Mechanochemical Synthesis of Hybrid Powder Nanocomposite Systems 274
8.4.1 Methods for Calculating the Electronic Structure of Molecules and Solids, Simulation and Quantum 274
8.4.2 Adsorption Complexes of Butanol and Urea on the Surface of Silica and Wollastonite 280

9. Mechanochemical Treatment and Modification of Metal Systems 295

9.1 Mechanochemical Treatment of Aluminum Powders with Organic Additives 298
9.2 Mechanochemical Activation of Aluminum Powders with Organic Modifiers in the Presence of Quartz 309
9.3 Study of Aluminum-Based Powders after MCT 313
9.4 Determination of the Activity of Modified Aluminum Powders 326

10. Fields of Application of Composite Materials Obtained Using MCT 335

10.1 Mechanochemical Synthesis of Multifunctional Sorbents Based on Nanostructured Composite Quartz-Containing Systems 335
10.1.1 Modified Quartz as a Sorbent Material for Water Purification from Various Types of Pollution 337

10.1.2 Magnetic Sorbents Obtained by Mechanochemical Treatment of Quartz-Containing Systems to Collect Oil from the Water Surface 343
10.2 Composite Systems with Fillers Modified by Mechanochemical and Ultrasonic Treatment 356
10.3 SH-Synthesis of Ceramic Materials Based on Pre-Activated and Modified Systems 375
10.3.1 The Main Macrokinetic Aspects of the Synthesis of SHS Systems and Methods of Process Control 375
10.3.2 Influence of Mechanochemical Treatment and Modification of Quartz, Calcite and Wollastonite on the Technological Combustion of Systems 380
10.3.4 SH-Synthesis of Composite Systems with Participation of Aluminum Modified during MCT 402
10.4 The Use of Energy-Intensive Powders Based on Aluminum, Obtained by Mechanochemical Treatment, in the Composition of Solid Rocket Fuels 415

Index 439

Introduction

Mechanochemical treatment is one of the modern promising directions of the chemical and technological processes of obtaining a new substance as a result of the transformation of mechanical energy into the chemical and physical processes of system restructuring. The peculiarity of the state of solid matter as a result of intensive mechanical action is determined not only by its destruction, i.e., dispersing and obtaining a powder material with a high new and active surface, but also by the accumulation of defects in the entire volume of particles, which increases their reactivity. The increase in the reactivity due to machining is one of the methods for obtaining solids in metastable, active form. In addition, in the process of deformation and destruction of the treated substance, various physical, chemical, optical, electrical, and other phenomena are observed, which, in turn, also affect the processed material. With an appropriate choice of modes of mechanochemical treatment, the material can be brought to the nanodispersed state, thereby increasing the solubility of hardly soluble substances, accelerating chemical reactions, and enhancing the catalytic properties. Much attention has been paid to obtaining alloys in the process of machining, as well as various composite systems.

Synthesis of new inorganic and organic compounds with the use of mechanical action makes it possible to obtain them in the regime of solid-phase processes (reactions) as a result of the large number of contacts between reagents and the possible release of heat in the local contact region, which substantially intensifies the interaction process. A sufficiently large number of scientific articles have been devoted to the problem of dispersion and activation of solid materials of various compositions (mineral systems, nonmetallic and metal compounds, polymer compositions). The results of these studies are summarized in various reviews, monographs, and textbooks. The creation of mechanochemical synthesis of composites, including both inorganic and organic ingredients, has received much less attention, although the possibilities of the use of such composites in terms of variation in their properties are very large.

This book presents the results of several years' research on the mechanochemical synthesis of composite systems, consisting of inorganic and organic components, obtained by the scientific team of the Institute of Combustion Problems, Kazakhstan. The book first presents the general ideas about the mechanochemical process and the phenomena accompanying it. The main provisions of structural rearrangement and modification of the surface of dispersible particles, as well as thermodynamics of mechanochemical treatment of materials, are then considered. Further, using the example of a number of natural minerals, it is shown how mechanochemical treatment can radically change their structure and properties, using various organic compounds as modifiers. Particular attention is paid to the use of ultrasonic treatment, as a kind of mechanical action in a liquid medium on a solid.

The book has been organized in 10 chapters. Chapter 1 presents the basic concepts and concepts of mechanochemistry and discusses the methods of mechanical action and physicochemical changes of inorganic and organic matter. In Chapter 2, the theoretical foundations of grinding and activation of solids, the main physicochemical processes of substance transformation, the features of structural rearrangement, and the modification of the surface of dispersible particles during mechanochemical treatment are described. The main models of mechanochemical processes are considered. Chapter 3 discusses some issues related to the course of mechanochemical processes, taking into account the shape of a small build and layer thickness, as well as the actual impact of a small build on theoretical and practical positions, which are necessary factors for assessing the quality of the material obtained.

Chapter 4 presents the results of the study of the features of the structural changes in silicon dioxide during mechanochemical treatment, modification, and radical polymerization of the surface of quartz particles, which results in the encapsulation of activated particles. In Chapter 5, the results pertaining to the synthesis of nanocomposite quartz particles encapsulated in carbon-containing shells are considered. The features of the state, structure, and properties of activated quartz are shown depending on the machining regimes. Particular attention is paid to the question of ferromagnetism of quartz powder induced as a result of mechanochemical treatment. In Chapter 6, the theoretical foundations for the creation of composite nanostructured systems based on quartz by mechanochemical synthesis are presented. The role of the piezoelectric effect of quartz is noted as a component of the process of the particle surface modification by organic compounds. In Chapter 7, the results on the mechanochemical synthesis of dispersed composite systems consisting of mineral particles (quartz, calcite, wollastonite) and a polymer shell are considered. Chapter 8 presents the results of the studies on mechanochemical treatment under the ultrasonic effect on powder systems in an aqueous medium. The role of the cavitation effect and patterns of changes in the structure of properties and modification of quartz, calcite, and wollastonite under the influence of ultrasound as a way of obtaining nanostructured systems are considered. Modeling and quantum chemical calculations of the formation of surface compounds in the mechanochemical synthesis of hybrid powder nanocomposite systems using various modifiers were carried out. Chapter 9 presents the examples of mechanochemical treatment of metal powders, in particular aluminum, to increase their reactivity. For this purpose, organic surfactants are used, which contribute to both grinding and modifying the surface of metal particles. It is shown how, as a result of mechanochemical treatment, structural changes occur both in the bulk and on the surface of metal particles and optimal treatment conditions are established to obtain maximum aluminum activity. In Chapter 10, some areas of application of composite materials obtained with the use of mechanochemical treatment are presented. The results of mechanochemical synthesis of sorbents for water purification from various types of pollution are considered. A method for obtaining magnetic sorbents by mechanochemical treatment of quartz-containing systems for collecting petroleum products from the surface of water is considered. Attention is paid to mechanochemical treatment as a method of preliminary activation and modification of systems designed for self-propagating high-temperature synthesis (SHS) of ceramic materials. The peculiarities of the influence of mechanochemical treatment and modification of quartz, calcite, and wollastonite in mechanical reactors and ultrasonic treatment on technological combustion of systems and the production of SHS composite systems are considered.

The book summarizes and systematizes the results of a fundamentally new complex approach to the creation of composite polymer-inorganic systems by mechanochemical treatment, both as a result of the traditional approach with the use of dynamic mills (mechanical reactors) and ultrasonic action on powder mixtures. It is shown that by selecting the processing conditions and the composition of the processed systems, composite systems with a nanostructured surface layer of particles can be obtained, which provides a directional change in the properties of the composition.

In addition, the book presents a chapter that focuses on the method of mechanochemical preparation of materials for high-energy systems, which are the most promising, since it allows the most productive management of the formation of new structural compositions that increase the energy intensity of the system. Stabilization of the increased activity of highly dispersed particles is provided by the surface layer modification. This method allows obtaining highly dispersed modified material for various types of fuels, including rocket technology.

Chapter 1

Basic Concepts and Representations of Mechanochemistry: Methods of Mechanical Action and Physicochemical Changes of the Substance

1.1 Historical Stages in the Development of Mechanochemistry as a Technological Process

Mechanochemistry, as a technological process, has its roots in the antiquity, when it was already noticed that mechanical ignition, in particular, friction, can cause ignition and detonation of solids as well as initiate many chemical reactions. A rather detailed analysis of the first serious applications of mechanochemistry in the preparation of metals and alloys is given in a monograph by Peter Balaz [1]. So, when obtaining mercury, cinnabar (mercury sulfide) was ground with vinegar in a mortar with a copper pestle. Vinegar was used to prevent side effects that accompany dry grinding in air, for example oxidation. Since the 16th century, the use of mechanical impact has received quite a serious application in the mining and metallurgical industry in the preparation of raw materials and increasing the degree of conversion and output

Mechanochemical Synthesis of Composite Materials
Zulkhair A. Mansurov, Nina N. Mofa, Tlek A. Ketegenov, and Bakhtiyar S. Sadykov

ISBN 978-981-4800-88-4 (Hardcover), 978-1-003-12081-0 (eBook)
www.jennystanford.com

of the final product. The efficiency of the process was determined by the surface area of the reacting solid particles obtained as a result of their preliminary grinding.

A significant contribution to the development of mechanochemistry was made by the English physicist Faraday in the first half of the 19th century. First, he drew attention to the fact that some hydrated salts are spontaneously dehydrated during mechanical processing (grinding). Second, it was found that silver chloride decomposed under the effect of mechanochemical action, and then studies were also carried out with other metals: tin, copper, iron. The mechanical action on the metal chloride system, for example that of zinc, leads to a strong self-sustaining exothermic reaction. All experiments were carried out under abrasion conditions, i.e., under the influence of shear stresses, when the state changes to a great extent, up to the melting of the surface of the reacting substances.

The rapid development of mechanochemistry took place in the 20th century with reference to various branches of technology. Particular attention was paid to the possibility of mechanically initiating explosion reactions in solids in the preparation of various explosive mixtures. Studies in this field were conducted in England, France, Germany, Russia, and other countries. Based on the results of the investigations, the occurrence and development of the explosion were associated with a local increase in temperature at the contact of two solids in the course of mechanical action. Much attention was also paid to the effect of mechanical action on metals. G. Tammann [2] showed that part of the spent mechanical energy remains in the metal in the form of potential energy and increases the thermodynamic potential of the solid, this accelerating interaction between different solid–phase systems, for example between quartz and calcite.

At the same time (i.e., the first 20 years of the 20th century), work on mechanochemistry of organic systems was started and rapidly developed. For the pulp and paper industry, the effect of grinding on the solubility of cellulose due to the destruction of macromolecules and creation of new hydrophilic groups at the point of splitting of valence bonds was shown. At the same time, work was carried out on mechanical destruction of polymers based on polystyrene and other systems, which developed rapidly in the middle and second half of the 20th century [3, 4]. In these works, not only attention was paid to shear deformation and destruction by abrasion, but grinding was carried out in mills of different actions using hydrostatic pressure.

The first deep analysis of the work on mechanochemistry was made by K. Peters in the article "Mechanochemical reaction" [5] in 1966 with reference to the works of K. Wenzel, V. Ostwald, G. Tammann, I. Hodwald, and others. He believed that K. Hess, E. Shtoyner, and Kh. Fromm were the first who introduced in 1942 the notion of "mechanochemistry" as the decomposition of carbonates, chlorides and other substances taking place mainly during conventional grinding processes in ball and vibrating mills; the formation of various substances, for example, sulfur compounds and silicates; the increase in the solubility of hardly soluble substances; acceleration of chemical reactions; strengthening of catalytic properties; improvement of physical and technical properties of artificial stones and polymers; the fall of temperature required for the response. The term "mechanochemistry" was introduced by V. Ostwald as a field of physical chemistry, in which the influence of mechanical energy on chemical reactions is considered.

A sufficiently detailed analysis of the various stages over 100 years (second half of the 19th and the first half of the 20th century) and some features in the development of mechanochemistry as a scientific direction is given in the book by G. Heinicke [6]. The author notes that the first systematic analysis of mechanochemical studies was given by M. Keri-Lee at the end of the 19th century considering an example of decomposition of silver, gold, platinum, and mercury halides under the effect of mechanical action. Moreover, M. Keri-Lee excluded local heating as the reason for initiating the decomposition reaction. G. Tammann emphasized that up to 15% of the spent mechanical energy is converted into the potential energy of the processed material and increases its thermodynamic potential, as a result, various processes are accelerated, in particular, dissolution of the crushed material.

In the fifties of the 20th century, K. Peters studied a number of inorganic reactions under the influence of mechanical energy and focused attention on the thermodynamic conditions of their flow. Heinicke believes that thermodynamics of irreversible processes is most effective. The preference is given to the "magma-plasma" model proposed by P. Thiessen, according to which the main role in the mechanical action is given to the high-energy excited states operating for a very short time.

A comprehensive analysis of numerous works made by H. Heinicke leads to the fact that the physical processes occurring under mechanical action are very numerous, and therefore the mechanisms initiating chemical reactions are very diverse and can have a multistage "hierarchical" character. The main feature of all the studies carried out is a more thorough and better knowledge of the physical, chemical, optical, electrical, and other phenomena occurring in the grinding process.

Mechanochemistry played a significant role in the construction industry. In this industry, a great contribution was made by the works of Johannes Hint and P. A. Rebinder [7–9]. As early as 1949, it was noted that when processing in a specially designed grinding machine—a disintegrator, where the particles of the material were subjected to fast successive strokes, at a maximum impact speed of 250 m/s, the lime–sand mixtures acquire completely new technological properties. The sand treated in the disintegrator, compared to the sand ground in a ball mill, imparts a great strength to the products from the calcic–sand mixture. And as a result of the joint processing of lime–sand mixtures in the disintegrator, a qualitatively new artificial stone with a unique structure, silicalcite, was obtained. In the course of the research, it was possible to develop a method—a formula for predicting the compression strength of a product, where, in addition to the amount of lime, product density and curing conditions, the magnitude of the new sand surface formed during the disintegration of the raw mix (the so-called specific surface of sand in the mixture) is an important factor. The strength of the samples prepared from sands ground in a ball mill and a vibratory mill was approximately the same, and the strength of samples prepared from sands crushed to the same dispersity in a disintegrator was 80% larger.

A. P. Rebinder associated the peculiarities of the state of the material after intensive mechanical treatment with its surface activity. He was the first to formulate the fundamental differences between mechanochemical transformations from classical chemical processes. In mechanochemical processes, the initial reagents are converted into final products as a result of the combined effect of elastic stresses and chemical forces. Elastic stresses increase the internal energy of the reagents. Accordingly, the thermodynamic parameters of the initial reagents, intermediates and final products differ from the standard values, and this leads to differences in the values of the equilibrium constant and the enthalpy of the reaction. The work done by elastic stresses is also spent on overcoming the energy barrier separating the initial and final states in elementary acts. Correspondingly, the kinetic parameters change and an additional contribution is made to the enthalpy of the reaction.

In addition, attention is drawn to the fact that a low temperature ($T < 0.5\ T_{melting}$) is favorable for mechanochemical transformations. Under these conditions, various forms of thermal motion, including translational diffusion, are inhibited and nonequilibrium states arising under mechanical influences are relatively stable and have time to manifest themselves in chemical reactions. The mobility required to carry out the reactions at a low temperature is ensured by the plastic deformation of the reactants or their mixtures. Thus, the equilibrium position, kinetics, route, and products of mechanochemical transformations obey their laws. These transformations are nonequilibrium, nonthermal, and diffusionless. Such indications distinguish them from the usual thermo-fluctuation chemical reactions. With mechanical action, not only long-lived disturbances of the chemical structure and crystal structure are formed. There appear vibrational and electronic excitations, ionized states, new forms of molecular mobility, and so on. All these changes are short-lived and are manifested directly during mechanical effects on the reagents and their mixtures. Therefore, mechanical transformations and, above all, synthesis reactions that are carried out directly during the mechanical treatment are also classified as mechanochemical processes. The energy supplied under these conditions is used much more completely.

At present, the concept of a free radical mechanism of bond splitting and the role of free radicals in subsequent changes in the chemical structure of substances that occur against the background of disordering of the supramolecular structure and static electrization are clearly formulated [10]. The peculiarity of deformation processes, which is important for mechanochemistry, consists in the fact that with deep disordering the formation of structural defects is accompanied by their migration and death. Splitting of covalent bonds is accompanied by recombination of free radicals, multiplication of dislocations leads to their unification into dislocation boundaries, dispersion is accompanied by aggregation, decay of the supramolecular structure and amorphization are limited by spontaneous return to the initial state, and so on. As a result, a significant part of the entire flow of the energy supplied passes through the substance. The energy level of such relaxing unstable systems is, according to estimates, tens of kJ/mol. It is important that the characteristic times of relaxation and death are greater than the times for establishing thermal equilibrium and the state of matter is determined not by temperature, but by the dynamics of relaxation processes. Since heat is released not only at the moment of impact but also during the subsequent relaxation of stresses, the temperature in the layer is also far from adiabatic. The actual level of heating depends on the relationship between the rates of relaxation and heat exchange with the environment.

The possibility of a global redistribution of atoms in a deformed substance, as well as the new role of deformation as a constructive and ordering rather than destructive factor, are of fundamental importance. To understand the constructive role of deformation, one should return to the notion of the existence of a dynamic equilibrium between the processes of accumulation and release (relaxation) of excess energy. In the mechanical treatment of individual substances, relaxation is completed by formation of deeply disordered unstable structures corresponding to the dispersion limit. In reaction mixtures, where chemical forces act additionally, relaxation leads to the formation of synthesis products.

Professor G. I. Distler (Institute of Crystallography, Academy of Sciences of the Russian Federation) showed that the difference in the methods of processing materials is associated with the amount and electrical activity of point defects arising in the near-surface layers. Moreover, processing of the material in such impact installations creates a greater number of electrically nonequilibrium charged centers in the material than under other mechanical influences, and destruction of the material proceeds to a greater extent along the boundaries of accumulations of impurities.

Describing the history of the development of mechanochemistry in Russia, V. V. Boldyrev [11, 12] notes a number of significant steps that determined the significance of the results and their contribution to world science in this area. Russian scientists, in particular in the Siberian Branch of the Academy of Sciences of the USSR and then the Russian Federation, paid much attention to the use of mechanochemistry in mineralogy to improve the efficiency of opening, enriching and processing mineral raw materials for the metallurgical industry and obtaining fertilizers from phosphate ores. Studies were carried out in the field of the theory of mechanochemical processes and the possibility of their application in various technologies. The results served as the basis for synthesis of catalysts, sorbents, intermetallides and hydrogen accumulators. With the development of nanotechnology, special attention was paid to the use of mechanochemistry of organic compounds and composite systems [13, 14], which contributes to the creation of environmentally friendly production of organic substances and nanocomposites, which form, in particular, the basis for the synthesis of medicinal and cosmetic forms.

Thus, at present, mechanochemistry as a scientific direction and as a technological device is one of the promising directions that ensure a high potential for the development of various industries.

1.2 Methods of Action and the Main Physicochemical Processes in the Mechanochemistry of Inorganic and Organic Substances

Mechanochemistry deals with chemical and physicochemical transformations of substances under mechanical influences (in mills, disintegrators, rollers, extruders, etc.), during deformation, friction and shock compression. Transformations due to friction are isolated in an independent section, called tribochemistry. The chemistry of ultrasound and the chemistry of shock waves are also considered as parts of mechanochemistry.

In the course of mechanical action takes place not only the change in the shape of a solid body and its destruction, but also accumulation in it of defects, amorphous regions, an increase in the area of grain boundaries and formation of new surfaces that change the physicochemical properties, including the reactivity of the substance. Accumulation of defects contributes to the acceleration of reactions involving solids, a decrease in the temperature of processes and other ways of intensifying chemical reactions in the solid phase [6, 10]. As a result of mechanical disruption of the atomic structure, solubility of the substance and dissolution rate increase, reactions with molecules of the medium and other solid bodies are facilitated, temperatures of solid-phase synthesis, thermal decomposition and sintering decrease by tens and hundreds of degrees. Mechanochemical methods are used to degrade polymers, synthesize intermetallic compounds and ferrites, produce amorphous alloys and activate powder materials. The mechanochemical synthesis of intermetallic compounds is usually called mechanical alloying [15]; its advantages over thermal synthesis are the possibility of producing powders of amorphous alloys (for example, nickel with titanium) and active catalysts (for example, nickel aluminide). All mechanochemical transformations are conditioned by transition of a substance to a metastable chemically active state as well as the intensification of mass transfer as a result of absorption of mechanical energy. The activity of solids during deformation, friction, or fracture is caused by the appearance of vibrationally and electronically excited states of interatomic bonds, mechanically strained and split bonds, including free radicals, ion radicals, coordinatively unsaturated atoms, decomposition of structural defects, and ionization of particles of a substance and stabilization of electrically charged centers.

Tribochemistry—the section of mechanochemistry—deals with the influence of mechanical energy on the reactions between solids and their structure. Under the influence of energy released during friction or impact, the elements of the disorder of the crystal structure arising due to thermal motion increase resulting in an active state [6]. Due to mechanochemical activation, significant adsorption effects are observed, while adsorbed components fill the submicroscopic pores and voids of deeper layers of the solid phase. Thus, when a jet of sand is shaken from a sandblaster for a short time (10-5-10-6 s), a high-energy state corresponding to a short-lived solid-state plasma is reached. It is characterized by electronic and light radiation (triboluminescence), charge transfer as well as high chemical activity.

The section of mechanochemistry using ultrasound in the treatment of liquids is called ***sonochemistry*** [16, 17]. It is associated with the appearance of cavitation in the absorption of ultrasound. Cavitation—the formation of a mass of pulsating bubbles in a liquid medium—leads to the appearance of micro-impact pressures up to 800 MPa, a local increase in temperatures to 7400 K (theoretically), the formation of electrical discharges and ionization. When the cavitation spaces collapse, there takes place the transfer of energy molecules of the vapor-gas mixture as they flow into the liquid flow and their dissociation. Under the influence of ultrasound, sonoluminescence occurs, which consists of the appearance of a bright, point source of bluish light due to formation of a standing spherical ultrasonic wave in water, i.e., the sound turns into light [10]. With ultrasonic irradiation, active particles develop in the liquid, which initiate chemical reactions.

Shock waves that occur under the action of a directed explosion also accelerate chemical reactions by several orders, for example vulcanization of rubber passes in fractions of a second. Using ultra-high pressures, it was possible to convert graphite into diamond and boron nitride into borazon.

A special place is given to the mechanochemistry of macromolecular compounds as an independent branch of the science of polymers. It originated from the fusion of two seemingly remote sciences—the mechanics and chemistry of polymers—and is one of the youngest branches of the chemistry of these compounds. A major contribution to its development was made by V. A. Kargin, N. K. Baramboym, and A. A. Berlin. Mechanochemistry includes all the features of the breaking of chain molecules under the action of stress [3, 4]. However, in a narrower sense, they speak of mechanochemical methods, if one means intentional mechanical degradation of (solid) polymers. The purpose of these methods is to grind or soften materials, as well as to produce large highly reactive surfaces to create permanent chemical bonds between different polymers. In mechanochemical methods, degrading solids undergo an indistinctly complicated complex loading that causes deformation, which always simultaneously includes forced elasticity, material flow and chain breaking. Thus, in unfilled and soot-filled elastomers under mechanical influence, there takes place splitting of C–C bonds with formation of free radicals and a series of reactions proceed which can lead to a profound change in the initial structure of the polymer. The method of mechanochemistry has been used to produce a high-strength material based on plasticized grades of a copolymer of styrene and nitrile rubber. Mechanochemistry is also used in the production of phenol-formaldehyde resins and the combination of resole resins with polyvinyl butyral or polyvinylformal, this allowing improve the adhesion properties and elasticity of polymers.

Any mechanical action is based on the absorption of mechanical energy, which initiates the decomposition of substances (including the destruction of polymers), polymorphic transformations, heterogeneous reactions of solids with gases and liquids, solid phase synthesis in powder mixtures and other reactions. Heterogeneous mechanochemical reactions of solid-gas and solid-liquid proceed on surfaces formed during destruction or participating in friction. The main contribution to the chemical activity of the surface is made by coordinatively unsaturated atoms. So, on the surface of silicon dioxide and germanium during the mechanical treatment of powders, the Si–O or Ge–O bond splits and free radicals, for example, = Si and = SiO, appear. Some of them quickly die and get stabilized on the surface up to 10^{17} m^{-2} radical and molecular active centers. The interaction of gases with short-lived and stable active centers that appeared on the surface of solids due to mechanical action is called mechanical chemisorption. Mechanical chemisorption, as well as polymerization of monomers on active sites, leads to surface modification and improved adhesion, when using powders as fillers. Heterogeneous mechanochemical reactions, starting on the surface, can develop into the depth of the material. This is how silicon dissolves in water during the mechanical treatment of suspensions: silicon dissolves completely to form H_2 and H_4SiO_4.

Areas of use of mechanochemistry. Mechanochemical research, first of all, makes a significant contribution to fundamental scientific knowledge and solves many practical problems of synthesis and modification of substances, improvement of production technology and mechanical processing of materials. Mechanochemical methods are used in many-tonnage production (plasticization of rubbers, synthesis of building and heat-resistant materials, preparation of drilling solutions, etc.) and in highly specialized fields (for example, ultrasonic preparation of vaccines). The achievements of mechanochemistry are important for the development of solid-state chemistry, solid-state reaction kinetics, strength and durability physics, the theory of polymers aging, physicochemical mechanics, and a number of problems of geochemistry, biochemistry, and biophysics.

Mechanochemical treatment is used as a means of accelerating technological processes or as a way of changing technological parameters and modes of processing of mineral raw materials. The general scheme of application of mechanochemical treatment in modern technological processes based on the results of [19] is presented in Fig. 1.1.

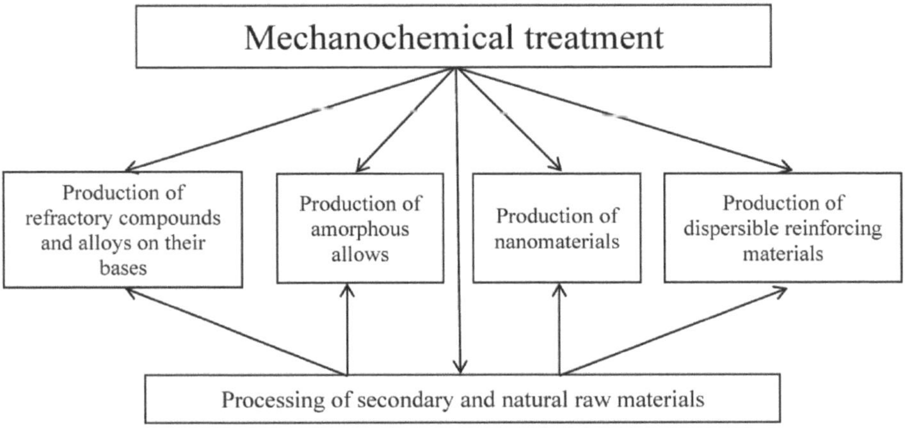

Figure 1.1 Application of mechanochemical technology in modern materials science [19].

Activation of minerals by grinding is successfully used in coal technology, for intensification of hydrometallurgical processes, in the production of fertilizers, building materials, composite mixtures, etc., and opens the prospect of secondary processing of mineral raw materials stored in dumps, and production wastes [20, 21]. As a result, the complex and rational use of mineral resources is increasing, and the harmful impact of industry on the environment is weakened.

Another promising direction of using mechanochemical treatment (activation) by grinding is the preparation of composite mixtures. Composite mixtures are widely used in a wide variety of industries. They are prepared in the form of charge before pyroprocesses; they are used in the preparation of pressure molds; and in the preparation of solid solutions for catalysts or other purposes; on their basis, the ceramic industry works; they are used in the preparation of molding lands, fluxes; to cover the electrodes; for stamping metal-ceramic parts, glue compositions, etc. Fillers (graphite, marble, wollastonite, etc.), phosphate fertilizers, natural and synthetic polymers and other materials are mechanically activated. The mechanical activation of moistened silica and some other oxides imparts them astringent properties and is the basis of the non-firing technology of heat-resistant materials. A significant role is given to mechanochemistry in the field of obtaining construction materials for various purposes, both on the basis of cement and gypsum binders, and in the preparation of polymer systems.

At present, mechanochemistry, as an area of investigation of chemical phenomena and chemical processes arising from all kinds of mechanical effects on substances, is a very promising and rapidly developing field of research in many countries. Particular attention is paid to the production of composite systems, with the combination of organo-inorganic constituents of the composition. In this case, both powdered and compacted systems are considered. In the preparation of such systems, the peculiarities of transformation characteristic of polymers and the reaction characteristics of solids from mixtures of powders in the presence of a gas and liquid medium are combined.

Questions

1. What were the main stages of the development of mechanochemistry in the 20th century in relation to various branches of technology?
2. What is the role of mechanochemistry in the creation of explosives?
3. What were the main directions of the development of mechanochemistry in Europe?
4. What were the main directions of the development of mechanochemistry in Russia?
5. What is the most promising contribution of mechanochemistry in the development of the construction industry, in the enrichment and processing of mineral raw materials for various industries?
6. What are the features of the structural and energy changes of a substance during friction and impact?
7. What is sonochemistry as a form of mechanochemistry?
8. Describe the features of mechanochemistry of macromolecular compounds as an independent branch of the science of polymers.
9. What are the fields that use mechanochemistry in solving fundamental and practical problems of synthesis of new materials?

References

1. Balaz P., *Mechanochemistry in Nanoscience and Mineral Engineering.* Springer, 2008. 422 p.
2. Tammann G. Der Einfluss der Kaltbeareitung auf die chemischen Eigenseliaften, insbesonbere von Metallen. *Z. Electrochem.*, 1929. Bd 35.–S.21–29.
3. Baramboym N. K. Mechanochemistry of macromolecular compounds. *Chemistry*, Moscow, 1971. 363 p. (In Russian).
4. Simionescu K., Oprea K. V. Mechanochemistry of macromolecular compounds. *Peace*. Moscow, 1970. 354 p. (In Russian).
5. Peters K. Mechanochemical reactions. In: *Proceedings of the European Meeting on Milling*. Stroiizdat, Moscow, 1966. pp. 80–97. (In Russian).
6. Heinicke G. *Tribochemistry.* Akademie Verlag, Berlin, 1984. 495 p.
7. Hint J. A. *Basics of Production of Silicalcitic Products.* Gosstroy, Moscow-Leningrad, 1962. 600 p. (In Russian).

8. Hint J. A. On the main problems of mechanical activation. Materials of the 5th symposium on mechanoemission and mechanochemistry of solids. *Tallinn*, 1975. vol. 1. pp. 12–23. (In Russian).
9. Rebinder P. A. *Selected Works. Surface Phenomena in Disperse Systems: Physicochemical Mechanics.* Science, Moscow, 1979. 384 p. (In Russian).
10. Avvakumov E. G. Mechanical methods of activation of chemical processes. *Science*, Novosibirsk, 1986. 304 p. (In Russian).
11. Boldyrev V. V. On the history of the development of mechanochemistry in Siberia. *Chemistry for Sustainable Development,* 2002. V. 10. pp. 3–12 (In Russian).
12. Boldyrev V. V. Mechanochemistry and mechanoactivation of solids. *Uspekhi Khimii*, 2006. vol. 75, no. 3. pp. 203–216 (In Russian).
13. Sebastian K. L. Mechanochemistry. The amazing viral DNA packaging molecular motor. *Resonance*, 2007. vol. 12, no. 5. pp. 48–59.
14. Lyakhov N. Z., Grigoryeva T. F., Barinova A. P., Vorsina I. A. Mechanochemical synthesis of organic compounds and composites with their participation. *Uspekhi Khimii,* 2010. vol. 79, no. 3. pp. 218–233 (In Russian).
15. Lu L., Lai M. O. *Mechanical Alloying*. Kluwer Academic Publishers, Norwell Massachusetts, USA. 276 p.
16. Margulis M. A. Fundamentals of sound chemistry. *Chemistry*, Moscow, 1984. 260 p. (In Russian).
17. Baranchikov A. E., Ivanov V. K., Tretyakov Yu. D. Sonochemical synthesis of inorganic materials. *Uspekhi Khimii,* 2007. vol. 76, no. 2. pp. 147–168 (In Russian).
18. Margulis M. A. Sound chemical reactions and sonoluminescence. *Chemistry,* Moscow, 1986. 288 p. (In Russian).
19. Popovich A. A. Prospects of development of mechanochemical technologies in solving modern problems of materials science. *Bulletin of the Far Eastern State Technical University*, 2009. no. 1(1) (In Russian).
20. Zhang X., Lu C., Liang M. Preparation of rubber composites from ground tire rubber reinforced with waste-tire fiber through mechanical milling. *J. Appl. Polym. Sci*, 2007. vol. 103, no. 6, p. 4087.
21. Lu C., Zhang X., Liang M. Mechanochemical recycling and processing of waste crosslinked polymers: Waste tire rubber and waste XLPE from cable scraps. *The 5th ISFR*, October 11–14, 2009, Chengdu, China. pp. 148–155.

Chapter 2

Mechanochemical Treatment, Activation, Synthesis, and Modification of the Surface of Particles of Inorganic Materials

2.1 Theoretical Basis of Crushing and Activation of Solids: Energy Intensity of Grinding Machines and Stored Energy of Materials Processed in Them

When preparing powder materials, various grinding devices are used, the action of which is based on different types of destruction: crushing, splitting, cutting, sawing, abrasion, impact, and various combinations of these methods [1]. The body in each of the above methods is crushed under the action of compressive, stretching, shearing, twisting forces of dynamic loading. To obtain a fine material, disintegrators of impact action are used (Fig. 2.1).

In addition, there are grinding machines in which abrasion is used in combination with crushing or impact. To evaluate the efficiency of grinding and to estimate the conditions for its determination, a number of assumptions are introduced, within which the grinding process is described [1, 2], depending on many

Mechanochemical Synthesis of Composite Materials
Zulkhair A. Mansurov, Nina N. Mofa, Tlek A. Ketegenov, and Bakhtiyar S. Sadykov

ISBN 978-981-4800-88-4 (Hardcover), 978-1-003-12081-0 (eBook)
www.jennystanford.com

factors, which provides a variety of known theories of the grinding process. The viability of hypotheses and theories is determined by how fully they reflect the physicochemical essence of the in all its details.

Figure 2.1 Multifunctional disintegrator of impact action, intended for fine grinding of materials.

Destruction occurs when the stress in the material exceeds the internal cohesive forces, i.e., ultimate strength. Determination of the energy consumed in this case is one of the main problems in the theory of grinding. From the foundations of the theory of elasticity it follows [3] that the energy u, and consequently the work (A) spent on grinding this material is directly proportional to its volume or mass, but it considerably depends on the degree of grinding:

$$A = (\sigma_p^2/2E)\,(D^3 - d^3) = (\sigma_p^2 d^3/2E)\,(i^3 - 1), \tag{2.1}$$

where σ_p is the breaking stress (ultimate strength), E is the modulus of elasticity, D^3 is the volume of the particles before grinding, d^3 is the volume of the particles after grinding, i is the degree of grinding.

This formula is valid only for course grinding. A more accurate formula is the formula of P. A. Rebinder [4], according to which the energy spent for grinding the material is the sum of the work spent on the deformation of a solid and formation of new surfaces:

$$A = \sigma_p^2 v/2E + k_r F, \tag{2.2}$$

where v is the volume of the deformed body, k_r is the proportionality factor, F is a newly formed surface when the body is destroyed.

The energy costs for grinding large and small particles are different. If we assume [4] that at each stage of destruction work is expended and that n steps are required to destroy a body of size D to particles of size d, the total work A spent on this operation is equal to

$$A = (\sigma_p^2 D^3/2E)\ (3\log a_o), \tag{2.3}$$

where a_o – the volume degree of single destruction of a given body (that is, for a single fracture, it is divided into particles a_o).

For the grinding devices of a particular action, the form of these dependencies varies. When grinding by impact, the material is crushed due to impact loads. The kinetic energy of the impacting body along the grinding material is determined [5] as

$$E_y = q_y w_y^2/2g, \tag{2.4}$$

where q_y is the weight of the striking body and w_y the velocity of the body at the moment of impact.

The destruction of the material consumes part of the energy E_y. In the case of an elastic impact, part of the energy E_p is returned to the striking body, which, after the impact, bounces off from the material being crushed. In this case, the energy transferred to the material being ground will be equal to

$$\Delta E = E_y - E_p \tag{2.5}$$

$$E_p = q w_p/2g, \tag{2.6}$$

where $w_p = \varepsilon w_y$ is the velocity of the body it has after the impact,

$$\Delta E = (q_y w_y^2/2g)\ (1 - \varepsilon^2) \tag{2.7}$$

To destroy the body being ground, the energy ΔE must be sufficient to overcome the internal cohesion forces between the body particles.

In addition to impact (compressive) action in grinding machines, abrasive (shear) stresses are of great importance. The latter facilitate the process of destruction and, to a greater extent, than the compressive stresses, change the structure of the dispersed particles, and, consequently, their properties. Shifting deformations and stresses play an extremely important role in the destruction of solids and formation of the structure and state of the fracture surface [6, 7]. It is the shearing stresses that ensure sliding and turning of individual parts of the solid, cause the bonds to break at all hierarchical levels of structures and determine the size of the zone at the top of the crack.

Among the shock-abrasive shredders are planetary mills (Fig. 2.2).

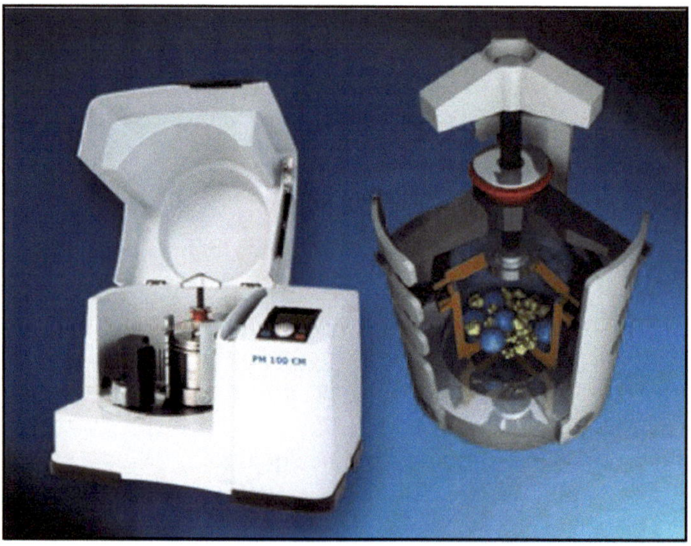

Figure 2.2 Laboratory high-energy planetary centrifugal mill.

Such mills operate on the principle of gravitational grinding, which is realized due to the interaction of two centrifugal fields [5]. To characterize the operating mode of planetary-centrifugal mills (PCM), the parameters *k* and *m* are introduced [5, 8]. The quantity $k = \bar{\omega}_2/\bar{\omega}_1$ ($\bar{\omega}_2$ is the number of revolutions around its own axis and $\bar{\omega}_1$ the number of revolutions around the common axis) is called the kinematic characteristic of the mill. The value $m = r_1/r_2$ (r_1, r_2 are the radii of the total and proper rotation) is the geometric characteristic of the PCM. By varying the values of k and m, it is possible to change the dynamic dispersion conditions over a wide range of modes.

The parameters influencing the dispersion and activation efficiency also include the density and size of balls, the volume filled with balls, the mass ratio of balls and the mass of the substance being treated, etc., It has been experimentally shown [9] that the greatest effect in grinding is achieved when the number of balls is 50 volume%, the diameter of the balls is 10 mm.

Thus, in order to estimate the energy costs for dispersing and activating the material, many factors must be taken into account. It is practically impossible to take into account all the processes of mechanical and physical interaction of grinding bodies and particles of the material being ground. All calculations are made with certain tolerances and limitations. In addition, part of the supplied mechanical energy is converted to thermal energy and dissipated, while the other part accumulates in the material being ground in the form of various defects (vacancies, dislocations, disclinations and their complexes, broken bonds and a modified surface). Evaluation of it is extremely important, as an indicator of mechanochemical activation of the material.

The value of the total energy accumulated in the dispersed substance depends on the deformation conditions: the intensity of the energy supply, the properties of the substance, the duration of the process, and so on. However, under the most favorable conditions, it does not exceed 25–30% of the energy supplied [10]. If the destruction is mainly related to elastic stresses, the stored energy is determined by the plastic deformation of the solid and is associated with the accumulation of defects and the transformation of the structure of the material, and, above all, of the surface layers. From the point of view of the chemistry of defective crystals, a solid body containing an increased amount of defects, as compared to an equilibrium one, has a higher reactivity. These structural changes take place in the crystal lattice during activation and facilitate the subsequent thermal decomposition of the substance.

The reactivity of a solid after deformation is determined both by the greater specific surface area of the particles with broken bonds and by the presence of defects in its volume. The surface is characterized by a relatively low binding energy density, and its values are uniquely related to the physicochemical properties of the substance [11]. Calculations of the surface energy of deformed particles [12] have shown that its values correspond to the melting heat level of inorganic substances (10–150 KJ/mol). The volume accumulation of energy due to the dislocation structure can be an order of magnitude higher. Since accumulation of defects is associated with plastic deformation, for activation during the dispersion process, it is necessary to realize such conditions under which the plastic flow of material will occur as quickly as possible occurs. This can be achieved by increasing the power and speed of loading, changing the loading scheme with the predominance of shear strains, or using smaller particles as initial particles, which can immediately transform into a plastic flow state, i.e., accumulation and redistribution of dislocations. The degree of structural disorder of the deformed material, as a rule, is estimated by X-ray diffraction analysis from the change in the lattice parameter and broadening of the X-ray lines [13]. From the values obtained, one can judge the degree of reactivity of the mechanically treated material [14].

Apart from the defects in the crystal structure accumulated in the volume of the particle reactivity of the dispersed material is significantly influenced by amorphization of the surface layer of the particle, which is a "loosened" structure of the material saturated, first of all, with vacancies and other defects, with great number of broken bonds, i.e., chemically active centers. Therefore, such a surface has an increased adsorption ability. The degree of mechanical activation of the material is also determined by the change in adsorption characteristics [15].

Since annealing and recombination of defects with heat generation occur during heating, the thermographic and calorimetric methods are used to estimate the total degree of activation (in terms of the total defect structure of the prototype) [16]. These methods make it possible to estimate the deformation energy stored in the material by the amount of heat released. For monolithic deformed samples, indeed, the released heat and stored energy are adequate. For activated powders, this indicator, like others, is quite conventional, since there is no complete annealing of the defects, especially in the surface and near-surface layers, where a thermal restructuring of the structure takes place with the formation of new active centers [17, 18].

Questions

1. How is the energy and work spent on grinding the material evaluated?
2. Name a distinctive feature of P. A. Rebinder's formula for the estimation of the energy spent on grinding the material.
3. What is the role of abrasive (shearing) stress in the process of grinding and the formation of the structure and state of the surface of destroyed solids.
4. Describe the principle of grinding and the characteristics of the operating mode of planetary-centrifugal mills.
5. What are the factors that determine the energy consumption for dispersing and activating the material when processing it in grinding machines.
6. What determines the reactivity of a solid after processing it in grinding machines?

2.2 Physicochemical Processes of Substance Transformation during Mechanochemical Treatment

2.2.1 Deformation, Destruction, and Activation are the Main Stages in the Transformation of Matter under Mechanical Action

In the previous section, we presented general ideas on the grinding of solids in various high-energy mills and on the assessment of the energy state of the material. However, a detailed, step-by-step consideration of grinding shows that this is a very complex process of multi-level change in the structure and state of the material being deformed. In the grinding process, all known stages and regularities of the formation of solids under force are focused, starting with elastic deformation, then plastic, due to the formation, movement, restructuring and annihilation of defects of different structural levels up to a complete transformation of the structure and destruction of the material. In all well-known monographs on mechanochemistry [16–21], a detailed analysis of the work on the process of grinding solid bodies in its multifaceted manifestation and the variety of physical and chemical phenomena accompanying deformation and destruction is given. As a result of the constantly growing large volume of experimental facts and the consistent development of theoretical ideas about the mechanism of the formation of solids under dynamic force effects, and, in spite of some disagreements among researchers in interpreting the results, a fairly clear idea of mechanochemical processes is obtained in the grinding of crystalline, amorphous and polymeric materials. The central object of research is the question of activation: What is it itself as a process and a state, and what is defined and how can it be described? As noted in the work of J. A. Hint [22], at first under mechanical activation different researchers understood different phenomena: an increase in catalytic activity, an increase in the rate of chemical reactions, an increase in strength and other properties of activated materials. Experience has shown that in all cases during machining there is a significant change in the structure of the material and a change in the energy state of the substance after its grinding.

At present, the concept of mechanical activation has become more capacious in meaning and content. This is a complex, multistage physicochemical process, at each stage of which some of the main processes and phenomena occur determining to some extent the change in the state of the material and its energy potential. Starting from the moment of elastic deformation to brittle failure, from the moment of accumulation of defects to complete structural rearrangement in a particle of ultra-small sizes, the energy state of the substance changes. The difference is only in the number of these energy changes, the possibility of their relaxation or long-term fixation due to fixed structural rearrangements of the material. Some authors [16, 20] take into account all stages of grinding and all structural changes in the surface and volume of particles of different degrees of dispersion as components of the process of mechanical activation. Others [21, 23] associate activation only with the final stage of a qualitative change in the substance as a result of its dispersion.

However, the change in the structure and, consequently, the state of the deformable material begins practically from the first moment of application of the load. The difference of all the stages of deformation is in the value of the accumulated energy caused by the depth of transformation of the structure of the material. Many researchers have shown experimentally that contribution of the surface energy in the grinding of solids is only 5–10% of the total free energy reserve. However, this does not mean that the role of the surface in changing the state of the crushed material should not be taken into account. The importance of the surface structure and the surface state of the particles is exceptionally great in the variation of their properties.

The main change in the internal energy of a substance during grinding is associated with a volumetric transformation of the structure, the appearance and relaxation of residual stresses caused by the shift of atoms from their normal positions to a state is described within the physics of the defective structure of the material and characterized by an increased entropy of matter after activation [20]. At this stage of activation, changes in the structure lead to significant changes in the property indicators. Moreover, periodicity is noted both in the restructuring of the structure and, correspondingly, in the properties of the material with the time of mechanical action.

The example of quartz (Fig. 2.3) shows [20] that the period of accumulation of distortions of the crystal lattice is replaced by a period of intensive growth of the specific surface. The density of quartz also decreases stepwise from 2.65 g/cm^3 (quartz) to 2.20 g/cm^3 (amorphous silica). Periodic accumulation of defects and stresses in the form of "frozen thermal vibrations" is replaced by discharging them to grinding and, consequently, the increase in the free surface and the decrease in the density of matter.

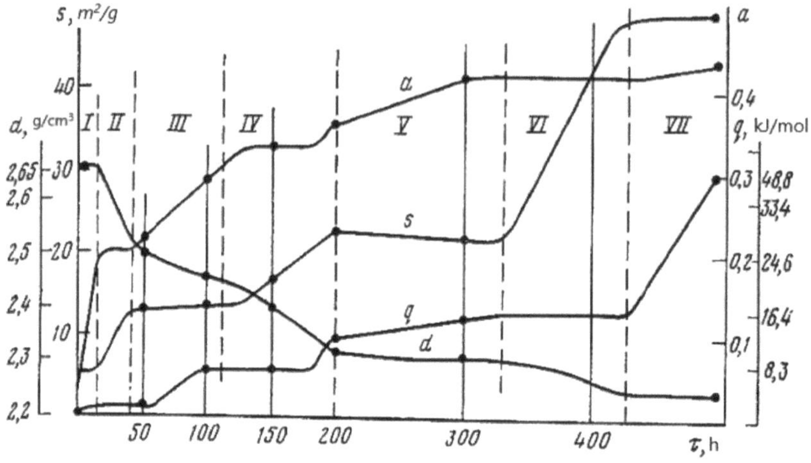

Figure 2.3 Dependence of the specific surface (s), density (d), lattice defects (a) and the heat of quartz dissolution (q) on the duration of its grinding (τ) in a vibrating mill [20].

This law of periodic accumulation—the discharge of the energy state of matter—takes place in all cases of grinding; only the degree of its manifestation depends on the method of action on the material to be crushed, which is clearly shown [19] by a wavy change in the chemical activity of crushed substances with the increase in the time of grinding. When the materials activated by grinding are exposed to air, their properties change. In particular, for quartz, the solubility of the powder decreased by 20% during the year, i.e., there is an aging effect due to the metastable state of the activated substance [16].

The highest level of activation is associated with the stage of ultrafine grinding, when the material practically transforms into a completely new substance with a new structure and properties at high force characteristics of the action. This stage is given preference in the works of M. V. Chaikina [21] in the classification as an actual activation stage when grinding with the highest energy indices of the substance with simultaneous stability of the state. Using apatite and quartz as an example, it was shown that in such a case the substance passes into a quasi-liquid state with a high reactivity. The intensity of such a transition is determined by voltages of 1.0–1.78 GPa (Fig. 2.4) and is associated with a change in the mechanism of matter transfer from the vacancy to the crowdion (interstitial) and stresses exceeding the theoretical tensile strength and/or dynamic yield strength. In the process of aging, such particles with a quasi-liquid structure assume the icosahedral and dodecahedral shape inherent in small particles in a non-crystalline state, with a specific electron diffraction pattern [21]. Such a transition in the plastic deformation of a solid under conditions of intense grinding is of independent interest as a new type of phase transition under conditions of a force action on a crystalline substance.

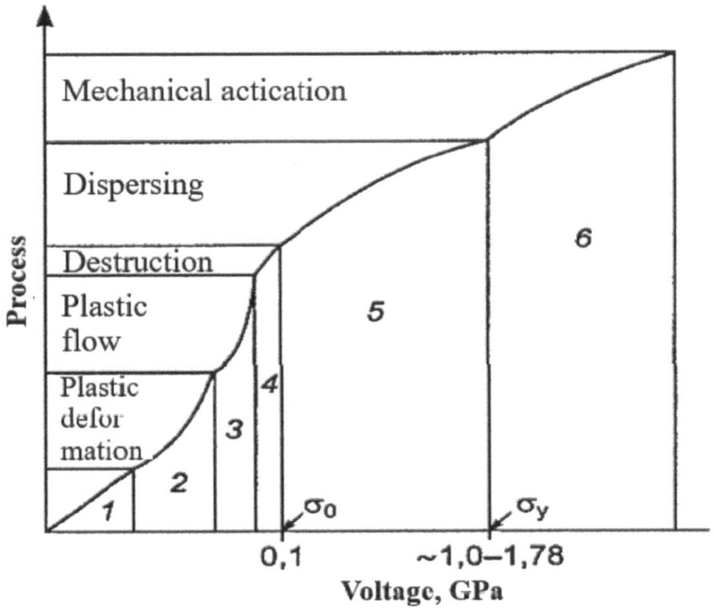

Figure 2.4 Change in the state of a solid body as a function of the stresses arising during dynamic mechanical impacts [21].

2.2.2 Models of Mechanochemical Processes: The Localization of Deformation and the Physicochemical Processes Accompanying It

The activation levels discussed above, their structural and energy characteristics, represent a static activation model, which practically reflects the result of mechanical action on the substance at various stages of its grinding. Dynamic models are directly related to the process of mechanical action itself. First of all, this is the thermal effect at the contact of friction, upon impact and at the mouth of a rapidly growing crack in the destruction of a solid body. However, thermal models cannot explain many mechanochemical processes, the direction of which is opposite to that of thermal treatment. In particular, after mechanochemical treatment with heating up to certain temperatures, an increase in the density of the substance is observed, while the density of a similar phase obtained during heat treatment is much lower [17]. Among the thermal models, the most popular is the tribo- or "magma-plasma" model proposed by P. Thiessen [24]. According to this model, friction or shock in the contact area of the colliding solids results in the instantaneous release of a large amount of heat, as a result of which the substance passes into the plasma state with the formation of short-lived highly excited ions, solid-state atoms responsible for chemical reactions (Fig. 2.5). The recombination of these active objects leads to gas-discharge emission, electron and ion emission, and also to their interaction. The physical picture of the processes is quite complicated.

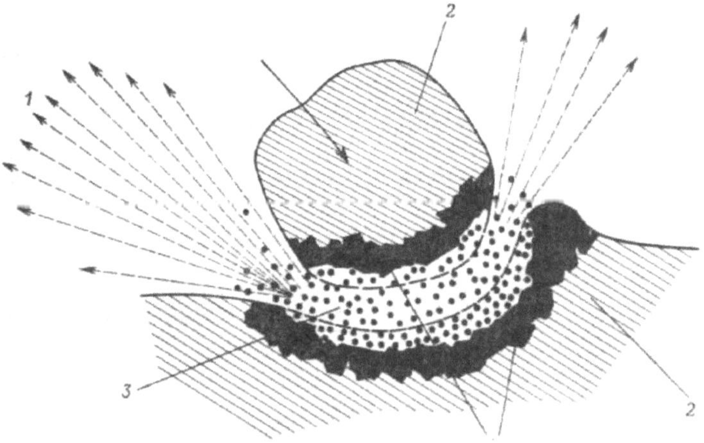

Figure 2.5 The model of "magma-plasma" [18]. (1) exoemission, (2) undistorted structure, (3) plasma, (4) disordered structure.

Much attention is paid to other areas of transformation and accumulation of mechanical energy and, above all, breaking of chemical bonds with formation of active centers on the newly formed surface. According to P. Yu. Butyagin [25] on a freshly formed surface with the destruction of a solid particle there is a significant number of active sites, the lifetime of which is 10^{-4}–10^{-7} s or more, depending on the conditions for carrying out mechanochemical treatment, including in air [26]. The active centers are well fixed by electron paramagnetic resonance (EPR). Their death is associated with the relaxation of excess energy due to rearrangement of chemical bonds or interaction with molecules of the environment. This process is exothermic and is accompanied by luminescence, as well as other phenomena associated with the emission of energy. The model of defect formation with mechanical activation of different quality and quantity at all structural levels includes a number of particular variants responsible for the different stages of transformation of a substance under mechanical action [17, 21, 27].

In recent years, the wave model of mechanochemical activation has gained the most recognition. It practically united all the models considered above by a single approach in the consideration and analysis of the physicochemical phenomena that determine and accompany the process of grinding and activation of solids. The wave approach in the analysis of the interpretation of mechanical energy transformations and its accumulation in a deformed material is shared by the authors of [20, 21, 28]. As shown in the monograph by M. V. Chaikina [21], under the conditions of dynamic mechanical action, the process of transformation of mechanical energy in a solid body has a complex wave character. Depending on the input power, and the direction of the applied forces, the wave process of energy transfer, the ways of its relaxation and the state of the solid will change (Table 2.1).

A single act (impact, compression, etc.), of an impact on a solid body does not always lead to destruction but is necessarily a source of forced oscillations. Elastic waves, propagating in accordance with physical laws inside a solid [29], reflected and refracted at the boundaries of inhomogeneities, lead to local stress concentrations. In these zones of stress concentration there is a disruption in the orderliness of the crystal structure and the loss of energy of the traveling elastic wave. The occurrence of extreme stresses is most likely when a traveling wave meets the inhomogeneities of the crystal structure (vacancies, impurity

Table 2.1 The nature of the transformation of mechanical energy and the channels of its accumulation, depending on the intensity of mechanical action on a solid body [21]

The change in the character of the process with the increase in mechanical stresses	Elastic region	Plastic deformation	Destruction	Grinding	Mechanical activation
	"Wave"		Autoacoustic excitations and superposition of waves		Autogeneration of "shock" waves of compression and unloading and superposition of waves
Transmission mechanism of energy and type of waves	Elastic waves	Plastic waves			
Ways of energy relaxation	Phonons, heat	Lattice defects, heat	Crack formation, heat	Formation of a new surface, heat	Excitation of vibrations of chemical bonds, structural and phase transformations, chemical interactions, heat.
Dynamics of the mass of matter and its primary forms	No	Motion of dislocations and point defects, twinning	Movement of vacancies, propagation of cracks		Atoms, ions, domains of molecules, molecules, intense diffusion of matter
Change in Macro Volume	No	No	No	No	Yes
Stored energy	Yes	$£_d = 5$ J/cm^3 for $p = 10^{11}$ cm^2	Surface energy for apatite = y_{s0}'' (0001) – 0.095 J/m^2, Y_{surf} (1010) – 0.48 J/m^2		Up to 100 kJ/mol and more
Laws, models (the boundaries of their actions)	Hooke (linearity of stresses and deformations)	Schmidt's law	Models: Zhurkov S. N. [1957], Bovenko V.N. [1983] [34]	Bovenko V.N. [1983] [34]; Bovenko V.N. [1986] [28] Ritterger, Kick, Bond, Sidenko P.M. [1] Khodakov G.S. [19]	The laws of conservation of mass, momentum, energy
	Debye's mode (harmonic approximation)				

atoms, dislocations). The propagation of an elastic wave is a change in the phases of compression and expansion. The addition of waves that coincide in phase can cause a local concentration of stresses exceeding the ultimate strength and leading to a discontinuity, i.e., to the rupture.

The addition of waves of direct and reflected waves, taking into account the fact that waves propagate from different points, will determine the level of activation, which is higher the higher, the frequency of mechanical action. The elastic energy transferred to the particles can be transformed into phonon energy, due to focusing to the symmetry center of the particle in the form of a converging spherical elastic wave [30]. In this case, the optical phonon modes are excited, this leading to creation of virtual photons which, propagating through the volume of the crystal, transmit energy to the symmetrically located section of the front of the spherical wave. This leads to an increase in the frequency of the optical phonon created on it, which in turn emits a new virtual photon with a higher frequency, etc., according to the scheme of chain reactions. Such a process, taking place repeatedly as the front of the elastic wave moves, will lead to a significant increase in the energy of the system as the phonon focusing proceeds.

In the light of the wave concept, of V. I. Molchanov [20] considers activation by grinding as a loosening of the crystalline structure of a solid under continuous pulsed excitation of elastic and thermal oscillations of elementary particles by analogy with autowave processes. These processes occur in samples of sufficient length under the influence of locally introduced thermal or mechanical excitation and, according to [31], are similar to chain combustion reactions. With an increase in the effective load and with the transition of the deformed body from the elastic to the plastic state, there take place formation, displacement, inhibition and interaction of dislocations occur accompanied by dissipation of energy. The periodic change in the configuration of the dislocation core and the nonuniformity of its motion lead to radiation by the dislocation of elastic waves [32]. The motion of the interstice-crowdion during the plastic deformation of solids is also considered as a soliton or crowdion wave [33].

Of particular interest are the works of V. N. Bovenko [34], in which the stage of pre-destruction of plastic deformation is considered from the standpoint of autoresonance. Frequent change of load and unloading with varying force action on a solid body leads to the appearance of large gradients of mechanical deformations and stresses. For large pressure amplitudes, the state of the solid on the boundary of the compression wave (in the head of the wave) and the unperturbed substance is mathematically described by discontinuous functions of the parameters of the medium. Such a mechanical action generates a strong plastic or conditionally "shock" wave [35].

The main distinguishing feature of plastic waves from elastic waves is that they possess an important property of mass transfer of the medium in which they move [36]. Moreover, plastic "fast" waves in a dynamically deformed medium are associated with the compressibility (change in the bulk density) of the medium and are single (disparate) waves, called solitons. The mechanism of mass transfer by such a wave is extremely complicated as a kind of mechanical motion of a deformed body and is not fully understood. At the moment, the very fact of the phenomenon under consideration is theoretically grounded and experimentally stated. In [21], the mass velocity of the matter moving in the activation zone is estimated as a wave process. It varies from 188 to 990 m/s. When such a "shock wave" emerges onto a free surface, the substance is discharged, at which the mass velocity doubles, which, according to [18], causes resonant excitation of lattice vibrations in matter with a frequency range of 400–500 MHz. Such resonant vibrations, excited by the energy of plastic waves under mechanical influences, can cause structural and phase changes [21].

The nature of these transformations will be determined by the parameters of the process. When the compression wave passes through the substance, the energy supplied during the restructuring of the structure will partly transform into the kinetic energy. Another part of it will transform into thermal or energy of various mechanochemical processes [37]. In this case, the substance will represent either a high-pressure phase or a metastable phase with a higher density than the original substance. In the unloading zone, this process can also be accompanied by ultradispersion. Such restructuring processes are called "deformation" processes.

If the structural rearrangement of the substance occurred not in the front of the "shock" wave, but during the unloading period, in particular on the surface of the particle, then the density of matter decreases, and the structure breaks down. It is in the zones of wave unloading that areas of liquid crystal type are formed in the volume of the activated material. The process of transition of a substance to a quasi-liquid state at the time of compression and unloading is not analogous to melting, since, according to [21], the input energy of a mechanical wave is an order of magnitude higher than the latent heat of melting and is close to the heat of evaporation. On the energy level, the substance "quenched" under these conditions must be in a state between the liquid and the gas and have a high reactivity.

All the processes of restructuring the structure and associated with the motion of defects under the influence of varying stresses in a solid are accompanied by generation of high-frequency phonons due to interaction of developing dislocations with other dislocations, vacancies, impurity elements and other defects. High-frequency phonons can initiate chemical reactions. Internal, local stresses are responsible [18] for the appearance of active radical centers both directly on the surface of the fracture, and in the surface layer of the particle, where the gradient of these stresses is particularly significant. In this case, mechanochemical activation and transformation of a substance are always accompanied by emission of energy in the form of light, electron flux or ultrasound. The authors of [38, 39] paid great attention to exoemission during deformation in general and to the grinding of solids, in particular. Emission of fast electrons with energies reaching tens of kiloelectronvolts and radio emission occur during the breakdown of crystalline dielectrics, amorphous bodies do not emit electrons upon destruction. Of the number of crystalline bodies, the most powerful electron emission is observed when crystals of piezoelectrics are split, for example, in quartz and Rochelle salt, the fused quartz does not. Thus, there is some parallelism between the emission of electrons and the presence of the piezoelectric effect of materials. Along with fast electrons, emission of slow electrons, which proceeds after destruction, is also observed. It is associated with the restructuring of surface layers or compensation for broken chemical bonds. The emission of electrons is accompanied by the luminescence of the substance [39], clearly demonstrating a change in its energy state.

The most effective exoemission proved itself in the mechanodestruction of polymers [40], facilitating the mechanocracking of organic compounds and self-organization in the process of creating new structural forms. For polymer mixtures, this effect of electron emission is enhanced by the destruction of the double electrical layer at the polymer interface and the formation of an uncompensated charge on fresh surfaces [41]. The presence of an uncompensated charge and an electric field can also lead to the rupture of macromolecules via the ionization of macromolecules as a result of electron tunneling and the formation of macroions. For metal–polymer mixtures, it is shown [42] that the stimulated emission (injection) of electrons promotes an increase in the degree of ordering of macromolecules. Practically all the processes accompanying destruction and grinding of solids (breaking of bonds and the formation of active sites, formation of radicals and their outlet to the surface, interaction of emitted electrons with the surface) lead to the electrization of the surface layer of the particle.

In all these cases, there is a local violation of the electrical neutrality of the double electric layer and formation of a free charge [43]. The local excess charge, mechanically induced on the surface of a solid, provides the appearance of electric fields with a strength of up to 10^7 V/cm in defective areas on the surface [18]. These areas are centers of polymerization and other chemical processes occurring on the surface of the material being crushed. Formation of charges of different signs on the surface is possible, i.e., formation of microelectrode pairs, leading to a change in the surface potential. In the impact zone, the highly disordered lattice structure of the crystal acquires an excess of energy, on account of which the electrochemical potential of the electrons changes (which is equivalent to a change in the Fermi level).

Between the disordered and undistorted regions, a galvanic potential arises, called a deformation potential in this case. The total potential of a real solid surface is the potential integrated over the entire surface of a solid body over all its local elements. Surface properties, in general, its sorption activity are determined to a large extent by both the level of this deformation potential and the reactivity of the system of local microelements. The regularities of interaction between the local and integral potential are determined in particular by the ability of the material to stress corrosion cracking and fatigue corrosion [44].

Local changes in the structure in the volume and on the surface of the deformed solid play a significant role in the overall change in the state of the material. The effect of localization of structural changes, according to [30], is due to the nonuniformity of the absorption of elastic energy in crystals and to the formation in the microregions of "supercondensates"—bundles of elastic energy of limiting density. It determines ultimately the concentration of defects, breaking of bonds and limiting possibilities of grinding the material.

2.2.3 Mechanisms of Initiation of Mechanochemical Reactions

The diversity and complexity of physical and chemical processes occurring under mechanical action determine the variety of approaches for explaining mechanochemical reactions.

Many studies have developed ideas about the mechanisms of initiation of reactions conditioned by the plastic flow of solids and dislocations coming to the surface in the course of mechanical processing [45–47]. It is stated by Bridgeman that application of shear during the action of high pressures leads to a sharp acceleration of solid-phase processes [48]. Such an impact is accompanied by different irreversible phenomena, including chemical reactions. For example, direct synthesis of copper sulfides, conversion of red phosphorus into crystalline black phosphorus, etc., When considering such processes, accelerated mass transfer under the influence of plastic deformation is associated with two reasons: the transfer of matter due to the existence of plastic flows and acceleration of actual diffusion due to formation of linear and point defects, as well as their displacement due to pressure gradient. Theoretically, the mobility of a dislocation is such that it can provide a body elongation of one third in a time of 10^{-9} s [18].

A. N. Dremin and O. V. Breusov [49] developed and proposed "model roller." According to it, if two layers of a substance are displaced relative to each other, the nucleation center of a new phase can be considered as a kind of motion roller. With the rapid displacement of two layers of the phase relative to each other, a multitude of atoms of different kinds pass by the nucleation center. According to calculations, with a plastic displacement for a distance of 1 mm (the size of one atom is 0.2 nm) and the time of mechanical action 10^{-5} s, the contact time of two atoms is 2×10^{-12} s. As the time of rearrangement of electron shells (from 10^{-13} to 10^{-14} s) is much less than the time of contact between atoms with each other, all atoms passing in the immediate vicinity of the nucleus have time to join it, and the formation of a new phase can occur.

According to P. Yu. Butyagin [37], deformation is the first response of a solid to the action of external mechanical forces. When solids get deformed, various forms of mobility are excited. These include both forms of thermal motion and deformation mobility. The movement of dislocations and grain boundaries refer to the deformation mobility. At a concentration of vacancies of $\sim 10^{17}$ g^{-1}, the solid is saturated with defects almost in the same way as during melting, when a vacant site accounts for 10^3–10^4 atoms in the lattice points. The difference from melting is that such a state is realized at a relatively low temperature, when atomic jumps are impeded. However, during the development of deformations, this restriction is removed, since internal stresses reduce the jump barrier.

In [50], it is shown that an effective way to obtain nanocrystalline materials with impurities is mechanochemical treatment, as a result of which there take place a convective transfer of impurity atoms by moving macrodislocations, i.e., by linear defects of the proper packaging of nanograins. As accumulation of impurities occurs periodically at the time of mechanical loading, and the accumulation of impurities in the particle is constantly increasing during the process of mechanical activation, there exists an effective mechanism for holding impurities in the period between the impacts.

The proposed mechanism is as follows. Mechanical loading leads to formation of macrodislocations in the bulk and surface layer of particles. Differences in the density of the medium lead to a change in pressure on the macrodislocation core. High pressure gradients are created around a moving dislocation, which lead to barrodiffusion of the substance along the polycrystal boundaries from the environment in the direction of the nucleus. If a macrodislocation is displayed on the medium through an external surface on which there are impurities, it captures them with its core and transfers them into the volume. When the macro-dislocation stops after the cessation of exposure or annihilation of macrodislocations of different signs, the pressure on the core drops, and impurities are easily dispersed throughout the volume of the medium. As a result, a crystal plastically deformed under mechanoactivation conditions with a well-developed intergranular slip will quickly become saturated with impurities along intercrystalline boundaries.

Much attention has also been paid to the formation of radicals during destruction of solids. So when typical homeopolar crystals (quartz, silicon, etc.), split, paramagnetic centers appear. It is stated [51] that when quartz, silicon, and silica gel break down, there appear paramagnetic centers of the same type, the spectrum of which is a narrow single line, and the g-factor is close to 2. The concentration of paramagnetic centers in all cases is close to $\sim 10^{17}$ g^{-1} [52]. The authors of [51, 52], came to the conclusion that the paramagnetic centers are located in a thin surface layer, and not on the surface of destruction. Centers are stable under vacuum at room temperature for months. Paramagnetic centers recorded by the EPR method are chemically inactive. The existence of paramagnetic centers makes a definite contribution to the energy accumulated by the material under the influence of mechanical processing. Regarding the mechanism of formation of paramagnetic centers, there are two theories: hemolytic and heterolytic decay.

For the rates of formation and consumption of centers in the process of mechanical destruction of solids, the following expressions were derived in [53]:

Center formation rate:

$$\varpi = \frac{dn(\tau)}{d\tau} = n\frac{ds}{d\tau} \tag{2.8}$$

Spontaneous death rate centers:

$$-\frac{dN(\tau)}{d\tau}=\frac{1}{\tau_0}N(\tau) \tag{2.9}$$

The rate of interaction with gas molecules:

$$-\frac{dN(\tau)}{d\tau}=\alpha kN(\tau)p' \tag{2.10}$$

where $N(\tau)$ is the concentration of centers at the moment of time τ (g^{-1}), s is the specific surface of the products of destruction (m^2/g), n is the concentration of centers on the freshly formed surface (m^{-2}), p is the gas pressure, kp is the frequency of collisions of gas molecules with active centers (s^{-1}), and α is the probability of a chemical reaction during collisions.

Thermal theory has witnessed the greatest development, according to which the sources responsible for initiating reactions can be local heating, which occurs when kinetic energy is converted into heat at the time of the collision of a grinding body with another body and friction of particles, as well as plastic flow of material [54]. Later, in the development of this theory, the researchers came to the conclusion that it is necessary to take into account the factors of time, pressure and temperature arising at the impact-friction contact.

The release of heat by mechanical action is associated with the conversion of kinetic energy into thermal energy at the time of the collision of the grinding body with the drum wall, grinding body, with particle friction, as well as with the plastic flow of the material. All of these sources may be responsible for the thermal initiation of reactions. The concepts of local heating during friction were developed by the school of F. P. Bowden [55]. As a result of research, it is stated that the area of hot spots is 10^{-2}–10^{-4} cm^2, the duration of their existence is 10^{-5} to 10^{-3} s, and the temperature jump can reach the melting point of the low-melting component, but not higher. For refractory substances, an instantaneous local increase in temperature up to 1300 K can be achieved. The bulk of the substance remains cold at this time. The time of existence of such temperature flashes on the area of 10^{-7} to 10^{-9} m^2 is about 10^{-4} s.

The calculation of the temperature peak at the contact of the rubbing bodies that take place in planetary mills was made in the work of F. Kh. Urakaev [56]. The main assumptions of his calculations are as follows. In contrast to thermal, radiation and other methods of conducting solid phase processes, the time factor (t), along with traditional pressure P (or mechanical stresses σ) and temperature (T), is decisive in the specifics of mechanochemical processes [57]. On this basis, the author [57] classified the grinding apparatus according to the type of mechanical effect on the substance. Grinding devices were divided into two classes. The first is the impact of grinding bodies and walls on the set of particles of the substance being treated. Such a "constrained blow" takes place in ball, rod, hammer and roller mills, attritors, and jaw crushers. This class is divided into three subclasses according to the nature of the impact of grinding bodies, where

(a) impact processing mode, which is characterized by a certain relative speed W of the impact of grinding bodies on the totality of the particles being processed, and is realized, for example, in ball vibration and hammer mills, jaw crushers, in shock mode of operation of ball and planetary-centrifugal mills;
(b) abrasive processing mode, which is characterized by the shear effect of the dynamic impulse force F or pressure P of the grinding bodies on a set of the particles being processed and is realized, for example, in roller mills, attritors, mortars and in the abrasive mode of operation of ball and planetary centrifugal mills;
(c) impact-abrasive processing mode, which is characterized by the ability to implement modes (a) and (b) in one unit and is implemented, for example, in rod mills and impact abrasive mode of operation of spherical planetary centrifugal mills.

In the second subclass, the mechanical action is carried out by the collision of a single particle of the treated substance with a grinding body. Such a "free kick" occurs when the particle interacts with the rotating fingers of the disintegrator or with the fixed wall of the jet mill. Some dispersing devices, for example a disintegrator, can combine both types of mechanical action:

"free kick"—with single fingers; "cramped punch"—between adjacent rows of fingers.

Both during the impact and under the dynamic impulse effect on the totality of the particles of the treated substance in the grinding apparatus, the impact force acting on the contact of the particles is decomposed into normal and tangential components. As a result, during a certain characteristic time of pulsed interaction, the process of deformation of particles with a shift takes place, leading to an increase in temperature (temperature pulse) at the contact. A special case of the described process is the shock-friction interaction of the particles being processed [58]. Therefore, the author of [56], based on the calculated models for all types of grinding machines, assumed theoretical positions of off-center impact of solids and their dynamic contact compression, which essentially boils down to a single model that allows determining parameters differing from each other (area-pressure-temperature) by contact of particles depending on the type and mode of operation of the grinding apparatus.

However, in the classical theory, the physical properties of the colliding bodies are not taken into account; therefore, its use for calculating grinding apparatus (mechanochemical reactors) did not give the desired results. On the other hand, the use of modern ideas about the collision of solids for the calculation of mechanochemical reactors, due to the complexity of the processes of mechanical interaction occurring in them, is almost impossible. This is possible only within the framework of the nonlinear elastic (or elastic-plastic [20]) theory of Hertz, therefore, its main provisions and results, which were used in constructing the model and calculating the grinding apparatus, are described below.

In the most convenient form, the final results of calculations for the most significant parameters are presented in [20, 59]:

- interaction time t_{ij} of colliding bodies:

$$t_{ij} = (\pi/4)(10\pi)^{0.4}\,\{w_{ij}^{-1}(\theta_i + \theta_i)^2(L_i + L_j)[\rho_i\rho_j/(\rho_i L_j^3 + \rho_j L_i^3)]^2\}^{0.2}, \quad (2.11)$$

where the numerical coefficient is determined from the elliptic integral through the Gamma function, $f(G[z]) \approx 3.0 \approx (\pi/4)(10\pi)^{0.4}$,

- maximum force f_{ij} interaction of bodies:

$$f_{ij} = (2/3)(10\pi)^{0.6}\,\{w_{ij}^{6}(\theta_i + \theta_i)^{-2}(L_i + L_j)^{-1}[\rho_i\rho_j/(\rho_i L_j^{0.3} + \rho_j L_i^{3})]^{3}\}^{0.2} \quad (2.12)$$

- maximum stress σ_{ij} in the center of the contact area of bodies:

$$\sigma_{ij} = 4(10/\pi^4)^{0.2}\,[\,w_{ij}^{2}(\theta_i + \theta_i)^{-4}(L_i + L_j)^{3}\rho_i\rho_j/(\rho_i L_j^{3} + \rho_j L_i^{3})]^{0.2} \quad (2.13)$$

- average mechanical stress $<\sigma_{ij}>$ on the contact of bodies:

$$<\sigma_{ij}> = f_{ij}/s_{ij} = (8/3\pi)(10\pi)^{0.2}\,[\,w_{ij}^{2}(\theta_i + \theta_i)^{-4}(L_i + L_j)^{3}\rho_i\rho_j/(\rho_i L_j^{3} + \rho_j L_i^{3})]^{0.2} \quad (2.14)$$

- maximum radius r_{ij} of the contact area of bodies:

$$r_{ij} = (1/2)[(3/2)\,f_{ij}\,(\theta_i + \theta_i)/(L_i + L_j)]^{1/3} \quad (2.15)$$

- maximum contact area s_{ij} of bodies:

$$s_{ij} = \pi r_{ij}^{2} = (\pi/4)\,[(3/2)\,f_{ij}\,(\theta_i + \theta_i)/(L_i + L_j)]^{2/3} \quad (2.16)$$

- maximum total deformation (approach) ε_{ij} of bodies:

$$\varepsilon_{ij} = \varepsilon_i + \varepsilon_j = (1/4)\,[(9/4)\,f_{ij}^{2}(\theta_i + \theta_i)^{2}(L_i + L_j)]^{1/3} \quad (2.17)$$

$$\varepsilon_i\theta_j = \varepsilon_j\theta_i$$

From (2.12), it follows:

$$\varepsilon_i = \varepsilon_{ij}\theta_i/(\theta_i + \theta_i);\ \varepsilon_j = \varepsilon_{ij}\theta_j/(\theta_i + \theta_i) \quad (2.18)$$

ρ_i and ρ_j are densities, $\theta_i = 4(1 - \nu_i^2)/E_i$ and $\theta_j = 4(1 - \nu_j^2)/E_j$ are compliance, ν_i and ν_j are Poisson's coefficients, E_i and E_j are Young's moduli of solid bodies *i* and *j*.

When applying Eqs. (2.11)–(2.14) to grinding devices, it is necessary to bear in mind that these equations are valid only for shock operation, and Eqs. (2.12)–(2.13) for any mode. For the abrasive mode of operation of the grinding apparatus, Eqs. (2.11)–(2.13) should be deduced from the corresponding models for the process of dynamic compression of the particles being processed, specially designed for this type of mill.

When calculating multicomponent systems (three and more), it should be borne in mind that at the same point only two homogeneous or dissimilar particles of the material being processed can contact, i.e., sequential calculations are performed for possible systems of two components that make up a multicomponent system. When studying mechanochemical processes (mechanolysis, activation) that occur during the processing of individual compounds in the mills, the calculation formulas are simplified [56].

The case considered establishes a quasistationary distribution of the particle dispersion, and the main role in heat generation is played by the deformation and friction of the particles against each other. The notation and scheme adopted by the author are shown in Fig. 2.6.

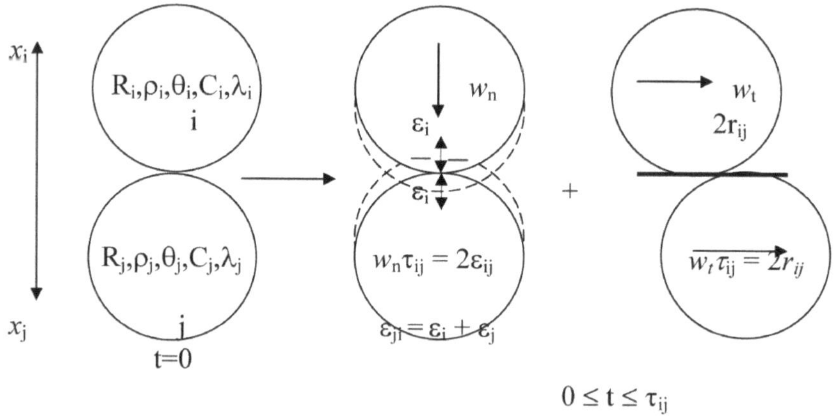

Figure 2.6 Diagram of the interaction of solid particles with each other at the time of the impact of the grinding body on them [56].

The indices i and j denote the interacting particles; R the particle radius; f_{ij} the particle compressive force; f the force shifting particles relative to each other; s_{ij} the particle contact area; r_{ij} the contact area radius; ε_{ij} the deformation of colliding elastic bodies; ρ substance density; $\Theta = \frac{4(1-\nu)}{E}$; ν Poisson's ratio; c the heat capacity; λ the thermal conductivity; w_{ij} the relative speed of interaction of elastic bodies at the moment of contact; and L the curvature of the particle surface at the point of contact.

Thus, the process of activation of a solid body as a result of mechanical action cannot be explained by one mechanism. On the other hand, all considered theories have common points. According to the dislocation theory, about 70% of surface atoms can be transferred to the active state, since they are in the zone of action of elastic distortions around dislocations that emerge to the surface during plastic deformation of a solid. This brings the dislocation theory closer to the theory of short-lived states, according to which all atoms of the surface are active at the moment of destruction.

The common point in the thermal and dislocation theory is the explanation of the mechanochemical decomposition: the existence of highly excited nonequilibrium states, caused in one case by the interaction of dislocations with each other and the generation of high-frequency phonons, and in the other by a local temperature rise on the surface of destruction, is allowed.

From the presented theories of mechanochemical initiation of solid-phase processes, at present only the theory of local temperature rise at the impact contact has been brought to full numerical calculations, which allow one to estimate the thermodynamic probability of the occurrence of mechanochemical processes.

2.2.4 Structural Rearrangement and Modification of the Surface of Dispersible Particles

As shown in the previous sections, the activated state of a dispersible solid is determined mainly by structural rearrangement in the volume of the particle. The surface component in the evaluation of the degree of activation is relatively small. However, this can be attributed to a greater extent to the static activation model, i.e., to the final result of the process. In the dynamics of the transformation of matter under the influence of mechanical loads during grinding, the role of the freshly formed surface is exceptionally great. First, the surface of destruction is the region of active centers, the presence of which is due to the breaking of bonds and the maximum defectiveness of the substance. According to the wave model of deformation and fracture of particles during grinding, the running, "shock" wave is discharged on the surface, which leads to a decrease in the density of matter and amorphization of the surface layer. The loosened substance of the surface actively interacts with the environment, primarily by absorbing water vapor and atmospheric gases [20, 21]. In addition, it is the surface that is the source of both fast and slow electrons, as well as other forms of radiated energy. The radiation generated by the deformation process in its turn has a re-activating effect on the material being processed. In the dynamics of grinding, the particle surface is constantly in an excited highly active state.

The degree of amorphization of the surface layer is determined both by the conditions of machining (the ratio of shearing and normal stresses; the material, the size and number of grinding balls, the time and the processing medium), and the type of the treated substance [16–21]. The study of the thermodynamics of surface states during grinding is the subject of special investigations, which makes it possible to predict the behavior of a substance under specific conditions [60, 61]. At the same time, some researchers, for example [62], pay more attention to the change in the electronic state of the surface, relating thermodynamic characteristics to an excessive concentration of hot electrons, i.e., electrons with an energy greater than the average energy of their equilibrium with the crystal lattice. With increasing thickness of the active surface, preference is given to Rayleigh waves of phonons heated by hot electrons.

Other researchers [17, 18, 25] emphasize the reaction centers that arise when bonds break at the surface of destruction. The presence of such active sites provides a chemical interaction of the material being ground with the surrounding medium. For example, numerous studies have stated [18] that during mechanical activation of quartz, reactive groups with unsaturated valences appear due to breaking of Si–O bonds. These centers react easily with water, as well as with organic substances with formation of surface groups. When machined in the presence of monomers, such as ethylene, butadiene, styrene, chlorovinyl, surface groups are grafted onto the below-mentioned reaction followed by polymerization.

$$\equiv \text{Si–O–Si} \equiv + \text{R–R}' \rightarrow \equiv \text{Si}^{\bullet} + \equiv \text{Si –O}^{\bullet} + \text{R}^{\bullet} + \text{R}'^{\bullet} \rightarrow$$
$$\rightarrow \equiv \text{Si–R} + \equiv \text{Si–O–R}', \qquad (2.19)$$

where R–R′ are radical groups of organic substances.

Thus, when treated in butadiene medium for a sufficiently long time, the amount of polymer reaches 30%. When processing inorganic materials, in particular quartz, with polymers, intensive polymer degradation and grafting of organic radicals to the surface of quartz particles occur. It is assumed [18] that the surface of silicon dioxide, formed in the course of processing, catalytically accelerates both the destruction of the polymer and the process of grafting to the surface. The final result is modification of the surface of inorganic particles.

The pattern of the physical processes occurring during grinding is quite complex and diverse depending on the specific conditions of their procedure, and the chemistry of creating a new structure of the surface layer in each case has its own specific features. A single generalizing factor is that the wave (periodic) nature of the mechanical action that causes a wave of elastic and plastic deformations in a solid and, accordingly, structural changes in it, leads to chemical reactions and transformations, in particular, on the surface of particles. These processes, both physical sorption and oxidation, carbidization, nitridation, hydrogenation, and grafting of polymer structures cause volumetric changes in the reactive surface layer. A consequence of this is formation of surface mechanical stresses, usually compressive and reaching several GPa [63]. The gradient of such stresses causes additional diffusion in the surface layer and, as a result, self-organization of the elements of the surface structure, energy reduction, and stabilization of the state. The final result of such chemical transformations is the formation of nanostructured layers on the surface of the particle being ground, formation of nanocomposite objects with a new complex of physical and chemical properties [64].

Questions

1. What processes and phenomena during machining determine the change in the state of the material and its energy capabilities, characterizing the degree of activation of the substance?
2. Describe the contribution of surface energy in the grinding of solids to the total supply of free energy of the material being crushed.
3. What determines the change in the internal energy of a substance during grinding in the framework of physics of a defective structure of a material and with a substance associated with increased entropy after activation?
4. How does the wavy change in the chemical activity of crushed substances with increasing grinding time and the effect of aging?
5. What is the static model of activation, as a result of mechanical action on the substance at various stages of its grinding?
6. Describe the dynamic and thermal models of mechanochemical activation processes.
7. Describe the thermal model of "magma-plasma" mechanochemical processes.
8. Describe the models of defect formation during mechanical activation of a substance.
8. What processes take place during mechanochemical activation?
9. What processes during plastic deformation ensure the initiation of reactions in the process of mechanical processing of solids?
10. What is the "roller model" of the transformation of the state of the material under mechanical action on it?
11. Describe the dislocation mechanism of deformational transformations (changes) in solids under mechanical action.
12. What is the role of paramagnetic centers in changing the energy state of solids during mechanical treatment?
13. What are the main provisions of the thermal theory of transformation of the state of matter and the initiation of mechanochemical reactions?

14. What are the main provisions of the Hertz theory when using it to assess the state of the ground substance and the calculation of mechanochemical reactors?
15. For which processing modes in grinding apparatus is Hertz's theory applicable?
16. Describe the exoemission in the deformation and grinding of solids, radio emission in the destruction of crystalline dielectrics.
17. What are the features of the state of the surface of particles in the process of grinding?
18. Describe the formation of reaction centers in the rupture of bonds on the surface of destruction.
19. What chemical reactions and transformations take place on the surface of particles, surface modification?
20. How does the self-organization of process take place of elements of surface structure, energy reduction, state stabilization and formation of nanostructured layers.

2.3 Kinetics and Thermodynamics of Mechanochemical Treatment of Inorganic Materials

Summarizing all known results of research on mechanical grinding and activation of solids, it should be noted that they are considered from two positions: the process itself and its final result. Depending on the approach, the thermodynamic characteristics, as well as the kinetics of mechanochemical processing of inorganic substances, are evaluated in different ways. But when choosing any variant of the evaluation of the energy state of the material, one must always have a sufficiently complete picture of the processes occurring in the material during machining. The sequence of processes and states that take place in a solid during machining, according to the ideas presented in [18], is given in Table 2.2.

The process of state change, especially at the last stages, develops rather slowly, so the activated state can be regarded as quasi-static and can be described with sufficient degree of approximation by classical thermodynamics [65]. When compared with ordinary heating, mechanically treated solids are much more reactive, especially in the low-temperature region.

Table 2.2 Processes and relaxation times of various excited states in solids during machining [18]

Excitation process	Relaxation time, s
The impact process	$>10^{-5}$
Triboplasma	$<10^{-7}$
Gas discharge	$\sim 10^{-7}$
Hot Spots	10^{-3}–10^{-4}
Electrification	10^{2}–10^{5}
Electronic emission	10^{-6}–10^{5}
Tribolumenismence	10^{-7}–10^{3} (fluorescence, phosphorescence)
Lattice defects	10^{-7}–$>10^{6}$
Motion of dislocations	$< 10^{5}$ cm/s
Lattice oscillations	10^{-9}–10^{-10}(at 0 K)
Crack formation	10–10^{3} cm/s (rate of destruction)
Fresh surface formation	1–100 at 1.3 10^{-4} Pa $< 10^{-6}$ at 10^{5} Pa
The lifetime of the excited metastable states	10^{-3}–10^{-2}

The activation energy E_A of reactions, carried out by thermal means, i.e., when energy is supplied solely by the supply of heat from the environment, is determined from the Arrhenius equation:

$$k = C\, e^{-EA/RT}, \tag{2.20}$$

where k – reaction rate constant; C – constant.

In the presence of elastic stresses (σ) in deformed particles of matter, equation (2.20) takes the form:

$$k = k_0 e^{-(\upsilon o\, -\, \gamma\sigma)/RT}, \tag{2.21}$$

where $-(\upsilon o - \gamma\sigma)$ represents the activation energy, taking into account the deformation energy due to internal stresses in the crystal lattice, created by various kinds of defects. These stresses cause weakening of bonds in the crystal lattice and, as a result, a decrease in the activation energy of various chemical processes.

For the thermodynamic description of active substances, the characteristic of excess free enthalpy is introduced, or it is also called the activity of a solid substance [18]

$$\Delta G = G_T^* - G_T, \tag{2.22}$$

where G_T^* – free enthalpy of the active solid, G_T is the free enthalpy of the same substance in the initial state at temperature T.

Since the excess free enthalpy affects both the equilibrium position and the rate of the solid-phase reaction, it is an important quantitative characteristic of the reactivity of the substance. One of the ways to determine ΔG experimentally is to consider the equilibrium of the chemical reaction and its shift in the transition from the equilibrium to the active state.

$$\Delta G = RT(\ln k_p^* - \ln k_p), \tag{2.23}$$

where k_p is the equilibrium constant for a non-mechanically activated system and k_p^* the equilibrium constant when using a mechanically treated substance.

The excess free enthalpy is determined by means of calorimetric and differential-thermal analysis methods. The excess enthalpy of a substance is the total property of all structure defects that are not in thermodynamic equilibrium.

The metastable states remaining in the solid after mechanical treatment have a sufficiently long lifetime at room temperature. The lifetime of a defective structure possessing excess energy can vary depending on the external conditions from 10^{-7} to more than 10^6 s. The lifetime of excited states in solids increases with decreasing temperature. This explains why machining often results in higher activities than in the preliminary chemical treatment of solids.

All processes of structural rearrangement and chemical interaction begin on the surface, and the rate of heterophase reactions, as is known [66], is proportional not to the amount of matter, but to the surface area of a solid. This surface is enriched with defects and radical centers, which, due to their high reactivity, enter various interactions with gaseous, liquid, and solid components resulting in modification of the surface and formation of composite, as a rule, metastable structures.

The metastable state of the obtained mechanochemical materials is fairly stable, and an explanation for this is given in terms of formation of dissipative structures [67]. The growth of energy dissipation with time ensures stability of a strongly nonequilibrium system. On the basis of these ideas, the authors of [68, 69] consider the mechanochemical process as a system of structural levels realized in the process of mechanical pumping of energy. At the same time, the intermediate structural levels act as accumulators of the supplied energy. Such intermediate levels, in particular, are amorphous compounds, the type and structure of which is determined by the specific conditions of mechanochemical treatment.

The restructuring of the structure under deformation always has a periodic character of the type: a chaotic defect structure ← → an ordered structure [54, 55]. This process has an oscillatory character and is described by a model of rotational instability of plastic deformation. Its essence boils down to the fact that the chaotic structure of dislocations of a deformed solid undergoes focal changes, in which part of the chaotically located dislocations is collected in the walls-nuclei of rotational deformation. As a result, the density of chaotically distributed dislocations and internal stresses decreases. The decrease in the density of chaotically distributed dislocations reduces the growth of the structure ordering. Again the chaotic structure begins to predominate. These oscillations in the structure are accompanied by fluctuations in the strength, plastic, and other properties of the material [69]. This approach to the process of deformation gives an idea of a solid and its crystal lattice, as a dissipative system capable of accumulating and dissipating the stored energy of "elastic distortions" during the deformation process [70].

Processes in dissipative systems are described by linear thermodynamics of irreversible processes, one of the most important provisions of which is the concept of constant states. The system in irreversible processes, with stabilization of certain forces in time, tends to a temporary constancy of the state, which, unlike thermostatic equilibrium, is denoted as a constant state of disequilibrium or simply a stationary state [71]. Stationary states can be of two types, when: (1) in a constant regime of external forces, a number of streams disappear; (2) the system introduces matter from the outside, which after a series of chemical transformations is again removed from the system, and the amount of material withdrawn per unit of time is equal to the amount of input. In accordance with this, each level of grinding should correspond to its level of mechanochemical equilibrium.

In [18], to describe mechanochemical processes, the thermodynamics of irreversible processes was used in the form

$$\Psi = TdS/dt = \Sigma^{n}_{1k=1} J_k X_k + \Sigma^{n2} j_{=1} J_k X_j, \quad (2.24)$$

where Ψ is energy dissipation, dS/dt entropy production per unit time, J_k X_k scalar generalized flows and forces, and $J_k X_j$ vector generalized flows and forces.

The left term is a "chemical term" that reflects the conjugation of scalar quantities, i.e., the reaction rates with the corresponding affinity values and scalar heat fluxes with scalar temperature changes. The first term relates the flows of matter with heat fluxes, which include the relations between dislocation flows by macroscopic and microscopic stresses, between heat fluxes and temperature gradients. To relate the reaction rate υ and affinity A at its small values, Heinicke G. (*Trlbochemistry*) suggests using the phenomenological equation of thermodynamics of irreversible processes

$$\upsilon = L_{i1} A + \Sigma^{n2} j_{=1} L_{1i} X_i, \quad (2.25)$$

where L_{i1} and L_{1i} are linear coefficients. In the stationary treatment mode, the second term of the equation is constant, so that the reaction rate will be limited mainly by the chemical stage of the transformation and will have the form

$$\upsilon = aA + b, \quad (2.26)$$

where a and b are constants.

The main factor for increasing the reactivity of mechanically activated solids is structural disturbances, which can be determined from the relative magnitude of microarrays of the crystal lattice $\Delta\alpha/\alpha$. Acceleration of processes can be calculated as

$$k = \alpha_2/\alpha_1 = \exp\left(E\Delta\alpha/\alpha V/(RT)\right), \tag{2.27}$$

where α_2 and α_1 denote mechanical activity, determined by broadening of X-ray lines.

Characteristics of X-ray diffraction of materials after mechanochemical treatment are used in other methods for calculating the nonequilibrium state from models of both thermodynamics [72] and macrokinetics [73] of processes of mechanochemical modification of various nanoscale systems. In these models, in addition to a general assessment of the change in the defect structure, a significant role is assigned to the surfaces of particles and the interphase surface. In [73, 74], a mathematical description of the macrokinetics of the chemical processes of interaction of the powder being ground with a gaseous medium consists of a joint examination of the heat balance, chemical transformation, grinding, dynamics of excess energy in the solid phase, determined by the intensity of the mechanical action and relaxation processes. Two channels of relaxation are taken into account: chemical transformations and normalization of the structure. The activation energy of these processes is determined by the excess energy of the system. The specific excess energy in the condensed phase φ is related to the specific excess energies in the reagent φ_r and in the product φ_α by the relation

$$\varphi = \nu\varphi_r + (1 - \nu)\,\varphi_\alpha, \tag{2.28}$$

where $\nu = V_o m_2 \rho_2 (1 - \alpha)/\rho_s$ the volume fraction of the reagent in the condensed phase.

Specific excess energy, in this or that phase, is considered [48] to be the ratio of excess energy to the volume of this phase. Its value is assumed to be proportional to the degree of structural imperfection of the material.

$$\varphi_r = A_r h_r, \quad \varphi_\alpha = A_\alpha h_\alpha, \tag{2.29}$$

where A_r, A_α are coefficients, and h_r, h_α the degree of structural imperfections in the reagent and the product, respectively. The values of h_r, h_α can be estimated from the relative broadening of X-ray diffraction lines.

The variation kinetics of φ_r and φ_α is estimated from the following dependencies:

For the reagent

$$d\varphi_r/dt = a_2\, W/V_T - \varphi_r m_o \exp\,[-(U - a_3\,\varphi_r)/RT] \qquad (2.30)$$

where m_o is the pre-exponential, U activation energy, a_2, a_3 coefficients, and $V_T = V_o m_2$ the volume of solid phase.

For the reaction product, the rate of change in excess energy will be

$$d\varphi_\alpha/dt = a_4\, W/V_T - \varphi\alpha n_o \exp\,[-(L - a_5\,\varphi_\alpha)/RT] + (j\varphi_r - \varphi_\alpha)/\alpha(1 + b\,\alpha) * d\alpha/dt \qquad (2.31)$$

The first term in Eqs. (2.30) and (2.31) indicates the change in the rate of excess energy due to grinding, the second term due to relaxation of the structure of matter, and the third in Eq. (2.31) due to the chemical reaction. The change in the total excess energy in the system is estimated as heat release, the rate of which, as a result of the change of φ, is defined as

$$I = -a_6 V_o d/dt\,[m_2\,\varphi], \qquad (2.32)$$

where a_6 is the coefficient.

A significant number of uncertain parameters make it difficult to use a macroscopic description of the process. For specific calculations, various simplified versions of evaluation of mechano-chemical transformations in inorganic and organic systems are adopted.

Questions

1. What is the sequence of processes and states in the mechanochemical treatment of solids?
2. How is the activation energy of the reactions carried out by the thermal path, i.e., due to the supply of heat?
3. How is the activation energy due to internal stresses in the crystal lattice created by various kinds of defects?
4. How is the excess free enthalpy of matter conditioned by structural defects not in thermodynamic equilibrium determined?

5. Describe the mechanochemical process, as a system of structural levels, realized in the process of mechanical pumping of energy.
6. Give a description of a solid body, like a dissipative system, capable of accumulating and dissipating the stored energy of elastic distortions during deformation.
7. What is the role of the surface of particles and the interphase surface in the macrokinetics of the chemical processes of interaction of the powder being ground with the gaseous medium?
8. Estimate the change in total excess energy in the system during mechanochemical transformations in inorganic and organic systems.

2.4 Features of Mechanochemistry of Organic Compounds and Systems of Inorganic and Organic Materials

A review of experimental and theoretical work, presented in the previous sections, gives a general idea of mechanochemical processes of changing the structure of solids and the physical and chemical phenomena accompanying them, with reference to both inorganic crystalline and amorphous and organic monomeric and polymeric compounds. Undoubtedly, depending on the structure, type and bonding force of elements in solid systems, certain phenomena will predominate in the process of grinding, activation and mechanochemical synthesis of powder materials.

Historically, priority in mechanochemistry belongs to natural minerals, which in complex are systems of crystalline compounds and inorganic polymers. However, mechanochemical phenomena were first theoretically substantiated in the study of high molecular organic materials [40]. It was found that under the action of various mechanical loads, the chemical bonds of polymer chains break up, i.e., mechanocracking takes place. It was precisely due to the study of mechanocracking that attention was paid to the emergence and the possibility of using highly active particles–free macroradicals formed during intensive mechanical action on polymer compositions–to initiate various chemical polymerization, copolymerization, and structuring processes at low temperatures [75]. Much later active long-life centers were registered on the surface of crystalline particles as a result of their grinding [17, 18]. Active particles in their variety are free radicals, free ions, radical ions, F, F′-centers (electrons, electron pairs trapped by the negative ion lattice vacancy), V centers (positive ion lattice vacancies), free electrons emitting when bodies disintegrated or contact with different electron densities. Both the above mentioned and other types of active centers most effectively and clearly manifest themselves in the mechanochemical processes of polymers.

It is on polymeric high-molecular systems the energy of the mechanical action at the time of its application is found to be localized not only on the main valence bonds in the chain, but to excite the whole system of bonds in a section the dimensions of which are determined by the conditions of mechanical action [75]. There is a kind of shake-up of the structure. The energy of the impact is consumed for increasing the interatomic distances, weakening of the bonds in the substituents, in the side groups, etc.

In such zones, there is a radical change in the chemical structure of the initial material, right up to mechanocracking. It was found that the number of free radicals can exceed more than 10 times the number of chain fragments, i.e., a polyradical state occurs within the same chain, and mechanocracking occurs over localized bonds, in particular, in quaternary carbon atoms. Moreover, this process proceeds according to the chain mechanism [30]. Ultimately, the emergence of a polyradical state characterizes the high activity of the material being processed. The development of the chain process occurs in different ways, depending on the type of the initial material.

Of the many options considered by N. K. Baramboym [40], of particular interest is the one that is characteristic for grinding of polyethylene, polystyrene and other similar polymers—this is a spontaneous isomerization of macroradicals with formation of secondary macroradicals. In this case, the unpaired electron formed at the extreme atom at the point of discontinuity during mechanocracking is not necessarily localized for a given atom. Due to conjugation, the electron delocalizes and migrates to the chain node, causing isomerization of the macroradical [76].

Along with the existing active radical fragments, a large number of inactive molecules can exist in the system, which facilitate chain transfer or initiated destruction. Free radicals formed during mechanocracking can cause initiated cracking of the neighboring chains with formation of a new free macroradical:

$$R^* + R' - R'' \rightarrow R - R' + R''. \quad (2.33)$$

The breaking of the chain and stabilization of radicals can be caused by various processes, in particular, disproportionation of macroradicals. But the most common cause of chain breakage is interaction of macroradicals with different acceptors, always present in technical polymers and in the environment (for example, O_2). Acceptance can also be associated with incomplete stabilization with the formation of end groups even of a free radical nature, but inactive so that they do not cause the development of chain processes and gradually finally get stabilized within the molecular rearrangement or other transformations of a non-chain nature. Breakage of the chain can also occur on the walls of the apparatus in which the mechanochemical treatment of the polymers is carried out. In this case, formation of organometallic derivatives is possible due to splitting atoms of the metal out by free radicals, which is most likely for peroxide radicals formed as a result of addition of oxygen to free macroradicals obtained in the process of mechanocracking.

In the course of mechanical action on the polymer until the moment of bond breaking, it can contain a continuous series of deformation states. These states are characterized by an increase in valence angles and interatomic distances. The weaker the bond, the lower the absorption frequency in the region of IR spectra, i.e., the absorption maximum characteristic for a given type of bonds decreases somewhat and shifts toward lower frequencies. This displacement is proportional to the applied voltage and leads to the emergence of a tense, active state. A gradual increase in interatomic distances up to the values corresponding to the rupture also reduces stability of the bonds to external influences. The process of mechanical destruction (mechanocracking) is followed by the stage of formation of new polymeric structures, most actively occurring in the presence of various activators. Thus, modified polymers are obtained with a purposeful property control [77].

As was shown earlier, in the case of mechanical dispersion of crystalline solids, depending on the nature of the substance and the type of crystal lattice, active centers of various types appear on the newly formed surfaces, which can play the role of electron "traps." Thus, during the crushing of ionic crystals, cation and anion vacancies can appear, which initiate the polymerization of monomers when crystals of ionic salts disperse. If the lattice is formed by covalent bonds, for example, in oxides (SiO_2), dispersion leads to the appearance of active centers on the atoms of the element, of which the oxide is composed, and on the atoms of oxygen [30].

Thus, mechanical dispersion of inorganic materials in the presence of monomers can initiate polymerization with grafting of particles of dispersible bodies on the surface or homopolymerization during the chain transfer. Thus, in particular, organomineral fillers chemically bound to the polymer are obtained [77]. Dispersion in the presence of polymers leads to similar results, due to recombination of macroradicals of the destructible polymer with active centers on the surface of particles. In most cases, the dispersible particles become polyradical and polymer fragments are grafted throughout the surface, forming a grafted polymer layer.

Initiation of mechanochemical processes in polymers with their simultaneous treatment with crystalline substances is associated not only with active centers on the surface of crystal destruction, but also with electric charges mosaically located on newly formed surfaces, with electron emission, with gas-discharge and other phenomena. Direct evidence of the presence of electron emission and its initiating effect was obtained by mechano-dispersion of quartz [40]. The authors of ref. [78] found the formation of chemical bonds between decomposed fresh surfaces and organic compounds during the crushing of quartz sand in the presence of butanol, styrene, vinyl chloride, methyl methacrylate and acrylonitrile and other organic substances. In all cases, formation of chemical bonds between the dispersed mass and organic additives was observed. For example [78], for methyl chloride,

$$\equiv Si - O - Si\equiv + CH_3Cl \rightarrow \equiv Si{-}OCH_3 + \equiv Si{-}Cl, \tag{2.34}$$

in the case of methyl methacrylate,

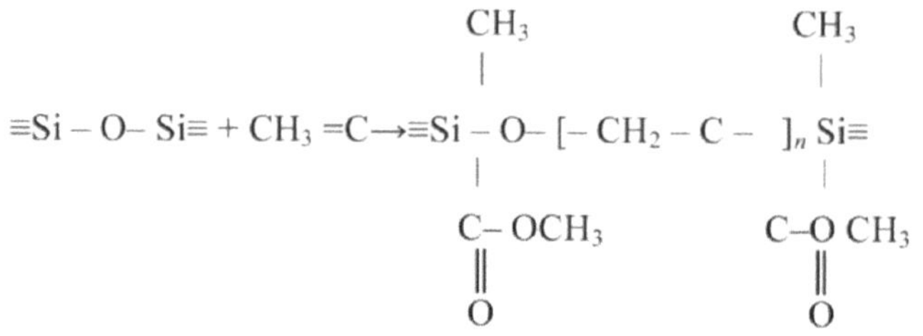

(2.35)

The results of grinding SiO_2 in the presence of alcohols also indicate their binding. It is assumed that in this case the mechanism of both the free radical process and the activated alcohololysis of Si–O–Si bonds is possible. The most widely used dispersed crystalline particles with a modified surface with organic compounds were obtained as fillers for composite polymeric materials, providing strength, thermal stability, electrical conductivity and chemical resistance [79, 80].

Questions

1. What is mechanocracking and the possibility of using free macro-radicals, formed during intensive mechanical action, for polymerization of systems at low temperatures?
2. What is the specificity of energy distribution upon impact in the structure change in the impact zone for organic and crystalline systems?
3. Describe the development of the chain process of structural changes in matter under mechanical action and the emergence of a polyradical state of the system.
4. What free radicals formed during mechanocracking, and initiation of polymer degradation?
5. Describe the process of polymer modification with purposeful regulation of system properties.
6. How does the dispersion of inorganic materials occur in the presence of monomers and production of organomineral fillers?

7. What are the electrophysical processes under mechanical action and a change in the state of matter on the example of mechanically dispersed quartz?

References

1. Sidenko P. M. *Crushing in the Chemical Industry*. Chemistry, Moscow, 1977. 368 p. (In Russian).
2. Guyo R. *The Problem of Grinding Materials and Its Development*. Stroiizdat, Moscow, 1964. 111 p. (In Russian).
3. Shermergor T. D. *The Theory of Elasticity of Microinhomogeneous Media*. Nauka, Moscow, 1977. 400 p. (In Russian).
4. Rebinder P. A. Physicochemical studies of the processes of deformation of solids. In: *On the Nature of the Friction of Solids*. Science and Technology, Minsk, 1971. pp. 8–20 (In Russian).
5. Golosov S. I., Molchanov V. I. Centrifugal planetary mill, its technical capabilities and application in the practice of geological research. In: *Physico-Chemical Changes in Minerals in the Process of Ultrafine Grinding*. Science, Novosibirsk, 1966. pp. 5–15 (In Russian).
6. Bacofen B. *Deformation Processes*. Metallurgy, Moscow, 1977. 325 p.
7. Shemyakin Ye I. About free destruction of solids. *Rep. Acad. Sci. USSR*, 1991. vol. 316, no. 6. pp. 1371–1373 (In Russian).
8. Bushuev L. P. On the design and application of planetary mills. *Izv. VUZ. Mountain Magaz.*, 1960. no. 32. pp. 17–20 (In Russian).
9. Boldyrev V. V., Avvakumov E. G., Logvinenko A. T. Efficiency of grinding devices for mechanical activation of solids. In: *Enrichment of Minerals*. Science, Novosibirsk, 1977. pp. 3–10 (In Russian).
10. Kolosov A. S. Some problems of modeling and estimating the energy efficiency of crushing processes for solids, *Izv. CO ANSSSR. Ser. Chem. Sci.*, 1985. vol. 2, no. 1. pp. 26–29 (In Russian).
11. Greene M. (ed.). *Surface Properties of Solids*, Mir, Moscow, 1972. 432 p. (In Russian).
12. Yakovlev V. M., Pertsov N. V., Alchagirov B. B. On the possibility of a quantitative description of the surface energy of metals in the solid state by their properties. *Colloid J.*, 1999. vol. 61, no. 5. pp. 705–708 (In Russian).
13. Korshunov A. B. Analytical method for determining the parameters of a fine crystalline structure by broadening X-ray lines. *Factory Lab. Diagn. Mater.*, 2004. vol. 70, no. 2. pp. 27–32 (In Russian).

14. Uybo L. Ya., Pae A. Ya. *Mechanical Activation of Chemical Reactions of Solids. Active Surface of Solids*. Publ. House VINITI, Moscow, 1976. pp. 220–229 (In Russian).
15. Kochegarov G. G. Technological aspects of mechanical activation of solids. In: *Abstracts of the All-Union Conference "Mechanochemistry of Inorganic Substances."* Novosibirsk, 1982. pp. 98–99 (In Russian).
16. Molchanov V. I., Yusupov T. S. *Physical and Chemical Properties of Finely Dispersed Minerals*. Nedra, Moscow, 1981. 170 p. (In Russian).
17. Avvakumov Ye. G. *Mechanical Methods of Activation of Chemical Processes*. Science, Sib. Branch, Novosibirsk, 1986. 300 p. (In Russian).
18. Heinicke G. *Tribochemistry*. Akademie Verlag, Berlin, 1984. 495 p.
19. Khodakov G. S. *Physics of Grinding*. Nauka, Moscow, 1972. 307 p. (In Russian).
20. Molchanov V. I., Selezneva O. G., Zhirnov E. N. *Activation of Minerals*. Nedra, Moscow, 1988. 208 p. (In Russian).
21. Chaikina M. V. *Mechanochemistry of Natural and Synthetic Apatites*. Izdatel'stvo SB RAS, Branch "Geo", Novosibirsk, 2002. 223 p. (In Russian).
22. Hint J. A. On the main problems of mechanical activation. *Materials of the V All-Union Symposium on Mechanoemission and Mechanochemistry of Solids*, Tallinn, October 1977. pp. 12–23 (In Russian).
23. Chaikina M. V., Kryukova G. N., Tatarintseva M. I. Simulation of the processes of mechanical activation and grinding by the method of normal tangential indentation of fluoroapitate monocrystals. *Izv. CO ANSSSR. Ser. Chem. Sci.* 1989. no. 5. pp. 134–144 (In Russian).
24. Thiessen K. P., Meyer G., Heinike K. *Grundlagen der Tribochemie.* Berlin, 1967. 194 p.
25. Butyagin P. Yu. Forced reactions in inorganic and organic chemistry. *Colloid J.*, 1999. vol. 61, no. 5. pp. 581–589 (In Russian).
26. Tipikin D. S., Lazarev G. G., Lebedev Ya. S. Mechanochemical generation of stable radical pairs. *J. Phys. Chem.*, 1993. vol. 67, no. 1. pp. 176–179 (In Russian).
27. Panin V. E., Egorushkin V. E., Khan Yu. A., Elsukov T. F. Atom vacant state in crystals. *Izv. High Schools. Phys.*, 1982. no. 2. pp. 5–28 (In Russian).
28. Bovenko V. N. Synergetic Effects in the Plastic Deformation and Destruction of Crystals. *Izv. Akad. Nauk SSSR. Ser. fiz.*, 1986. no. 3. pp. 509–512 (In Russian).

29. Vinogradova M. B., Rudenko O. V., Sukhorukova A. P. *Theory of Waves*. Nauka, Moscow, 1990. 432 p. (In Russian).

30. Kuznetsov V. A., Lipson A. G., Sakov D. M. On the Limit of grinding crystals. *J. Phys. Chem.*, 1993. vol. 67, no. 4. pp. 782–786 (In Russian).

31. Zanin A. M., Kiryukhin D. P., Barelko V. V., et al., On the pulsating nature of the distribution of low-temperature chemical transformations of irradiated solid systems initiated by brittle fracture, *Dokl.*, 1983. vol. 268, no. 35. pp. 1146–1149 (In Russian).

32. Alshits V. I., Indenbom V. L. Dynamic braking of dislocations. Dynamics of dislocations. *Science. Dumka, Kiev*, 1975. pp. 274–323 (In Russian).

33. Pushkarev D. I. *Defects in Crystals. Method of Quasiparticles in the Quantum Theory of Defects*. Nauka, Moscow, 1993. 135 p. (In Russian).

34. Bovenko V. N. Relation of autoacoustic emission to the pre-destructive state of a crystal. *Dokl. AN SSSR.*, 1983. vol. 271, no. 5. pp. 1086–1090 (In Russian).

35. Zel'dovich Ya. B., Ryizer Yu. P. *Physics of Shock Waves and High-Temperature Hydrodynamic Phenomena*. Nauka, Moscow, 1966. 640 p. (In Russian).

36. Dobrolyubov A. I. *Wave Transfer of Matter*. Belarusian Science, Minsk, 1996. 304 p. (In Russian).

37. Butyagin P. Yu. Disordering of structure and mechanochemical reactions in solids. *Uspekhi Khimii.*, 1984. vol. LIII, vol. 11. pp. 1769–1789 (In Russian).

38. Deryagin B. V., Krotova N. A., Smilga V. P. *Adhesion of Solids*. Nauka, Moscow, 1973. 280 p. (In Russian).

39. Aman S., Tomas J. *Mechanoluminescence of Quartz Particles in Stirred Media Mill/Fourth International Conference on Mechanochemistry and Mechanical Alloying*. Braunschweig, Germany, 2003. pp. 56–57.

40. Baramboym N. K. *Mechanochemistry of Macromolecular Compounds*. Chemistry, Moscow, 1978. 384 p. (In Russian).

41. Polukhina L. M., Khrustalev Yu. A. Electrical phenomena during mechanical action on a mixture of polymers. *J. Phys. Chem.*, 1993. vol. 67, no. 4. pp. 795–797 (In Russian).

42. Berlin Yu. A., Bershenko S. I., Zhorin V. A., Yenikolopyan I. S. Stimulated by pressure injection of electrons into thin layers of dielectrics. *Rep. Acad. Sci. USSR.*, 1960. vol. 132, no. 5. pp. 1140–1143 (In Russian).

43. Kochurova N. N., Rusanov A. I., Myrzakhmetova N. O. The Jones-Ray effect and surface electrification. *Rep. Acad. Sci. USSR.*, 1991. vol. 31, no. 6. pp. 1425–1427 (In Russian).
44. Gutman E. M. *Mechanochemistry of Metals and Corrosion Protection.* Metallurgy, Moscow, 1981. 299 p. (In Russian).
45. Koloberdin V. I. Influence of relaxation of internal stresses on the material grinding process. *News Univ. Chem. Chem. Technol.*, 2002. vol. 45, no. 2, pp. 67–68 (In Russian).
46. Alaeva S. S., Bobkov S. P., Talanova V. A. Investigation of the distribution of energy accumulated by the body during mechanical activation within the framework of a spatial model. *Izvestiya VUZ. Ser. Chem. Chem. Technol.*, 1999. vol. 42, no. 2. pp. 129–130 (In Russian).
47. Hamanov M. Kh. The mechanism of the collective motion of dislocations during plastic deformation of crystals. *Phys. Chem. Mater. Proc.*, 1998. no. 4. p. 61–66 (In Russian).
48. Bridgman P. V. *The Study of Large Plastic Deformation and Rupture.* IL, Moscow, 1955. 444 p. (In Russian).
49. A. Dremin, O. Breusov Processes occurring in solids under the action of strong shock waves. *Uspekhi khimii.*, 1968. vol. 37, no. 5. p. 898–916 (In Russian).
50. Vasiliev L. S., Lomaeva S. F. The mechanism of saturation of nanocrystalline powders with interstitial impurities during mechanical dispersion. *Colloid J.*, 2003. vol. 65, no. 5. pp. 697–705 (In Russian).
51. Butyagin P. Yu., Berlin A. A. On the formation of macroradicals during mechanochemical destruction of vitrified polymers. *High Mol. Compounds.*, 1959. vol. 1. pp. 865–869 (In Russian).
52. Ahmed-Zade K. A., Baptismansky V. V. On the mechanism of mechanodistruction of silicone polymers. *Macromol. Compounds.*, 1972. vol. 14. pp. 1360–1364 (In Russian).
53. Butyagin P. Yu., Bystrikov A. V. On the initiation of chemical reactions in the destruction of solids. *Mat. 5th All-Union Symposium on Mechanical Emission and Mechanics of Solids.* Tallinn, 1977. Part 1. pp. 63–78 (In Russian).
54. Rusanov A. I. Thermal effects in mechanochemistry. *J. Gen. Chem.*, 2002. vol. 72, no. 3. pp. 353–372 (In Russian).
55. Bowden F. P., Taybor L. *Friction and Lubrication of Solids.* Mashgiz, Moscow, 1960. 2002 p. (In Russian).

56. Urakaev F. Kh. Theoretical estimation of pressure and temperature pulses at the contact of friction particles in dispersing apparatus. *Izv. SOAN USSR. Ser. Chem. Sci.*, 1978. no. 7, vol. 3. pp. 5–10 (In Russian).

57. Boldyrev V. V. On the kinetic factors that determine the specificity of mechanochemical reactions in inorganic systems. *Kinet. Catal.*, 1972. vol. 13, no. 6. pp. 1411–1427 (In Russian).

58. Panovko Ya. G. *Introduction to the Theory of Mechanical Shock*. Science, Moscow, 1977. p. 244 (In Russian).

59. Dinnik A. N. Impact and compression of elastic bodies. *Kiev: USSR Acad. Sci.*, 1952. vol. 2. pp. 25–81 (In Russian).

60. Rusanov A. I. *Phase Equilibria and Surface Phenomena*. Chemistry, Leningrad, 1967. 270 p. (In Russian).

61. Volkov V. M., Shorshorov M. Kh., Morokhov I. D. *Theoretical Examination of the Limit of the Existence of the Crystalline and Amorphous Phases During Dispersion. Physicochemistry of Ultra-dispersed Systems*. Nauka, Moscow, 1987. pp. 41–45 (In Russian).

62. Zaitsev L. A. Influence of a layer of hot electrons near the surface on the thermodynamic energy of semibounded solids. *Chem. Phys.*, 2002. vol. 21, no. 2. pp. 15–25 (In Russian).

63. Eremeev V. S. *Diffusion and Tension*. Energoatomizdat, Moscow, 1984. 184 p. (In Russian).

64. Yezhovsky Yu. K. Chemical assembly of surface nanostructures. *Chem. Phys.*, 2005. vol. 24, no. 4. pp. 36–57 (In Russian).

65. Glazov V. M., Pavlova L. M. *Chemical Thermodynamics and Phase Transformations*. Metallurgy, Moscow, 1988. 560 p. (In Russian).

66. Ivanov E. Yu. Solid-phase reactions during mechanical alloying of metals. In: *Mechanochemical Synthesis in Inorganic Chemistry*. Science, Novosibirsk. Sib. department., 1991. pp. 190–204 (In Russian).

67. Carey J. *Order and Disorder in the Structure of Matter*. Mir, Moscow, 1985. 228 p. (In Russian).

68. Shapkin V. A. Role of the intermediate amorphous phase in the reactions of solid-state mechanochemical synthesis. In: *Mechanochemical Synthesis in Inorganic Chemistry*. Science, Novosibirsk. Sib. Department, 1991. pp. 237–242 (In Russian).

69. Barakhtin B. K., Vladimirov V. I., Ivanov S. A., Ovidko I. A., Romanov A. E. Periodicity of structural changes during rotational plastic deformation, Fiz. meth. and metal expert. 1987. vol. 63, no. 6. pp. 1185–1191 (In Russian).

70. Panin V. E., Likhachev V. A., Grintsev Yu. V. *Structural Levels of Deformation of Solids*. Science, Novosibirsk, 1985. 243 p. (In Russian).
71. Prigogine I., Stengers I. *Order from Chaos*. Progress, Moscow, 1986. 432 p. (In Russian).
72. Dorofeev G. A., Yelsukov E. P. Thermodynamic modeling of mechanical alloying reactions in the Fe-Sn system. *Inorg. Mater.*, 2000. vol. 36, no. 12. pp. 1460–1466 (In Russian).
73. Smolyakov V. K., Lapshin O. V., Maksimov Yu. M. Nonisothermal interaction of powders with an active gaseous medium during grinding. *Phys. Combustion Explosion*, 2003. vol. 39, no. 6. pp. 77–84 (In Russian).
74. Smolyakov V. K., Lapshin O. V. Modelling of mechanosynthesis macrokinetics in "solid-gas" system. *International Conference "Mechanochemical Synthesis and Sintering."* Novosibirsk, Russia, 2004. pp. 165–172.
75. Radtsik V. A. Mechanochemistry of polymers through the eyes of the EPR method (Review). *Chem Phys.*, 2004. vol. 23, no. 23. pp. 70–100 (In Russian).
76. Simionsku K., Oprea K. *Mechanochemistry of High-Molecular Compounds*. Mir, Moscow, 1970. 354 p. (In Russian).
77. Kochnev A. M., Galibeev S. S. Modification of the structure and properties of polymers. Izvestiya VUZov. *Chem. Chem. Technol.*, 2003. vol. 46, no. 4. pp. 3–10 (In Russian).
78. Ivanchev S. S., Dmitrenko A. V. Polymerization filling by the method of radical polymerization as a method of obtaining composite materials. *Uspekhi Khimii.*, 1982. vol. LI, no. 7. pp. 1178–1200 (In Russian).
79. Berlin A. A., Wolfson S. A., Enikolopyan N. S. *Kinetics of Polymerization Processes*. Chemistry, Moscow, 1978. 240 p. (In Russian).
80. Berlin A. A., Wolfson S. A., Oshmyan V. G., Enikolopyan N. S. *Principles of Creating Composite Polymer Materials*. Chemistry, Moscow, 1990. 237 p. (In Russian).

Chapter 3

Principles of Mechanochemical Activation in Technological Processes

Mechanochemical processing is effectively used in various technological processes. In particular, in the technology of processing minerals, which largely determine both the completeness of extraction of minerals and elements from rocks, and the kinetics of various heterogeneous processes involving solid substances in a finely disperse state. With an increase in dispersion, leading to an increase in the activity of a substance, it is possible to intensify many technological processes. A significant role of mechanochemical processing is given to the creation of the production technology of various composite ceramic materials from low-grade raw materials and industrial waste. The quality of the product is largely dependent on the technological conditions of the mechanochemical processing of raw materials.

3.1 The Influence of the Form of Grinding Bodies on the Parameters of Mechanochemical Treatment

As a rule, in industrial conditions at the stage of preliminary preparation of raw materials, conventional ball rotary mills are use, in which ball, cylpebs [1], or mixed loads are used as grinding

Mechanochemical Synthesis of Composite Materials
Zulkhair A. Mansurov, Nina N. Mofa, Tlek A. Ketegenov, and Bakhtiyar S. Sadykov

ISBN 978-981-4800-88-4 (Hardcover), 978-1-003-12081-0 (eBook)
www.jennystanford.com

bodies, which provide the necessary level of material dispersion. In some cases, the degree of material activation in such mills is commensurate with the activation in laboratory vibratory and planetary centrifugal mills, and is associated both with significant drum sizes of industrial mills and with changes in the shape of grinding bodies.

In the process of wear, cilpebs acquire an ellipsoidal shape (the figure of rotation of an ellipse around a large a or small b semiaxis). Therefore, it was important to assess the impact of changes in the shape of grinding bodies on the parameters of the machining of materials, in particular quartz [2]. Using the example of transformation of a ball into an ellipsoid of rotation, the effect of changing the shape of grinding bodies on the parameters of shock–friction activation of solids in mechanochemical reactors was estimated. Numerical calculations were carried out for treating quartz with cylpebs in an industrial rotating drum mill (Figs. 3.1 and 3.2).

For simulation of mechanical activation, the existing theories of the shock interaction of solids, described in [3–5], were used. They are divided into three groups: the classical theory of Newton [5, 6] with the recovery coefficient $0 \leq \gamma \leq 1$; wave theories used primarily to study the collision of beams and rods; Hertz theory [5–7], applicable to the description of shock processes with arbitrary values of the curvature $L_k = 1/R_k$ (R_k are the radii of curvature) of solids at the point of contact.

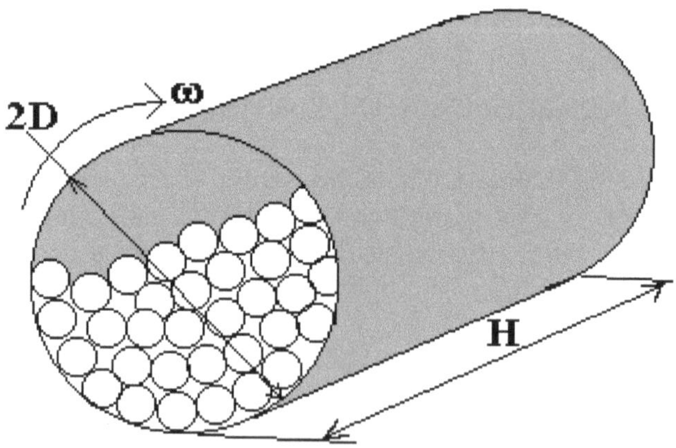

Figure 3.1 Diagram of a conventional rotating ball mill.

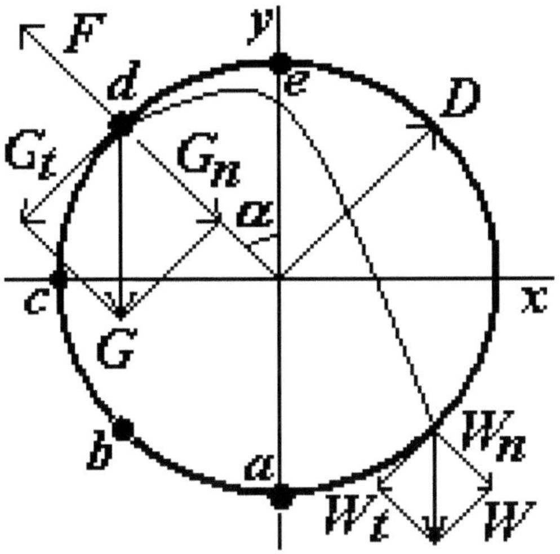

Figure 3.2 Scheme for calculating the conditions for the separation of the grinding bodies from the mill wall and the velocity vector **W** of their fall along a parabolic curve.

Modern trends in the synthesis of these theories are carried out on the basis of the Hertz theory in [6] and the Rauss "ξ-hypothesis" [5] for noncentral (oblique) collisions, according to which the relationship between the values of tangent (I_t) and normal (I_n) pulses upon impact is formulated like the law of friction: $I_t \leq \xi I_n$ (the inequality sign refers to the cases when I_t is so small that no body slippage occurs, and an equals sign must be taken when slipping), where ξ is the dynamic coefficient of friction. In Hertz nonlinear elastic theory, the following relation is assumed between the force f_{ij} and the deformation $\varepsilon_{ij} = \varepsilon_i + \varepsilon_j$ of the colliding bodies i and j: $f_{ij} = B\varepsilon_{ij}3/2$. The coefficient B depends on the properties of materials of bodies and L_k at the point of contact. The deviations in the calculations for this theory increase with an increase in the normal velocity $\mathbf{W}_n$ of a collision due to the appearance of plastic deformations and losses due to internal friction. More accurate values (up to $\mathbf{W}_n \sim 50$ m/s) are given by a semi-empirical nonlinear elastoplastic model, in which $f_{ij} = b\varepsilon_{ij}^n$ is adopted, where b and n are constants determined from experiment. We present the results of calculations in the framework of the Hertz theory, which is valid in some cases up to $\mathbf{W}_n \sim 10$ m/s [2, 6–9], since for ball mills the speed of the relative impact of grinding bodies is in the indicated value [10].

The most significant parameter for mechanical activation is the relative speed **W** of interaction of grinding bodies. However, for a conventional rotating drum mill (CRDM), the correct calculation of **W** was not found. This can be done using the example of a specific industrial mill with a diameter of $2D$ = 2.8 m and a length of H = 14 m (Fig. 3.2).

Let n [s^{-1}] be the rotational speed of the CRDM; g = 9.8 m/s^2 is acceleration of gravity; $G = mg$ is gravity; F is the centrifugal force of inertia; $v = D\omega$ is the circumferential velocity, $|\omega| = 2\pi n$; α is the angle from point d to point e. Then, the conditions for separation of grinding media $G_n = G \cos \alpha = F = m\,|\omega|^2 D = m\,|v|^2/D = 4\pi^2 n^2$ mD; critical conditions (α = 0, grinding bodies do not perform useful work) $\omega_c = (g/D)^{0.5}$, $v_c = (gD)^{0.5}$, $n_c = (g/D)^{0.5}/2\pi$; optimal conditions for dry grinding (α = 54°40′ or cos 54°40′ = 0.5784 determines the maximum trajectory of the grinding media) $v_o \approx 2.8$ m/s, $n_o \approx 0.33$ s^{-1}. Knowing the optimal parameters, you can calculate the coordinates of the point of intersection of the parabola and the circle: the height of the fall of the grinding body h, the direction and speed at the moment of impact $|\mathbf{W}| = (2gh)^{0.5}$.

If we place the origin of coordinates at the point of maximum lift height (at the top of the parabola), then the equation of the parabola is $y = kx^2$, and the circles are: $(x - x_0)^2 + (y - y_0)^2 = D^2$, where (x_0, y_0) are the coordinates of the center of the circle. Solving a system of two equations, we find the intersection point of a parabola and a circle. Numerically, for the CRDM we find the following:

$$h \approx 2.4 \text{ м}; |\mathbf{W}| \approx 6.9 \text{ m/s} \tag{3.1}$$

$$\mathbf{W}_t(\perp\mathbf{D}) = \mathbf{W} \cos\varphi = 2.02 \text{ m/s}, \mathbf{W}_n\,(||\mathbf{D}) = \mathbf{W} \sin\varphi = 6.62 \text{ m/s}, \tag{3.2}$$

where φ = 73° is the angle between **W** and $\mathbf{W}_t$. We see that the conditions of impact rather than abrasion are preferable for CRDM.

For a numerical assessment of the influence of the shape of the grinding bodies on the process of mechanical activation in the CRDM (Fig. 3.3): (1) the volume of the ellipsoid of rotation is equal to the volume of the ball; (2) the radius of the ball is $R = 0.05$ m << $D = 1.4$ m; (3) the ellipsoid semi-axes are related by $a = 5b$ and $R^3 = ab^2$ and vice versa. Then it turns out that the dimensions of ellipsoids equal to a large ball of that model cilpebs are: elongated (draw) $a = 14.62$ cm, $b = 2.924$ cm; flattened (flattening) $a = 1.71$ cm, $b = 8.55$ cm. The ball curvature is constant $L_b = 1/R = 0.2$ cm^{-1}. The average curvature $\langle L_e(x) \rangle$ in the vicinity of any point on the surface of the ellipsoid (it is recommended to use the geometric mean [5]) is a combination of the curvature of the ellipse $L_e(x)$ and the circles $L_b(x)$ perpendicular to the x axis: $\langle L_e(x) \rangle = [L_e(x) \times L_b(x)]^{0.5}$.

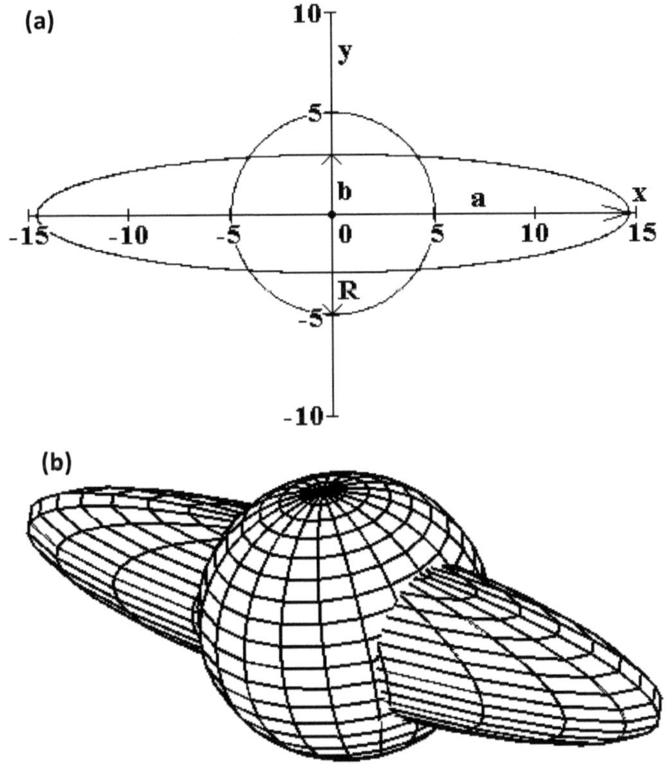

Figure 3.3 Transformation of a ball into an ellipsoid: (a) in a plane; (b) in volume.

If the ellipse is given by the equation $x^2/a^2 + y^2/b^2 = 1$, the curvature is given by the formula $L_e(x) = a^4b^4/(b^4x^2 + a^4y^2)$ 1.5, where $y = b(1 - x^2/a^2)^{0.5}$. The curvature of the ellipse at the points of intersection with the x axis ($y = 0$) is a/b^2, at the points of intersection with the y axis ($x = 0$) it is b/a^2. The curvature of the circles is defined by the formula $L_b(x) = 1/y = (1 - x^2/a^2)^{0.5}/b$. The average curvature of the ellipsoid at the points of intersection with the x axis is also equal to a/b^2.

Using the formulas for $L_e(x)$ and $L_b(x)$, for both types of ellipsoids (figures of rotation of ellipses) we will obtain:

$$<L_e(x)> = |\{a^4b^4/[b^4x^2 + a^4b^2(1 - x^2/a^2)]^{1.5}\} \times [(1 - x^2/a^2)^{0.5}/b]|^{0.5} \quad (3.3)$$

According to the condition $R << D$, the curvature of the wall of the mill $L_v = 1/D$ is small compared to the curvature of the grinding bodies $L = 1/R$. Therefore, when calculating the parameters of the impact of the grinding bodies, the wall of the mill is modeled by a plane with zero curvature.

Hertz's collision theory for bodies of arbitrary shape was numerically and experimentally tested only for the central collision of a ball and a ball with a plane [7]. In the application of the design formulas for such interactions to the collision of equal-sized balls of rotation ellipsoids with the flat wall of the mill drum, it is necessary to make changes. Each point on the surface of ellipsoids in this case can only be simulated by a ball with a radius $R_e = 1/<L_e(x)>$. When simulating the collision of an ellipsoid surface with an infinite number of balls with radii R_e, dimensionless similarity criteria must be introduced into the corresponding equations of collision of a ball with a flat wall $k_n = [R \times <L_e(x)>]^n = (R \times L)^n$. The change in the mass of such balls during collisions will be described by the criterion $m = [R \times <L_e(x)>]^3 = (R \times L)^3$, therefore, the total criterion is defined as mk_n. Now you can determine the t–s–T conditions on the shock–friction contact of the grinding body with the curvature L and the flat layer of quartz particles on the surface of the mill wall, taking into account the generalized criterion mk_n:

- Interaction time $t(L) \approx (R \times L)^2$ 3, 1 $L^{-1} \rho^{0.4}(\theta + \theta_1)^{0.4} \mathbf{W}_n^{-0.2} \sim L$, as for a ball $t \sim R$; therefore, $mk_n = (R \times L)^3(R \times L)^{-1} = (R \times L)^2$.

- Contact area $s\ (L) \approx (R \times L)\ 3.1\ L^{-2}\ [\rho(\theta + \theta_1)\mathbf{W}_n^2]^{0,4} \sim L^{-1}$, since for the ball $s \sim R^2$; consequently, $mk_n = (R \times L)^3(R \times L)^{-2} = (R \times L)$.
- The maximum value of the temperature pulse at the contact of the grinding body and the MA of quartz particles is $\Delta T(0, t) \approx (R \times L)^{-1}\ 2q\ (cc_1\lambda\lambda_1\rho\rho_1)^{0.25}\ t^{0.5}\ iErfc\ [0] \sim L^{-0.5}$ is a derivative only from the collision time t with $m = 1$ and $k_n = (R \times L)^{-1}$.

Consequently, $\Delta T(L) \sim (R \times L)^{-1}\ t^{0,5} \sim (R \times L)^{-1}\ L^{0,5} \sim L^{-0,5}$, since the density of the heat source $2q = \xi <P> W_t$ does not depend on the curvature of the grinding bodies.

With an increase in the curvature of the L surfaces at the point of contact, only the collision time $t(L)$ increases linearly, since the distance traveled by the compression wave increases. The hyperbolic $s(L)$ and quasi-hyperbolic $\Delta T(L)$ functions decrease with increasing L and are antibat with respect to the function $t\ (L)$.

Density ρ, compliance θ, Poisson's v, Young's modulus E, dynamic coefficient of friction ξ, specific heat capacity c and thermal conductivity λ for estimating t–s–T conditions for the interaction of steel grinding bodies (no index) and quartz particles (index 1) at 2]: $\rho = 7.86$ and $\rho_1 = 2.59$ g/cm^3; $\theta = 4(1 - v^2)/E = 1.65 \times 10^{-12}$ and $\theta_1 = 4(1 - v_1^2)/E_1 = 3.95 \times 10^{-12}$ cm^2/din; $v = 0.85$ and $v_1 = 0.302$; $E = 2.23 \times 10^{12}$ and $E_1 = 0.983 \times 10^{12}$ din/cm^2; $\xi \approx 0.65$ for steel friction on quartz; $iErfc\ [0] = 0.642$; $c = 66.7 \times 10^5$ and $c_1 = 125 \times 10^5$ erg/g·K; $\lambda = 75.3 \times 10^5$ and $\lambda_1 = 7.7 \times 10^5$ erg/cm·K·s.

Figures 3.4–3.6 show the results of the calculations performed for the t–s–T conditions in a CRDM for selected ellipsoid sizes.

In conclusion, it should be noted that the surface area of the ball is $S_b = 4\pi R^2 = 314$ cm^2, while the surface areas of the ellipsoids are $S_e = 2\pi\ [b^2 + (ab \arcsin \varepsilon)/\varepsilon]$, where $\varepsilon = [1 - (b^2/a)]^{0.5}$ will be as follows: draw (draw) $S_{ed} = 432$ cm^2; flattened (flattening) $S_{ed} = 503$ cm^2.

The curvature of the ellipsoids becomes less than the curvature of the ball 0.2 cm^{-1}:

At $x = 12$ cm for an elongated ellipsoid, which corresponds to the surface area of the ellipsoid 387 cm^2, the main working area used for mechanical activation is in the interval of the elongated ellipsoid surface curvature 0.07–0.2 and in the interval of curvature 0.2–1.71, only near-focal surfaces of the ellipsoid with an area of

45 cm^2 are used, which provide conditions smaller than for the ball s–T at the shock–friction contact (Figs. 3.5, 3.6) for longer collision times t (Fig. 3.5), but large t–P conditions at the shock-friction contact. Thus, with this ratio of ellipsoid semiaxes, cylpebs become more effective when used in a conventional rotating drum mill (CRDM) than the equivalent ball load.

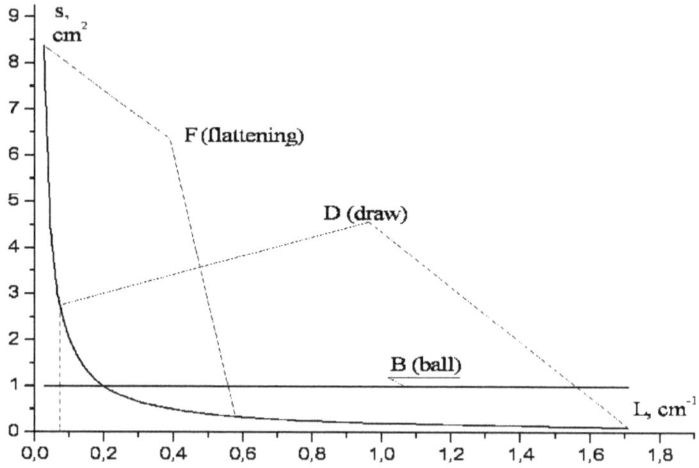

Figure 3.4 The dependence of the curvature of the surface of the grinding bodies on the radius of curvature.

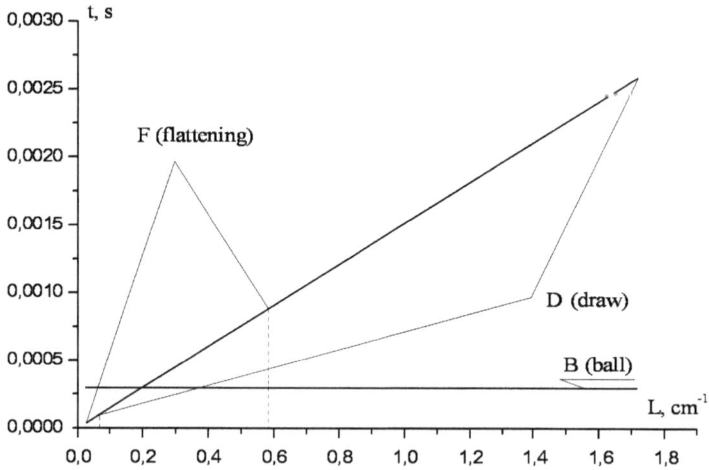

Figure 3.5 Dependence of contact time on the surface curvature of grinding media.

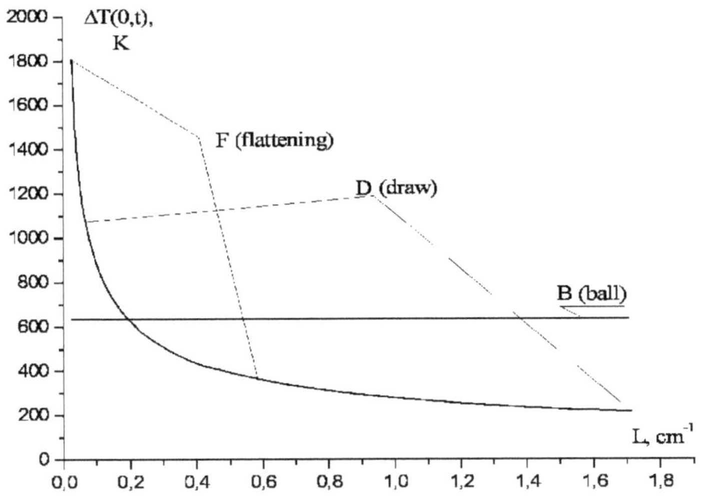

Figure 3.6 The calculated temperature dependence at the shock–friction contact on the curvature of the surface of the grinding bodies.

At x = 0.614 cm for a flattened ellipsoid, which corresponds to the surface area of the ellipsoid 414 cm^2, the main working area used for mechanical activation is in the interval of curvature of the flattened ellipsoid surface of 0.024–0.2 and in the curvature interval of 0.2–0.587, only near-focal ellipsoid surfaces with an area of 89 cm^2 work, which provide similar elongated ellipsoid *t–s–T* conditions at shock–friction contact (Figs. 3.4–3.6).

Thus, given the ratio of the semiaxes of a flattened ellipsoid, it becomes relatively more effective when used in a conventional rotating drum mill than an equivalent ball load.

Questions

1. What types and shape of grinding media are used in ball mills for processing solids?
2. What is the relationship between the values of tangential and normal pulses in the collision of grinding bodies in the framework of the Hertz theory?
3. Evaluate the relative velocity **W** of the interaction of grinding bodies for a conventional rotating drum mill.
4. Evaluate the influence of the shape of grinding bodies on the process of mechanical activation.

5. At what ratio of ellipsoid semiaxes (the shape of the grinding body of cylpebs) is the most efficient use of such grinding bodies in comparison with ball loading?

3.2 Influence of the Coated Layer Thickness on the Parameters and Kinetics of Mechanical Activation

One of the well-known factors, which hold back wide introduction of mechanical activation methods, is the absence of a visual quality monitoring of the behavior of materials inside mill reactors. So, the need for the development of the theoretical models which can adequately describe mill or disintegrator conditions and materials activation is very real today. In real conditions, the output of mechanochemical synthesis products is only a few percent. One of the reasons of such low output, taking into consideration also dissipation energy, may be formation of a ground material layer on the disintegrator surface or milling bodies. The quartz processing in a planet centrifugal mill is a good example of either distribution of pressure or the temperature impulse related to it on the layer depth during the impact moment of milling body.

It is known [11, 12] that processes in mechanochemical reactors (MR) in a number of cases do not proceed completely. For example, in the process of growing shallow (breakage), the sizes of the mechanically activated (MA) particles reach dynamic equilibrium [13–15], which is a part of well studied concept of "mechanochemical equilibrium" [11, 12, 16, 17]. The presence of dynamic equilibrium between the processes of destruction and aggregation of particles serves as the basis for self-fettling of the surface of the grindings bodies in MR (Fig. 3.7a) with the thickness of the steel-lined layer δ [18–20].

In principle, it is near to the phenomena of compaction and agglutination of powders [21], the difference is only in mechanochemical realization of these processes in MR. So, for MA particles of quartz in a roll planetary mill [22, 23]:

$$\delta = m_1/\rho_1 (1 - p) (\Pi_v + \Pi_b) \approx 4m_1/\pi\rho_1 (\Pi_v + 4\pi R - N), \quad (3.4)$$

where $\rho_1 \approx 2590$ kg/m^3 is the solidness (density) of quartz, $p \approx 1 - \pi/4$ the sponginess of layer, Π_v & Π_b the areas of drum walls and surface of rolls, and R and N the radius and the number of rolls, respectively.

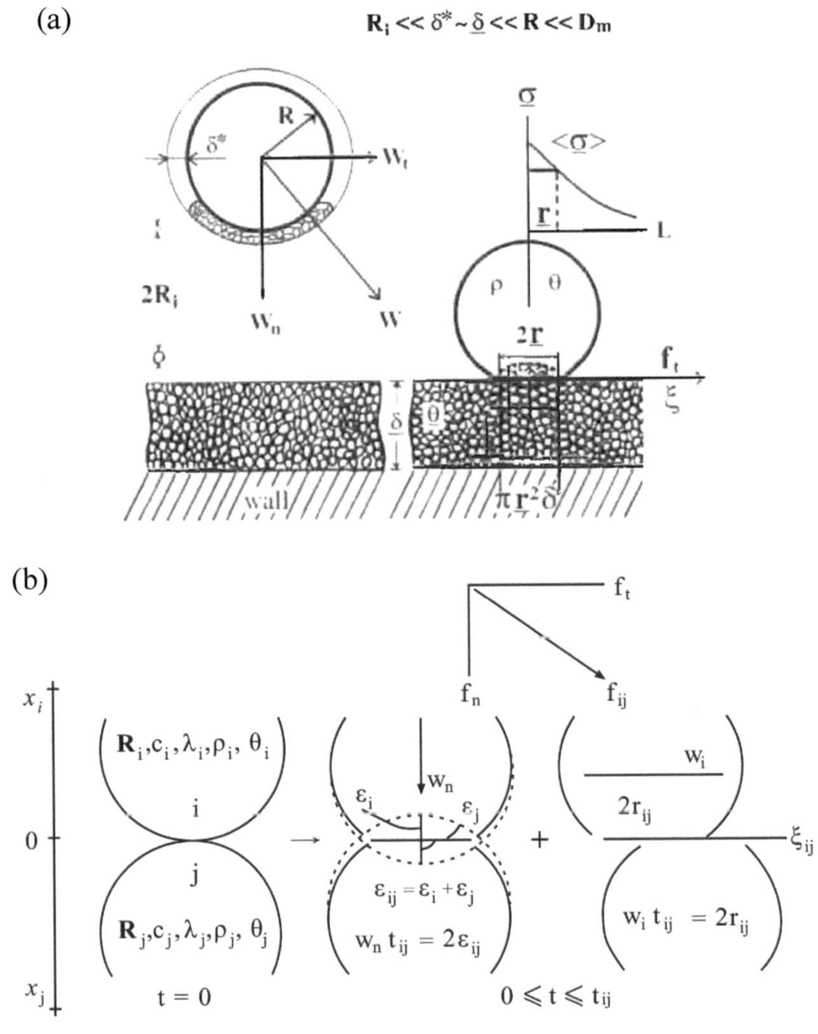

Figure 3.7 Chart of impact–frictional interactions of grindings bodies and MA of quartz particles: (a) grindings bodies; (b)—any contacting particles i and j from the selected volume $\pi r^2\delta$.

On the other hand, in practice the study of MA processes in MR, primary application refers to the net balance of mass m_b of moving grindings bodies (roll loading for all types of roll mills) and a hinge-plate m_1 of MA material [22, 24, 25]: $m_b/m_1 \gg 1$. As a rule, this figure is about 50 (there is about 2 g of the processed material per 100 g of the roll loading). Just the same choice of interaction m_b/m_1 is related to that at its considerable increase, the processes of MA in MR do not proceed completely or require unjustified long time of MA, caused by the necessity in the process of mass transition in MR, required to complete MA [26]. With an increase in m_b/m_1, consumption of mechanical energy becomes unjustified per unit of mass m_1 of MA material. As $m_b = 4\pi R^3\rho N/3$, where ρ is the solidness (density) of roll, expression (3.4) is written down as

$$\delta = m_1/\rho_1 (1 - p) [\Pi_v + (3m_b/\rho R)] \quad (3.4a)$$

$$\approx 4m_1/m_b \pi\rho_1 [(\Pi_v/m_b) + (3/pR)] \sim m_1/m_b.$$

Expression (3.1a) shows inversely proportional dependence of ratio δ: and m_b/m_1. In [22, 23], the process of MA of quartz was studied in steel roll 3 drums of the planetary mill, produced by NPO "Mekhanobr," in drums with capacity of 0.48 liter (0.00048 m^3). Affecting of the roll loading on quartz particles is carried out through a layer δ; with relative speed **W** of co-impact of grindings bodies (Fig. 3.7a):

$$|\mathbf{W}| = 27\pi\omega D_1 [(\kappa + 1)^2 + \Gamma^2 - 2\Gamma(\kappa - 1) \cos\varphi + (\Gamma + 1)^2]^{0.5}$$
$$= 16.7 \text{ m/s} \quad (3.5)$$

$$\mathbf{W}_n = |\mathbf{W}| \sin\varphi = 15.8 \text{ m/s, and}$$

$$W_t = |\mathbf{W}| \cos\varphi = 5.1 \text{ m/s,} \quad (3.6)$$

where $D = 0.115$ m and $D_1 = 0.05$ m are led and drums radii; $\omega = 11.7$ s $^{-1}$ and $\omega_1 = 20$ s^{-1} the frequencies of the opposite rotation of led and drums; $G = D/D_l = 2.3$ the geometrical factor; $\kappa = \omega_1/\omega = 1.7$ the kinematics factor; φ the angle of roll breaking away; cos $\varphi = -(1 + k)/G$; and $\mathbf{W}_n$ and $\mathbf{W}_t$ the normal and tangential constituents **W**. Relation of m_b to the mass of MA of quartz m_1 was equal to 4 at $m_b + m_t = 0.48 + 0.12 = 0.6$ kg.

The radius of rolls made up $R = 0.005$ m, solidness (density) $\rho = 7800$ kg/m^3, number $N = 120$, total surface $\Pi_b = 4\pi R^2 N = 0.037$ m^2, cylindrical ("working") constituent of drum surface area $\Pi_v = 2\pi D_1 h \approx 017$ m^2, where $h = 0.055$ m is the height of the drum. Calculation of (Eq. (3.4)) gives $\delta \approx 10^{-3}$ m, this is an order of size more than for traditional values m_b/m_1 or $\delta = 10^{-4}$–10^{-5} m accepted at the calculations of kinetics of mechanochemical processes [10]. Theoretical description of MA of quartz is presented in [22, 23]. Calculations were conducted for "equilibration" radius of quartz particles $R_1 = 3 \times 10^{-7}$ m [13] except the influence of fettling layer thickness δ of the processed matters on the parameters and kinetics of MA. Therefore, for the traditionally accepted periods of MA (to 90 min) the reaction of iron oxidization of abrasive wear of grindings bodies with subsequent formation of silicates of iron on-the-spot quartz particles proceeds incompletely—only ~10% [22, 23].

Thus, it is possible to state the possibility of substantial influence of value δ on parameters and kinetics of MA processes of quartz at m_1 in relation to near to m_b. Below we made the first attempt to introduce the ratio δ for the theoretical estimation of parameters of MA matters on the example of treating quartz according to the above presented method.

It is known [27] from the of powders' pressing theory, that the key parameter—mechanical tension σ on the impact-friction contacts of grindings bodies and MA particles—must depend on the thickness of the pressed powder-like material:

$$\sigma(h) = \sigma_0 \exp(-2\eta\xi h/r), \tag{3.7}$$

where $0 \le h \le \delta$ is variable h on the height of pressing δ, σ_0 the push of pressing, r the pressing radius, η the coefficient of lateral pressure, and ξ the coefficient of external friction.

The effect of tension of pressing σ_0 on the surface ($h = 0$) of roll impact with the flat layer (curvature $L_1 = 0$) of MA quartz particles on the wall of a drum is described by the expression [20]

$$\sigma_0 = (8/3\pi)(10\pi)^{0.2}\,[\rho(\theta + \theta_1)^{-4}\mathbf{W}_n^2]^{0.2} \tag{3.8}$$

The force of interaction of grindings bodies F and radius r of the contact surface are determined by expressions

$$f = (2/3)\,(10\pi)^{0.6}\,R^2\rho^{0.6}\,(\theta + \theta_1)^{-0.4}\mathbf{W}_n^{1.2} \quad (3.9)$$

$$r = (1/2)\,[(3/2)\,f(\theta + \theta_1)R]^{1.3} \quad (3.10)$$

The calculations presented below were performed using the program Mathcad (version 13.0/2005). The calculation results in equations (3.7)–(3.10) at $\eta = 1$ (at the lateral friction coefficient for a polycrystalline material [27] it was designed in [22] and is a flat layer [18, 19]), as well as at $\xi = 0$ (at the coefficient of external friction of iron on quartz [28]) as illustrated in Fig. 3.8.

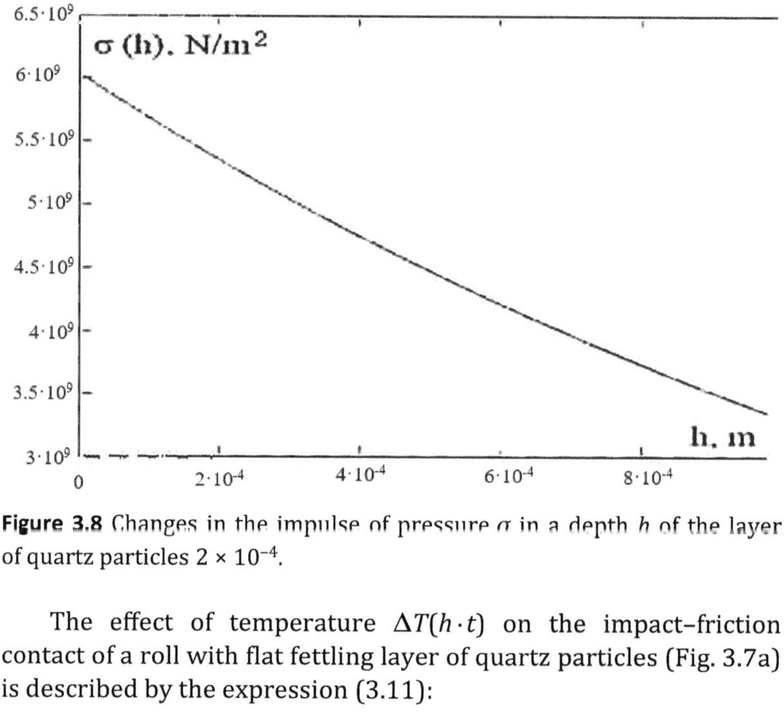

Figure 3.8 Changes in the impulse of pressure σ in a depth h of the layer of quartz particles 2×10^{-4}.

The effect of temperature $\Delta T(h \cdot t)$ on the impact–friction contact of a roll with flat fettling layer of quartz particles (Fig. 3.7a) is described by the expression (3.11):

$$\Delta T(h, t) = \xi\sigma(h)\,\mathrm{W_t}\,(cc_1\lambda\lambda_1\rho\rho_j)^{0.25}\{t^{0.5}\,iErfc[h/2(a_1t)^{0.5}]$$

$$- (t - \tau)^{0.5}\,iErfc\,[h/2a_1^{0.5}S(t - \tau)^{0.5}]\} \quad (3.11)$$

Here $\tau = (\pi/4)\ [10\pi\rho(\theta + \theta_1)R\mathbf{W}_n^{-0.2}$ is the time of impact–friction interaction. In particular, in Fig. 3.9 they show a cut ΔT $(h \cdot t = \tau)$—the change of maximal value of temperature in the depth of flat layer of quartz particles. It is evident that a sharp fall of temperature takes place along the depth of layer (close to the surface, and particles are only warmed up).

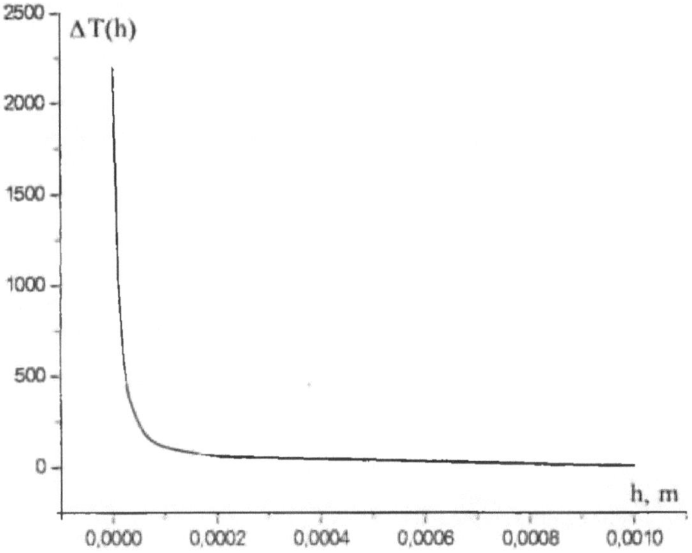

Figure 3.9 The effect of temperature impact distribution ΔT on the thickness h of quartz particles distribution layer.

Local warming up $\Delta T(x, h, t)$ of quartz particles in the volume of fettling layer takes place on impact-frictional contacts as a result of impulse of pressure $\sigma(h)$ on a fettling layer. In the calculations we do not taken into account distortion x of the contact flatness deep into the quartz particles (see Fig. 3.7b), i.e., it was accepted as $x = \theta$. To derive the formula for $\Delta T(h, t)$, it is necessary to revise the formulas derived for parameters of MA of particles in the volume of fettling layer [20, 22, 23] and bind them with $\sigma(h)$. Power f_1 of interaction of quartz particles now will be determined by expression $f_1(h) = (f/s) = S_1(h)\ \sigma(h)$, where $S_1 = \pi r_1^2$ is impact area of contacted particles, and $s = \pi r^2$. Then $s_1(h)$ is derived from expression $s_1(h) = (\pi/4)\ [(3/2)\ \sigma(h)\ s_1(h)\ \theta_1 R_1]^{23}$

$$S_1(h) = (\pi/4)\ x[(3\pi/8)\ \sigma(h)\ \theta_1\ R_1]^2 = \pi < r_1(h) >^2 \quad (3.12)$$

Knowing $S_1(h)$, we find $f_1(h) = [\sigma(h)\ \pi/4]^3 \times [(3/2)\ \sigma_1 R_1]^2$ and other parameters of MA of particles:

Total deformation (rapprochement) of contacting particles of quartz (Fig. 3.7b):

$$2\varepsilon_1(h) = 2R_1\ [\sigma(h)\theta_1 3\pi/8]^2 \quad (3.13)$$

The time τ_1 of particles' interaction is determined by expression $\tau(h) = 2\varepsilon_1(h)/w_n = 2r_1(h)/w_t$, where we will find normal speed w_n from

$$2\varepsilon_1\ (h)/w_n = (\pi/4)(10\pi)^{0.4}\ [\mathbf{w}_a^{-1}\ 2(\rho_1 R_1 \theta_1)^2]^{0.2}$$

and determine:

$$w_n = (2^{0.5}\ 10\pi\rho_1\ R_1^{2.5}\theta_1)^{-0.5} \times [8\varepsilon_1(h)/\pi]^{1.25}$$

$$= (9\pi)^{1.25}\theta_1^2 < \sigma\ (h)>^{2.5}/16(10\pi\rho_1)^{0.5} \quad (3.14)$$

Placing (3.9)–(3.11), we find $\tau_1(h)$ and tangent constituent $w_t(h)$ of particles (Figs. 3.10 and 3.11):

$$\tau_1(h) = 2\varepsilon_1(h)/w_n = [5\pi^{2.5}p[R_1^2/2^{0.5}3\sigma(h)]]^{0.5} \quad (3.15)$$

$$w_t(h) = 2r_1(h)/\tau_1(h) = (3/8) \times [2^{0.5}\ 3 <\sigma(h)>^3\ \theta_1^2/5\pi^{0.5}\rho_1]^{0.5} \quad (3.16)$$

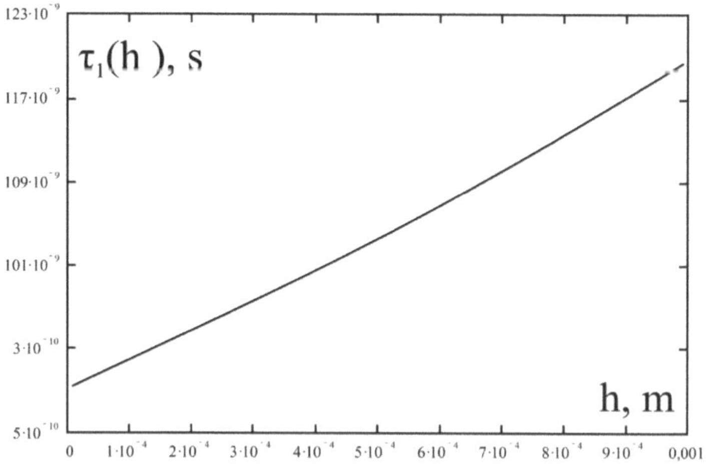

Figure 3.10 The change in the time of interaction of quartz particles along the depth of flat layer.

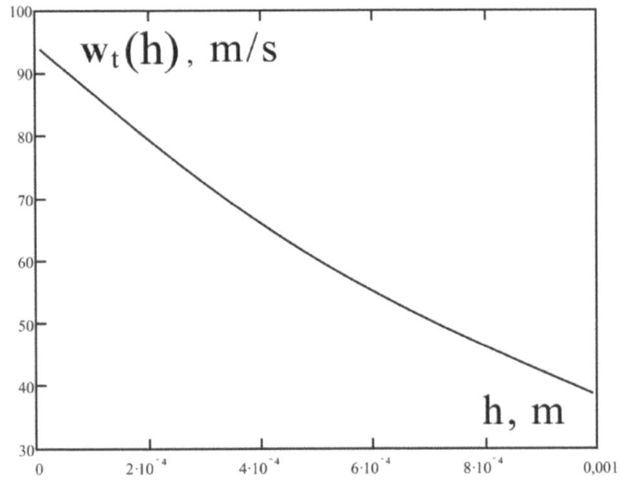

Figure 3.11 Changes in tangent speed of interaction of particles along the depth *h*.

Using Eqs. (3.15) and (3.16), we define the desired impulse of temperature on the surface of the particles:

$$\Delta T(h, t) = 0.5642\xi_1 \sigma(h)\, w_t(h)\, (c_1\lambda_1\rho_1)^{-0.5}\{t^{0.5} - [t - \tau_1(h)]^{0.5}\}$$

$$= 0.5642\, \xi_1\, (3/8)\times[2^{0.5}\, 3 <\sigma(h)>^5 \theta_1^2/5\pi^{0.5}]^{0.5}$$

$$(c_1\lambda_1\rho_1^2)^{-0.5}\, |t^{0.5} - \{t\, [5\pi^{2.5}\rho_1 R_1^2/2^{0.5}\, 3\sigma(h)]^{0.5}\}^{0.5}|$$

Here 0.5642 = *iErfc*, ξ_1 = 0.65 is coefficient of dynamic friction between quartz particles [29, 30]. Calculations are illustrated in Fig. 3.12, which state the most meaningful sections of impulse of temperature $\Delta T(h, t)$ on the friction contact of quartz particles:

- On the lower plane sections, one finds the values of depth *h* and corresponding values of τ_1 (Fig. 3.10);
- On a right lateral plane, a section $\Delta T[h, \tau_1(h)]$ is given on a depth *h* of fettling layer; on the left plane, a section $\Delta T[h(\tau_1), \tau_1]$ is given at time τ_1 of interaction of quartz particles.

Thus, the performed calculations specify a substantial influence of thickness of fettling layer on the parameters of MA of the matters processed in a planetary mill and, as a result on kinetics of mechanochemical processes. We should mark that according to expressions (3.4), the thickness of fettling layer of the processed material is totally determined by the size of hinge plates of load and by the area of surface of grindings bodies (walls of drum and roll loading).

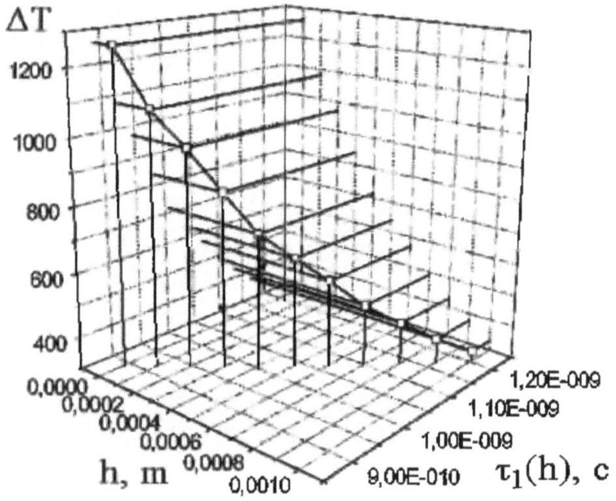

Figure 3.12 Changes in the impulse of temperature $\Delta T(h,\ t)$ along the depth h within time τ_1.

Questions

1. What is the role of the presence of dynamic equilibrium between the processes of destruction and aggregation of particles for the occurrence of the self-extinguishing surface of grinding bodies?
2. Give examples of calculating the mass balance of grinding media and the weight of the material being processed during mechanical activation (MA).
3. Give calculations of the influence of the thickness of the enveloping layer of the substances being processed using the example of quartz on the parameters and kinetics of MA? How does the thickness of the ground layer of the material being processed depend on the magnitude of the load and the surface area of the grinding bodies?
4. Calculate the mechanical tension on the contacts of impact–friction of grinding bodies and particles during mechanical activation, depending on the thickness of the compressed powder material.

5. Give examples of calculating the temperature impulse ΔT on the shock–friction contact of the roll with a flat layer of ground quartz particles depending on the pressure impulse. How does the time of interaction of quartz particles change over the depth of a flat layer?

3.3 The Phenomenon of Abrasive-Reactive Wear in Mechanochemical Processes

When analyzing mechanochemical processes, mainly physico-chemical phenomena [31–34] occurring only at impact–friction contact of the treated particles are considered, and the process of abrasive wear of grinding bodies and the role of the new fine-dispersed phase formed during this process do not receive due attention. At the same time, participation of the material of grinding bodies in mechanochemical processes plays a significant role in the qualitative change in the properties of the material obtained, which was clearly shown by the example of quartz processing in a planetary centrifugal mill. Quartz materials are widely used in researches on the mechanochemical synthesis of inorganic materials [35, 36].

It is known that during mechanical actions on crystalline silicon dioxide, phase transitions take place, as a result of which phases with denser packing are formed [37, 38]. The authors of these works associate formation of a glassy (amorphous) phase on the surface of quartz particles only with shear and dislocation mechanisms. At the same time, it is known from [39, 40] that high pressures and temperatures occur on the contact surface of friction particles, which can also influence the kinetic parameters of phase transition processes. To confirm expressed in [39, 40] assumptions, a numerical estimate was made of the pressure and temperature pulse in the system based on silicon dioxide, for which the following initial data were taken. As a mechanochemical reactor, we considered a steel ball three-drum planetary centrifugal mill with the following characteristics [33, 34]: geometric factor $m = l_1/l_2 = 2.3$, where $l_1 = 11.5$ cm and $l_2 = 5.0$ cm are the radii of the carrier and drum of the mill, respectively; kinetic factor $k = \omega_2/\omega_1 = -1.7$, where $\omega_2 = 20$ cm^{-1} and $\omega_1 = 11.7$ cm^{-1} are the number of revolutions of the drum and the opposite number of revolutions of the carrier, respectively; drum volume $V = \pi l_2^2 h \approx 450$ cm^3, where $h = 5.5$ cm is the height, and the cylindrical component of the drum surface $\Pi = 2\pi l_2 h \approx 170$ cm^2 (the effect of ball loading on the material being processed is mainly carried out on this surface). The ratio of the mass of ball loading to the mass of processed quartz was taken as 4: 1, and the total mass of ball loading (M) and quartz (M_1) was $M + M_1 = 480 + 120 = 600$ g. The radius of the balls is $R \approx 0.5$ cm, the density is $\rho = 7.8$ g/cm^3, the number of balls $N = M/\rho V = 3M/4\pi R^3\rho \approx 120$, total surface $\Pi_b = 4\pi R^2 N \approx 370$ cm^2.

Compliance (θ), Poisson's ratio (ν), Young's modulus (E), specific heat (c), thermal conductivity (λ) and thermal diffusivity (a) for steel balls are as follows [33, 34]: $\theta = 4(1 - \nu^2)/E = 1.65 \times 10^{-12}$ din/cm^2; $\nu = 0.85$; $E = 2.23 \times 10^{12}$ din/cm^2; $c = 66.7 \times 10^5$ erg/g·K; $\lambda = 75.3 \times 10^5$ erg/cm·K·s; $a = \lambda/\rho c = 0.144$ cm^2/s. Similar data for the quartz particles being processed are as follows: $R_{10} \approx 0.1$ cm – original size (average radius); $\rho_1 = 2.59$ g/cm^3 – density; $\nu_1 = 0.302$; $\theta = 4(1 - \nu_1^2)/E_1 = 3.95 \times 10^{-12}$ cm^2/din; $E_1 = 0.983 \times 10^{12}$ din/cm^2; $c_1 = 125 \times 10^5$ erg/g·K; $\lambda_1 = 7.7 \times 10^5$ erg/cm·K·s; $a_1 = \lambda/\rho c$ = 0.024 cm^2/s.

When processing particles of quartz, the phenomenon of self-lining steel fittings is not observed. Therefore, the mechanical effect of ball loading on quartz particles is carried out through a moving layer of the material being processed with a calculated thickness δ [33]:

$$\delta = M_1/\rho_1(\Pi + \Pi_b)(1 - p) \approx 4M_1/\rho_1(\Pi + \Pi_b)\pi, \tag{3.17}$$

where the porosity of the moving layer p is taken as equal as for the closest packing of spherical particles ($p \approx 1 - \pi/4$). The estimate gives $\delta \approx 0.11$ cm.

For the relative velocity **W** of the collision of the ball with the mill drum wall, simulated by a flat layer ($R_1 << l_2$), we have [33]:

$$\mathbf{W} = 2\pi\omega_1 l_2 \, [(k + 1)^2 + m^2 - 2m(k - 1)\cos\varphi + (m + 1)^2]^{0.5} = 1660 \text{ cm/s}, \tag{3.18}$$

where $\cos\varphi = -(1 + k)/m \approx 0.3$ determines the angle of separation of the ball from the wall. In contrast to [33], to analyze the process of interaction of a ball with a moving flat layer of processed quartz particles, it is necessary to take into account both the normal component $\mathbf{W}$ – $\mathbf{W}_n$ (determines the conditions for the interaction of quartz particles in the layer) and the tangential component $\mathbf{W}$ – W_t (determines the conditions for occurrence and mechanochemical processes on the friction surface of the contact of a ball with quartz particles):

$$\mathbf{W}_n = \mathbf{W} \cos\varphi \approx 500 \text{ cm/s};$$

$$\mathbf{W}_t = \mathbf{W} \sin\varphi \approx 1580 \text{ cm/s} \quad (3.19)$$

According to expression (3.19), it is clear that abrasion conditions are more preferable for a given mill compared to a shock. On this basis, it is possible to determine the *t–P–T* conditions on the shock–friction contact of the ball and the layer of quartz particles being processed:

Time (t) of interaction of colliding bodies

$$\underline{t} \approx 3.1\, R\rho^{0.4} (\theta + \theta_1)^{0.4} \mathbf{W}_n^{-0.2} \approx 32 \times 10^{-6} \text{ s} \quad (3.20)$$

Maximum normal P_n and average tangential $\langle \underline{P}_t \rangle$ pressure at the shock–friction contact of the grinding bodies:

$$P_n = 2.5\, \rho^{0.2} (\theta + \theta_1)^{-0.8}\, \mathbf{W}_n^{0.4} \approx 4.5 \times 10^{10} \text{ din/cm}^2 \quad (3.21)$$

$$\langle \underline{P}_t \rangle = 0.68 \underline{P}_n (\mathbf{W}_t/\mathbf{W}_n)^{0.4} = 1.7\, \rho^{0.2} (\theta + \theta_1)^{-0.8}\, \mathbf{W}_t^{0.4}$$
$$\approx 4.9 \times 10^{10} \text{ din/cm}^2 \quad (3.22)$$

The maximum value $\Delta T(x, t) = \Delta T(0, \underline{t})$ on the friction contact will be

$$\Delta T(0, \underline{t}) = \xi \langle \underline{P}_t \rangle \mathbf{W}_t\, (c\lambda\rho)^{-0.5}\, \underline{t}^{0.5}\, iErfc[0] = B \cdot \underline{t}^{0.5}\, iErfc[0], \quad (3.23)$$

where *iErfc* [0] ≈ 0.5642, if we take $\xi \approx 0.3$ for the dynamic coefficient of friction of steel over a quartz crystal, then $B \approx 11.7 \times 10^5$ K/c$^{0.5}$. The full expression for $\Delta T(x, t)$ in the vicinity of x of the contact of the ball with a flat layer of SiO_2 particles is

$$\Delta T(x,t) = B\cdot\{t^{0.5}\, iErfc[x/2(at)^{0.5}] - (t-\underline{t})^{0.5}\, iErfc[x/2a^{0.5}(t-\underline{t})^{0.5}]\} \tag{3.24}$$

Considering (3.23) and (3.24), it is easy to see that the stresses developed at the shock–friction contact of steel balls with a flat layer of quartz particles are close to the values of the Young's modulus for these materials. Since it is known [41] that the destruction processes due to the imperfection of the crystal structure occur at stresses of the order of (0.1–0.01) E, the abrasive wear of steel balls and grinding of quartz particles inevitably occurs. The experiment states the occurrence of these phenomena.

Due to the uncertainty of the ξ value, it is difficult to estimate from (3.23) the maximum [33, 34] temperature value of steel balls on the frictional contact surface $T = T_0 + \Delta T\ (0, t)$, where $T_0 \approx 350$ is the background temperature in the current mill, but at the accepted ξ it significantly exceeds the melting temperature of steel $T_m = 1812$ K. It is known [34] that the temperature at the impact–friction contact cannot exceed the melting temperature of the low-melting component itself, i.e., steel. Using (3.24) in the form

$$\Delta T_m = \Delta T(x=0, t=\underline{t}_m) = \Delta T(x=0, t=t'_m) = T_m - T_0 \approx 1460 \text{ K} \tag{3.25}$$

The lifetime of the molten zone on the surface of the steel balls:

$$t'_m - t_m \approx 33 \times 10^{-6} \text{ s}, \tag{3.26}$$

where $\underline{t}_m = (1460/B\cdot iErfc[0])^2 \approx 4.9 \times 10^{-6}$ c^5, $t'_m = (\underline{t} - \underline{t}_m)^2/4\underline{t}_m \approx 38 \times 10^{-6}$ s are the corresponding roots of Eq. (3.24). It was also shown in [34] that the thickness of the molten zone $x = x_m$ is close to the near-contact deformation value of the steel ball ε during shock interaction [33, 34]:

$$\underline{\varepsilon}(\text{Fe}) = 1.0\, R\rho^{0.4}\, \theta(\theta + \theta_1)^{-0.6} W_n^{0.8} \approx 6.3 \times 10^{-4} \text{ cm} \tag{3.27}$$

In this near-contact zone, various short-term ($t \sim t \sim 10^{-5}$ s) physical and chemical phenomena with a steel ball occur: melting–crystallization; chemical reactions; abrasive wear. The value of ε can be used to estimate the size of the broken iron particles, and the iron crystallites formed during the crystallization of the molten zone [34] will determine the size of individual iron particles in the process of abrasive wear of grinding media (balls and walls of the mill).

The high value of ΔT (0, t) facilitates chemical reactions between the material of the grinding bodies, the substance being treated and the surrounding gaseous medium. Since the reaction between Fe and SiO_2 is thermodynamically forbidden, it is only possible to oxidize steel balls with atmospheric oxygen according to the following scheme:

$$4Fe + 3O_2 \Leftrightarrow 2Fe_2O_3 \Leftrightarrow nFe_3O_4 \Leftrightarrow mFeO$$

Then, as a result of abrasive wear, FeO and Fe leave the surface of steel grinding bodies in an ultra-dimensional form and are adsorbed on much larger particles of quartz. Since the amount of FeO and Fe is insignificant, their presence does not affect the elastic properties of SiO_2 particles. This makes it possible to determine similar t–P–T conditions for the reaction of FeO with SiO_2 particles and oxidation of iron particles, already occurring as a result of impact–friction interaction of quartz particles in the δ layer [33, 34]:

$$t_1 \approx 6.6 \times 10^{-8}\ \text{s}$$

$$P_n \approx 3.6 \times 10^{10}\ \text{din/cm}^2$$

$$P_t \approx 7.9 \times 10^{10}\ \text{din/cm}^2$$

$$\Delta T_1(0, t) \approx B_1 \cdot t_1^{0.5} iErf[0]$$

$$\Delta T_1(x, t) = B_1 \cdot [t^{0.5}\ iErfc(x/2(a_1 t)^{0.5}] - (t - t_1)^{0.5}\ iErfc[x/2a_1^{0.5}(t - t_1)^{0.5}];$$

$$\varepsilon_1 \approx 2.6 \times 10^7 \tag{3.28}$$

If we take $\xi_1 \approx 0.8$, then $B_1 \approx 1.7 \times 10^7$ K/c$^{0.5}$, and the value $\Delta T_1(0, t_1)$ will exceed the melting point of quartz.

Thus, according to the presented model, the mechanism of formation of the amorphized zone is due to contact melting of quartz particles, and the total value of quartz particle deformation

ε_1, can serve as an estimate of the thickness of the amorphized layer on the surface of quartz particles. And as a result of the reaction of FeO and SiO_2 iron silicates can form on the surface of particles [42]. Assuming that iron meta-silicate formed by the $FeO + SiO_2 = FeSiO_3$ reaction [43] uniformly covers the amorphized surfaces of quartz particles, the thickness of this layer is $\partial \sim 2M_{FeO}/M_1S(\tau = 900\text{ s}) = 20\text{ nm} \sim \varepsilon_1$.

Experiments were carried out in a steel ball (radius ~ 0.5 cm), in a three-drum (volume ~ 450 cm^3) and planetary centrifugal mill, with a ratio of the mass of ball loading to the mass of quartz being processed 4:1, with a total mass of 600 g. The time of mechanical treatment varied from 5 to 90 min, with the normal and tangential components of the relative velocity of the collision of the balls with the drum wall ~5 m/s and ~15 m/s, respectively. The initial size of quartz particles did not exceed ~ 0.2 cm. To determine the possible forms of the presence of iron inside the quartz particles, the activated material was washed from the metallic iron phase, and the samples were examined by X-ray, electron microscopic, nuclear gamma resonance spectroscopy (NGRS) and other methods of analysis.

Table 3.1 The content of iron in a quartz mixture

Mechanical treatment time, min	The content of iron, % wt.		
	After mechanical activation	After etching	
		X-ray analysis	Microprobe analysis
0	0.06	—	—
5	1.43	0.1	—
15	3.74	0.24	0.23
30	3.9	0.41	0.35
60	4.5	0.52	—
90	5.14	0.64	0.40

In the process of grinding quartz, instrumental iron from the drum walls of the grinding chamber and metal balls enters the volume of the material being dispersed and can reach up to ~5% wt. (Table 3.1), however, the dependence of the amount of iron on the grinding time is not linear. The deviation from linearity is associated with the phenomenon of self-lining of grinding bodies [44], the degree of which increases with an increase in the specific surface of the material being processed and provides partial protection of grinding bodies from wear.

The bulk of iron in quartz is present in the acid-soluble (metallic and magnetic) phase [45], and the washed samples are represented only by particles of quartz (Fig. 3.13) with a very low content of iron in acid-insoluble form (up to ~0.6%). The amount of iron after etching was determined by the methods of emission spectroscopy and microprobe analysis (Table 3.1).

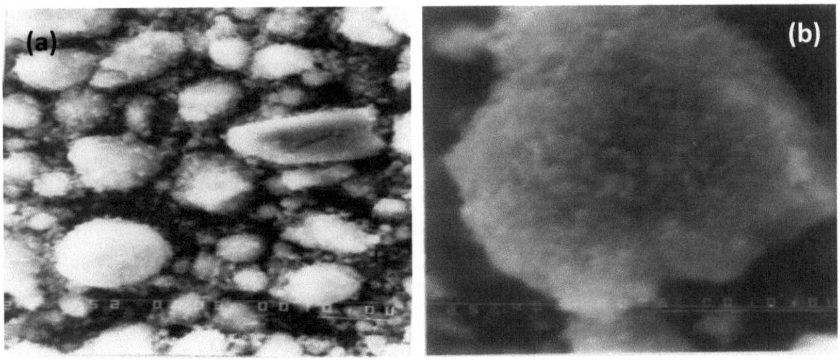

Figure 3.13 Electron-microscopic picture of a quartz particle after mechanical treatment: (a) general view (b) view of an individual particle.

It is known [46] that, during the process of mechanical activation of quartz, some surface layer is amorphized, and the internal volume retains the crystal structure. Both the amorphization mechanism and the thickness values of the amorphized layer are the subject of a long discussion. In this aspect, the question of the form of the iron content in activated quartz particles is of interest. In this regard, nuclear gamma resonance spectroscopy has determined the valence state of iron atoms in the quartz lattice (Figs. 3.14a,b) depending on the activation time.

The studies were conducted on a MS1104EM spectrometer at room temperature in the mode of constant accelerations. ^{57}Co was used as a source of gamma quanta in a rhodium matrix with an activity of 60 mCi. Registration of gamma radiation was carried out using a scintillation counter with a NaI (Te) crystal.

It is known that the Mössbauer spectra of iron atoms in Fe_2O_3 and Fe_3O_4 are characteristic sextets of magnetic splitting of nuclear levels [47]. However, additional lines appear in the spectra of the samples studied, indicating that the nature of the bond between the iron atoms and the environment has changed. The internal part of the spectrum is well described by a quadrupole doublet, the values of the parameters of the hyperfine interactions of which indicate that iron occupies two nonequivalent positions in the D1 and D2 quartz lattice.

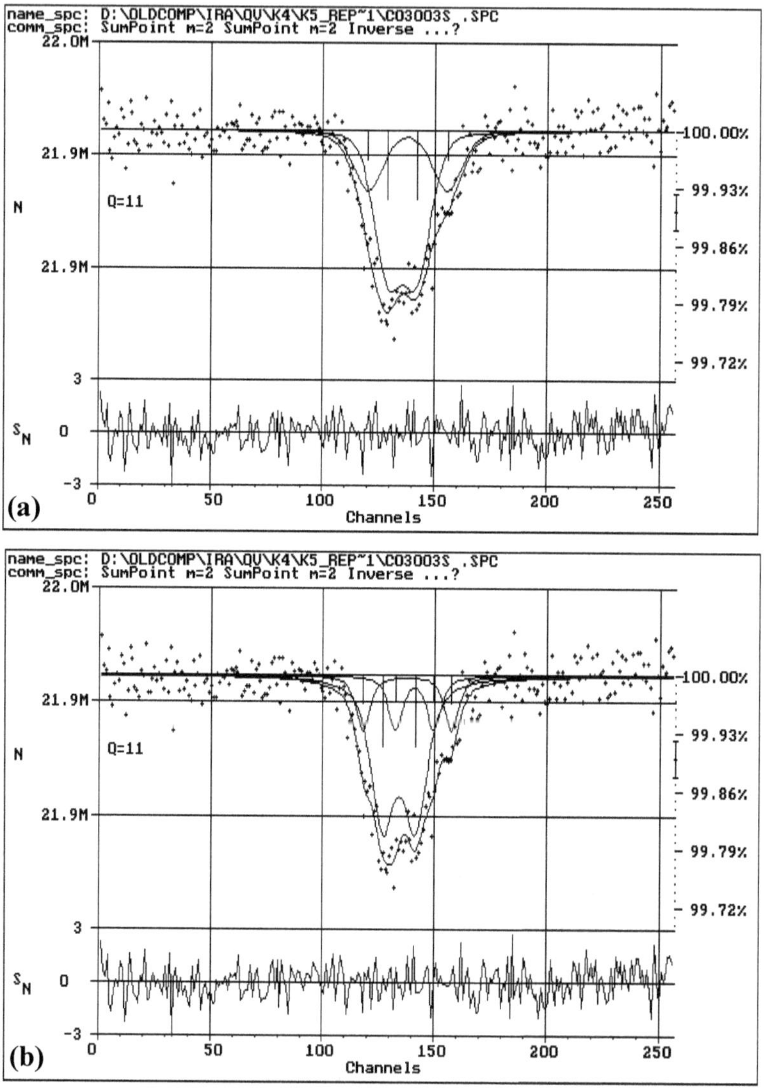

Figure 3.14 Mössbauer spectra using ^{57}Fe source for the systems of quartz particles treated 30 (a) and 90 (b) min.

For the first doublet, the hyperfine parameters of the isomer shift ($\delta \sim 0.648$ mm/s) and quadrupole splitting ($\varepsilon \sim 1.054$ mm/s) indicate that iron is most likely in octahedral oxygen coordination. Such atoms can belong to oxides that are in a superparamagnetic state.

The values of the hyperfine parameters of the second doublet ($\delta \sim 0.835$ mm/s and $\varepsilon \sim 2.8$ mm/s) can be attributed to the part of iron that is in the divalent high-spin state also in the octahedral oxygen coordination. From a comparison of Figs. 3.14a,b, the population of the positions with iron ions varies depending on the time of the mechanical action. According to [48], this position of iron ions in the quartz lattice can be explained by the formation of both meta- and iron ortho-silicates. From a thermodynamic point of view, in the system iron oxides – silica –formation of iron silicates in the temperature range of 1500–1700 K – is preferable from wustite (FeO) than from hematite (Fe_2O_3) and magnetite (Fe_3O_4). Moreover, both meta- and ortho-silicates which are resistant to mineral acids can be formed.

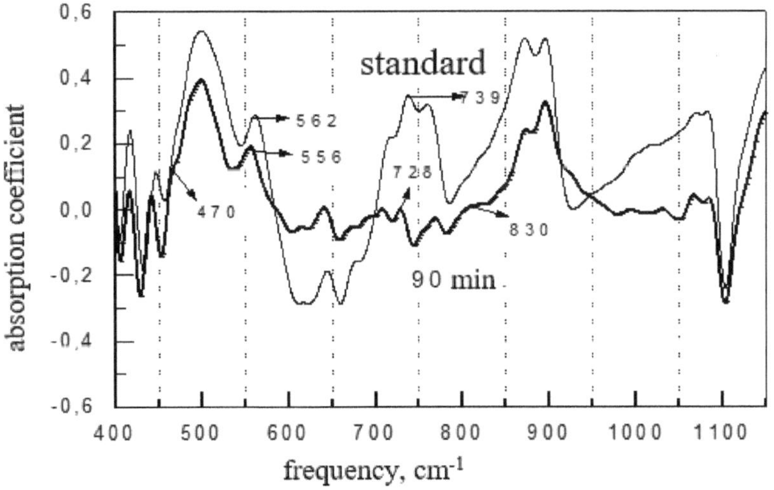

Figure 3.15 Infrared spectra of standard and activated quartz.

Formation of iron silicates is indirectly confirmed by the data of IR spectroscopy of the initial and last 10 min mechanical processing of quartz samples (Fig. 3.15). The differences in the bands of the Si–O–Si and O–Si–O bonds for the compared samples were not found. However, in the activated sample, a ~ 833 cm^{-1} band appears, which is attributed to Si – O – Me bonds, and the band corresponding to the Si – OH bond (~881 cm^{-1}) is attenuated.

The data presented are direct evidence of the formation of an iron-containing phase (iron silicates) on the surface of a quartz particle during its mechanical treatment, as a result of the abrasive reaction wear of grinding bodies during the treatment of oxide systems.

Questions

1. What are the main changes in the state of a solid surface under mechanical action that are responsible for abrasive-reactive wear?
2. What is the thickness of the moving layer of the material being processed at a mechanical impact?
3. How can one determine (*t*–*P*–*T*) conditions on the shock–friction contact of the ball and the particle layer being processed?
4. What is the mechanism of formation of the amorphized zone and the assessment of the thickness of the amorphized layer on the surface of quartz particles?
5. What is the dependence of the amount of rubbed iron with abrasive wear on the grinding time and the specific surface of the material being processed?
6. How can the amount and form of iron content in activated quartz particles be estimated due to abrasive wear of grinding media?

3.4 The Effectiveness of the Implementation of the Abrasive-Reactive Wear in the Mechanochemical Processing of Mineral Raw Materials

Mechanochemical treatment has found practical application in the technological scheme of enrichment and processing of mineral raw materials, as well as industrial wastes. At the same time, it is the fact of abrasive-reactionary wear of grinding media that has become an effective component of the process of mineral treatment. The abrasive-mechanochemical method was successfully implemented for the reduction of tenorite and galena by their joint processing with quartz material in a planetary mill with steel fittings [49].

The work was carried out using a steel two-drum ball (volume of each V = 140 cm^3, number of balls N = 400, radius R = 0.2 cm), water-cooled (water flow 1 cm^3/s, control of the outlet water temperature) as a mechanochemical reactor and the AGO-2 planetary mill with the speed of the relative impact of grinding bodies $\mathbf{W}_a$ = 11 m/s. For more efficient sample processing, all four possible orientations of the mill axis were used. The products of mechanical activation were investigated by X-ray phase analysis (XRD). Mechanically activated samples were isothermally annealed for 2 h at ~700°C in an argon flow (~1 cm^3/s) in alundum crucibles with graphite plugs and with a titanium sponge placed in front of the crucibles in the heated area of the quartz tube.

To 3 g crushed amorphous (fused) quartz, 1.5 g of tenorite or galena was added. The samples were pre-ground and homogenized for 1 h in a Fritsch Pulverisette mill equipped with agate fittings. Figure 3.16 presents the XRD result of one of such samples based on natural galena, indicating not only the presence of galena, but also some impurities (mainly crystalline quartz) inherent in this mineral. The original tenorite was obtained by thermal decomposition of malachite at ~250°C for 7 h. Figure 3.17 shows the result of the X-ray phase analysis of this powder.

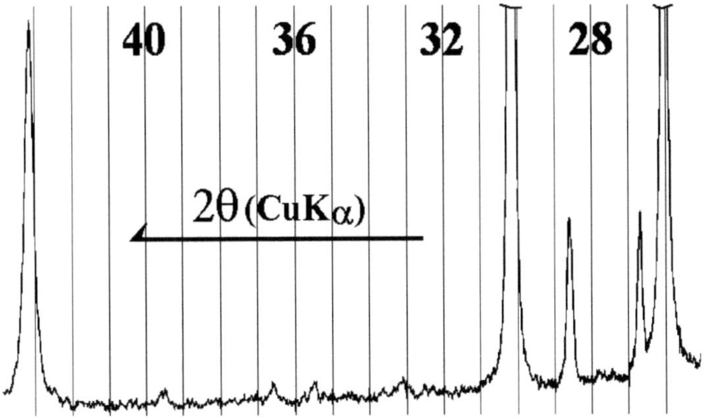

Figure 3.16 Data from the X-ray phase analysis of the system fused quartz–galena.

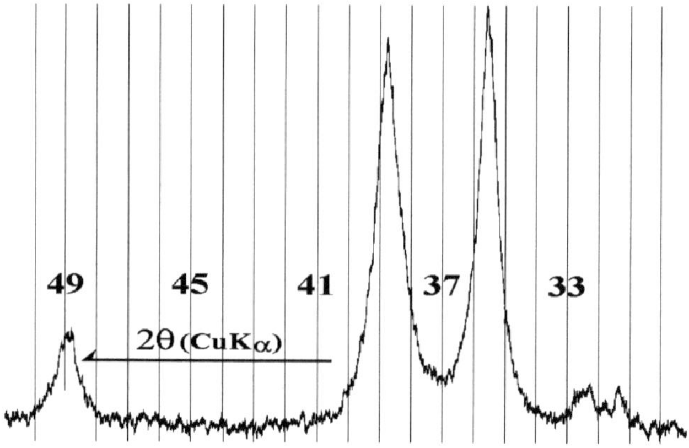

Figure 3.17 X-ray phase analysis of the original tenorite.

Preliminary experiments showed both similarity and a significant difference in the course of mechanochemical processes involving galena and tenorite. In both systems, abrasive wear of steel grinding media occurs, more pronounced on the system with galena. After a short-term (5–15 min) processing of the system with tenorite, there is a strong self-lining of the grinding bodies with the material being processed, which begins to change color from black to completely green after just 1 h of processing. This color, like self-lining, is preserved and enhanced during further processing. On the contrary, in the system with galena, the phenomenon of self-lining is completely absent, and after a short processing time a black and homogeneous "oily" powder is formed.

The results of the study of mechanical activation products (MA) by X-ray phase analysis and isothermal annealing showed the following processes: abrasive wear of steel grinding media, amorphization, recovery of copper from tenorite and lead from galena, the absence of crystalline phases of iron oxides and sulfides—other products of reduction reactions, for example, CuO + Fe = Cu + FeO ($\Delta rG = -28.0$ kcal/mol).

Figure 3.18 shows the formation of an X-ray amorphous product in a system with tenorite and the phenomenon of abrasive wear of the steel accessories of the mill (peak in the angle range 44–45). Moreover, the degree of amorphization with the increase in processing time "increases"—even the shape of the peak of iron becomes wider and more difficult to distinguish. The product of annealing of this sample in argon atmosphere at ~700°C (the value of this temperature corresponds to the maximum adopted for the crystallization of MA products) is only metallic copper with no reflections of other possible products of the exchange reaction, iron oxides (Fig. 3.19).

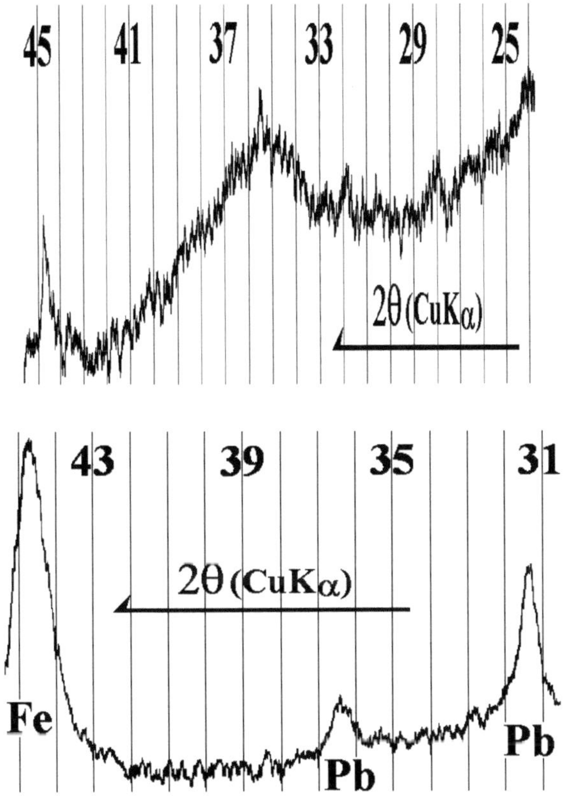

Figure 3.18 Data of the X-ray phase analysis of the system fused quartz–tenorite; MA time = 1 h.

Similar phenomena, but already in the process of mechanochemical activation of the system with galena, are shown in Fig. 3.20. Here you should also pay attention to the lack of reflexes of other possible products of the reducing exchange reaction—iron sulfides. An increase in the duration of mechanical treatment up to 2 h does not lead to significant changes in the spectrum, with the exception of the disappearance (amorphization) of the reflections of the admixture of crystalline quartz in natural galena. Annealing of these products does not change the situation either: the result is only insignificant changes in relative intensities and a significant narrowing of the corresponding reflexes. A further increase in MA time leads to the prevalence of abrasive wear of iron compared to all other processes (Fig. 3.21).

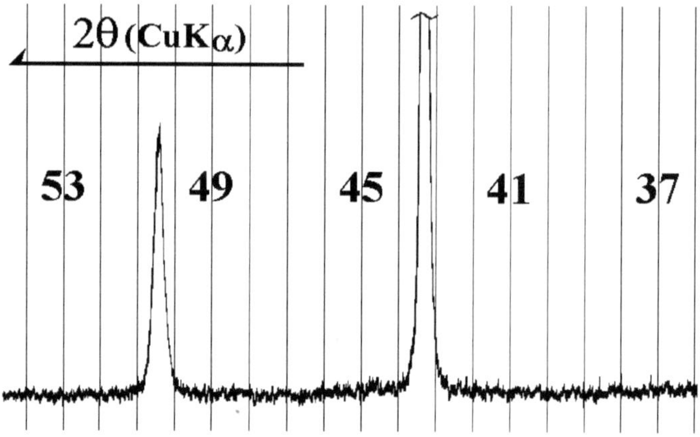

Figure 3.19 Data of the X-ray phase analysis of the system fused quartz–tenorite, MA time 1 h, annealing 2 h in argon at 700°C.

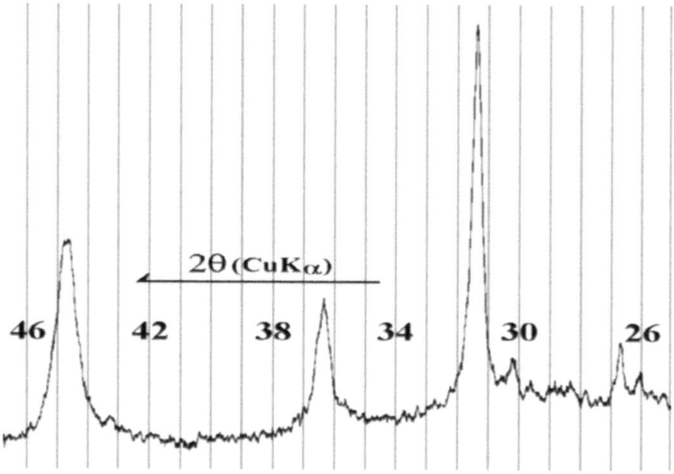

Figure 3.20 Data of the X-ray phase analysis of the system fused quartz–galena, MA time 1 h.

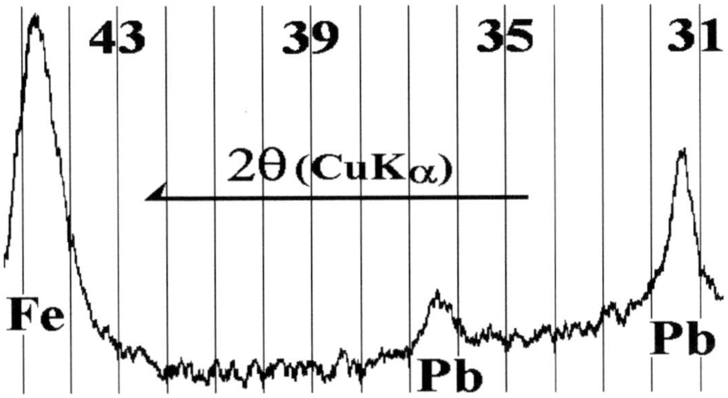

Figure 3.21 Data from the X-ray phase analysis of the system fused quartz–galena, MA time = 3.5 h.

The data obtained unequivocally confirm the fact that the phenomenon of self-lining of grinding bodies [50] prevents their abrasive wear and, as a result, the reducing reaction of tenorite with the material of grinding bodies in the process of mechanical activation. The complete absence of self-lining of the grinding bodies in the system with galena, on the contrary, leads to a quick breakdown of the galena with the steel material of the grinding bodies.

Without a comprehensive study of the mechanochemical activation of natural quartz with the participation of the material of steel grinding bodies [50], the disappearance of reflections of iron oxides (sulfides) after reduction of tenorite (galena) it would be difficult to give an unambiguous explanation. However, using the presented results, it can be argued that the surface of quartz particles can be modified not only by formation of amorphous iron silicates (after annealing of mechanically activated tenorite samples), but also by iron sulfides after mechanical activation of galena together with amorphous quartz particles. Moreover, in case of galena, the nanoscale surface layer on quartz particles should correspond to some amorphous compound in the Fe_xS_y–SiO_2 system. An even more complex compound of green color, formed after mechanical treatment of tenorite and containing nano-sized particles of abrasive wear of steel accessories of the mill, corresponds to a certain amorphous compound $Cu_xFe_yO_z$–SiO_2 on the surface of SiO_2 particles.

The next aspect is related to dimensional effects. With the traditional method of mechanochemical reduction, for example, copper sulfides [51–54], with the use of metallic iron powders with initial particle sizes of ~ 10–100 μm, the mechanical activation times are ~10–100 h. In the proposed abrasive-reducing method of breaking down, for example, galena, the size of the initial particles of abrasive wear of iron is close to 10 nm, i.e., orders of magnitude smaller than with the traditional method of carrying out an exchange mechanochemical reaction. Therefore, the reaction rate of using the abrasive exchange method with participation of steel grinding bodies is higher and, as a result, the time of mechanical processing of breaking down mineral raw materials by this method becomes 1–2 orders of magnitude smaller.

Thus, in the process of mechanochemical activation of natural minerals (tenorite and galena) mixed with abrasive (fused quartz) in the AGO-2 mill with steel fittings, nanocomposite metal-oxide-sulfide powders based on a quartz matrix were obtained. So, abrasive effects in mechanochemical processing play a significant role in the processes of obtaining new materials.

Questions

1. What are the similarities and differences between abrasive and reaction wear during the mechanochemical processing of tenorite and galena?
2. What are the features of the formation of X-ray amorphous product in systems with tenorite and galena in the process of mechanochemical processing in the presence of abrasive wear of grinding media?
3. What compounds are formed on the surface of quartz particles during joint mechanochemical treatment with galena and tenorite?
4. Discuss the role of the size effect in the abrasive-reducing method of breaking down mineral raw materials.
5. What is the role of self-grinding of grinding bodies in the process of mechanical activation of various mineral raw materials?

References

1. Ignatov V. A., Soleny V. K., Yarmolenko A. I., Zhuk V. L., Tuyakhov A. I. Selection and development of a rational form of the grinding body

for grinding raw materials and materials. *Metal Casting Ukraine.*, 2002. no. 1–2. pp. 23–26 (In Russian).

2. Urakaev F. Kh., Ketegenov T. A., Tyumentseva O. A., Boldyrev V. V. Simulation of the reaction of the material of grinding bodies with the processed substance on the example of quartz processing in a mill with steel fittings. *J. Phys. Chem.*, 2004. vol. 78, no. 5. pp. 828–834 (In Russian).
3. Zegzhda S. A. Theories of Hertz, Saint-Venant and Sears. In *The Book: Impact of Elastic Bodies*. St. Peterburg: Izd. St. Petersburg University, 1997. pp. 11–35 (In Russian).
4. Kilchevsky N. A. *Dynamic Contact Compression of Solids*. Naukova Dumka, Kiev, 1976. 240 p. (In Russian).
5. Zegzhda S. A. *Impact of Elastic Solids*. Science, Moscow, 1997. 310 p. (In Russian).
6. Robotnov Yu. N. Hertz's contact problem. In *The Book: Mechanics of a Deformable Body*. Moscow, 1988. pp. 378–385 (In Russian).
7. Urakaev F. Kh. Thermodynamic interpretation of mechanochemical reactions at the frictional contact of the treated particles in ball mills. *Friction Wear.*, 1980. vol. 1, no. 6. pp. 1078–1088 (In Russian).
8. Urakaev F. Kh., Zhogin I. L., Goldberg E. L. Description of the processing of particles in the disintegrator. *News Soan USSR. Ser. Chem. Sci.*, 1985. no. 8. vol. 3. pp. 124–131 (In Russian).
9. Batuev, G. S., Golubkov, Yu. V., Efremov, A. K., Fedosov, A. A. *Engineering Methods for Studying Shock Processes*. Mashinostroenie, Moscow, 1969. 356 p. (In Russian).
10. Urakaev F. Kh., Boldyrev V. V. Mechanism and kinetics of mechanochemical processes in comminuting devices. Theory. *Powder Technol.*, 2000. vol. 107, no. 1–2. pp. 93–107 (In Russian).
11. Avvakumov E. G. *Mechanical Methods of Activation of Chemical Processes*. 2 nd book, revised and add. Nauka, Novosibirsk, 1986. 306 p. (In Russian).
12. Heineke G. *Thribiochemistry. English Translation* Mir, Moscow, 1987. 584 p. (In Russian).
13. Schneider U. Makroskopische und mikroskopische Eigenschaftsanderungen von Feststoffpulvem infolge starker mechanischer Beanspruchung in Miihlen.
14. Okczky Z., Famady M. F. Kinetics and equilibrium of grinding. *Epitonyag (Hungary)*, 1982. vol. 34, no. 12. pp. 441–448.

15. Yokagama T. Pulverisation and mechanochemistry. Pulverisation limit of impact mills, Kagaku Kogaku Zasshi. *J. Chem. Eng.*, 1986. vol. 50, no. 7. pp. 467–469.
16. Zyraynov V. V. Mechanochemical balance in synthesis of PbMo05, *Izv. SOAS USSR. Ser. Chem. Sci.*, 1990. vol. 2. pp. 101–106.
17. Goldberg E. L., Shapkin V. L., Vibration instability and "mechanochemical balance". *Siberian J.*, 1991. vol. 6. pp. 120–126.
18. Kobayashi K. Formation on coating film on milling rolls for mechanical alloying. *Materials Trans.*, 1995. vol. 36. pp. 2134–137.
19. Belyaev E. Yu., Lomovsky O. l., Ancharov A. I., Tolochko B. P. Investigation of the reaction zone structure under mechanochemical synthesis of metal disilicides by a method of local diffractometry. *Nuclear Instr. Meth. Phys. Res. Sec. A*, 1998. vol. 405. pp. 435–439.
20. Urakaev F. Kh. Theoretical bases of mechanochemical processes in roll milling. *Eur. Chem. Technol. J.*, 2004. vol. 6, no. 4. pp. 239–254.
21. Cocks A. C. F., Constitutive modelling of powder compaction and sintering. *Prog. Mater. Sci.*, 2001. vol. 46, no. 3–4. pp. 201–229.
22. Urakayev F. Kh, Ketegenov T. A., Petrushin E. I., Savintsev U. P., Tumentseva O. A., Tchupakhin A. P., Shevchenko V. S., Yusupov T. S., Boldyrev V. V. Complex studies of abrasive-reactive modification of quartz particles surfaces by amorphous components of iron in mills with steel accessories. *Phys. Tech. Probl. Mineral Deposits*, 2003. vol. 3, pp. 110–122.
23. Urakayev F. Kh., Ketegenov T. A., Tumentseva O. A., Boldyrev V. V. Modeling the reactions of material in grinding charge with processed substance on the example of processed quartz in the mills with steel furniture. *J. Phys. Chem.*, 2004. vol. 78, no. 5. pp. 828–834.
24. Suryanarayana C. Mechanical alloying and milling. *Prog. Mater. Sci.*, 2001. vol. 46, no. 1–2. pp. 1–184.
25. Takacs L. Self-sustaining reactions induced by roll milling. *Prog. Mater. Sci.*, 2002. vol. 47, no. 4. pp. 355–414.
26. Urakayev F. Kh., Shevchenko V. S. The studies of mechanochemical synthesis nano-particles T1C1 through dissolving with final product, *J. Phys. Chem.*, 2006. vol. 80, no. 2. pp. 218–225.
27. Zhdanovich G. M. *Theory of Metal Powders Pressing*. Metalurgia, Moscow, 1969. 264c p. 129.
28. Van Nierop M. A., Glover G., Hinde A. L, Moys M. H. A discrete element method investigation of the charge motion and power draw of

an experimental twodimensional mill. *Int. J. Mineral Process.*, 2001. vol. 61 no. 2. pp. 77–92.

29. Zhang D., Whiten W. J. Contact modeling for discrete element modeling of roll mills. *Minerals Eng.*, 2001. vol. 11, no. 8. pp. 689–698.
30. Cleary P. W. Charge behaviour and power consumption in roll mill: Sensitivity to mill operating conditions, liner geometry and charge composition. *Int. J. Mineral Process.*, 2001. vol. 63, no. 2. pp. 79–114.
31. Courney T. H. Process modeling of mechanical alloying (Overview). *Mater. Trans.*, 1995. vol. 36, no. 2. pp. 110–122.
32. Zyryanov V. V. Model of the reaction zone under mechanical loading of powders in a planetary mill. *Inorg. Mater.*, 1998. vol. 34, no. 12. pp. 1525–1534 (In Russian).
33. Urakaev F. Kh., Boldyrev V. V. Mechanism and kinetics of mechanochemical processes in comminuting devices. Theory. *Powder Technol.*, 2000. vol. 107, no. 1–2. pp. 93–107.
34. Urakaev F. Kh., Boldyrev V. V. Mechanism and kinetics of mechanochemical processes in comminuting devices. Applications of the theory. Experiment. *Powder Technol.*, 2000. vol. 107, no. 3. pp. 197–206.
35. Streletsky A. N., Leonov A. V., Butyagin P. Yu. Amorphization of silicon during the machining of its powders. *Kinet. Process. Colloid J.*, 2001. vol. 63, no. 5. pp. 630–634 (In Russian).
36. Avvakumov E. G., Pushnyakova V. A. Mechanochemical synthesis of complex oxides. *Chem. Technol.*, 2002. no. 5. pp. 6–17 (In Russian).
37. Butyagin P. Yu. Diffusion and deformation models of mechano-chemical synthesis. *Colloid J.*, 2003. vol. 65, no. 5. pp. 706–709 (In Russian).
38. Steinike U. Mechanically induced reactivity of quartz and its relation to the real structure. *News Siberian Branch USSR Acad. Sci. Chem. Ser.*, 1985. vol. 3, no. 8. pp. 40–47 (In Russian).
39. Bowden F. P., Taybor L. *Friction and Lubrication of Solids.* Mashgiz, Moscow, 1960. 543 p. (In Russian).
40. Urakaev F. Kh. Theoretical estimation of pressure and temperature pulses at the contact of friction particles in dispersing apparatus. *Izv. SOAN USSR. Ser. Chem. Sci.*, 1978. no. 7. vol. 3. pp. 5–10 (In Russian).
41. Regel V. R., Slutsker A. I., Tomashevsky E. E. *Kinetic Concept of Solid Strength.* Science, Moscow, 1974. 560 p. (In Russian).

42. Babushkin V. I. *Thermodynamics of Silicates*. Science, Moscow, 1986. 408 p. (In Russian).
43. Urakaev F., Ketegenov T., Petrushin E., and others. Complex investigation into the abrasion-reaction modification of quartz particle surface by amorphous iron compounds. *J. Mining Sci.*, 2003. vol. 39, no. 3. pp. 304–314.
44. Urakaev F. Kh., Boldyrev V. V. Calculation of physico-chemical parameters of reactors for mechanochemical processes. *Inorg. Mater.*, 1999. vol. 35, no. 2. pp. 248–256 (In Russian).
45. Pae A. Ya., Pae P. I., Realo E. H., Uibo L. Ya. Study of mechanochemical reactions and activation of quartz using the Mössbauer effect. *Rep. Acad. Sci. USSR*, 1971. T. 200, no. 5. pp. 215–219.
46. Khodakov G. *Physics of Grinding*. Science, Moscow, 1972. 308 p. (In Russian).
47. *Chemical Applications of Mössbauer Spectroscopy/Trans. from English*; under. ed. V. I. Goldansky. Mir, Moscow, 1970. 502 p.
48. Vereshchak M. F., Zhetbaev A. K., Satpayev K. K. Mössbauer effect on 57Fe impurity atoms in quartz single crystals. *Solid State Phys.*, 1972. T. 14. pp. 3082–3083 (In Russian).
49. Urakaev F. Kh., Shevchenko V. S., Ketegenov T. A. New mechano-chemical method of producing pyrite. *News Natil. Acad. Sci. Kazakhstan Chem. Ser.*, 2003. vol. 3, no. 2. pp. 114–118 (In Russian).
50. Ketegenov, T. A., Urakaev, F. Kh., Tyumentseva, O. A., Mofa, N. N., Mansurov, Z. A. Synthesis of iron silicates on the surface of quartz particles during their machining. *Rep. Natl. Acad. Sci. Kazakhstan.*, 2003. no. 2. pp. 66–72 (In Russian).
51. Balaz P., Takacs L., Jiang J. Z., Soika V., Luxova M. Mechanochemical reduction of copper sulfide. *Mater. Sci. Forum.*, 2002. vol. 386–388. pp. 257–262.
52. Balaz P., Godocikova E., Boldizarova E., Luxova M., Bastl Z., Jiang J. Characterization of nanocrystalline products prepared by mechanochemical reduction of copper sulphide. *Czech. J. Phys.*, 2002. vol. 52. pp. A65–A68.
53. Mulak W., Balaz P., Chojnacka M. Chemical and morphological changes of millerite by mechanical activation. *Int. J. Miner. Process*, 2002. vol. 66, no. 1–4. pp. 233–240.
54. Bakhshai A., Pragani R., Takacs L. Self-propagating reaction induced by ball milling in a mixture of Cu_2O and Al powders. *Metall. Mater. Trans. A*, 2002. vol. 33, no. 11. pp. 3521–3526.

Chapter 4

Structural Changes of Silicon Dioxide under Thermal and Mechanical Impact

4.1 The Variety of Structural Forms and Features of the Surface Layers of Silicon Dioxide

Among the many organic and inorganic materials studied from the standpoint of mechanochemical transformations, special attention was paid to silicon oxides and, in particular, to quartz. It is on quartz that all structural and radical transformations in the process of mechanochemical action have been quite consistently studied [1, 2]. The researchers involved in mechanochemical processes took an intense interest in silicon dioxide primarily due to two reasons: First, it is one of the main rock-forming minerals (58.3% of SiO_2 is contained in the lithosphere); second, the diversity and characteristics of the structural forms and modifications of silicon dioxide determined its active participation in the formation of both inorganic, polymeric and organoelement compounds [3]. In a number of works, a more capacious term is used—silica, which unites the concept of "silicon dioxide" in all its crystalline, amorphous, hydrated, or hydroxylated forms. The term "silica" indicates that the content of silicon in the analysis

Mechanochemical Synthesis of Composite Materials
Zulkhair A. Mansurov, Nina N. Mofa, Tlek A. Ketegenov, and Bakhtiyar S. Sadykov

ISBN 978-981-4800-88-4 (Hardcover), 978-1-003-12081-0 (eBook)
www.jennystanford.com

corresponds to the mass ratio in SiO_2, regardless of the form in which "silica," as a compound containing hydroxyl groups, is actually present. The main crystalline modifications of anhydrous silica or silicon dioxide are α-quartz (up to 573°C), β-quartz (up to 867°C), tridymite (up to 1470°C) and cristobalite (up to 1723°C), which then transforms to quartz glass.

Of all the modifications, quartz with a trigonal symmetry is most common. Its density is 2.65 g/cm^3, hardness is 7, the average temperature coefficient of linear expansion is 12.3×10^{-6}°C^{-1}. Quartz is a dielectric with a specific electrical resistance along the *c* axis and in the direction perpendicular to it, 0.1×10^{15} and 20×10^{15} Ohm/cm^3, respectively. At 573°C, quartz acquires a hexagonal form with a lower density (2.52 g/cm^3). The transition to tridymite occurs mainly in the presence of certain impurities and mineralizers. Quartz is usually preserved up to 1000–1100°C, then it irreversibly passes into cristobalite (density 2.32 g/cm^3, cubic symmetry) [4]. The structure of quartz consists of silica-tetrahedra [5], in which there are four oxygen atoms at approximately the same distance around each silicon atom, and each oxygen atom connects two tetrahedra. The tetrahedra are interconnected by vertices, forming a common skeleton motif of the structure (Fig. 4.1).

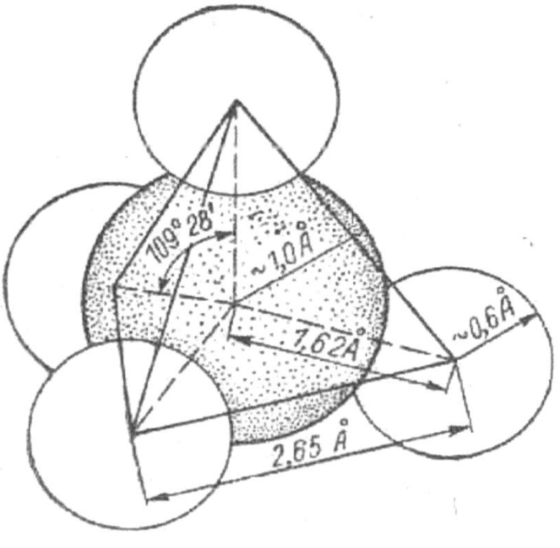

Figure 4.1 Scheme of the covalent model of the SiO_4 elementary tetrahedron [5].

A distinctive feature of quartz is that it is a piezoelectric. When exposed to external forces, the crystal lattice changes its state. Deformation of the lattice caused by mechanical tension results in the redistribution of electric charges in the presence of polar directions: on the negative edges of the zone of hexagonal prisms there appear negative electric charges, and on the positive edges, positive ones (Fig. 4.2).

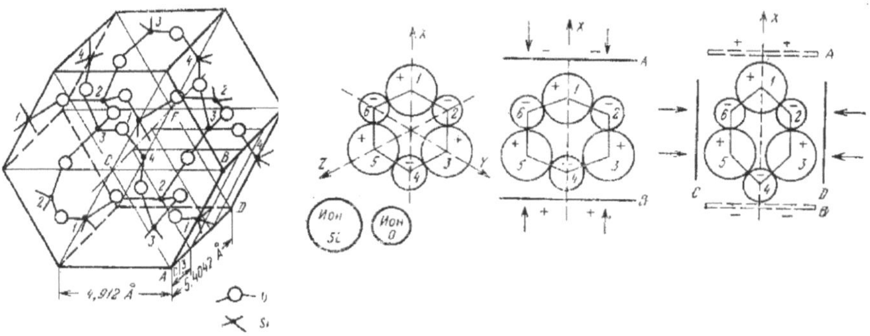

Figure 4.2 A structural cell of quartz (a) and formation of the piezoelectric effect (b–d) [6].

From a molecular point of view, this is associated with a helical chain of SiO_4 tetrahedra, which are located along a third-order symmetry axis (along the optical axis) [6]. There is also a reverse piezoelectric effect, that is internal stresses occur in an electric field in crystals that are proportional to the field strength.

Another fundamentally important feature of all forms of silicon dioxide, including quartz, is that the Si–O–Si bond angle is not 90°, as it should be, according to general ideas about the binding of tetrahedral [SiO_4] groups via common oxygen atoms by an unpaired electron on 2p-orbitals. It varies in a fairly wide range from 120 to 180°C [3, 7]. For low-temperature quartz, the bond angle is 143–147°C. This fact is very important, because the bond angle determines the mutual arrangement of tetrahedral [SiO_4] groups in space and, therefore, the structure and properties of various compounds involving silicon dioxide in its various crystalline and amorphous modifications. The formation of bonds by an oxygen atom with two silicon atoms with a bond angle that differs significantly from 90° is explained [7] by the presence of different hybridization options for the s and p orbitals of the oxygen atom. The variability of the Si–O–Si siloxane bond angle suggests its known flexibility, which determines the variety of silicate structures in which this bond is realized.

The third important fact of the structural features of any form of silicon dioxide, including quartz, is the presence of hydroxyl (silanol) groups on its surface. It is the silanol groups that are the sorption centers [8, 9]. They absorb water molecules and organic compounds having polar groups by hydrogen bonds. The concentration of SiOH groups on the surface of crystalline and amorphous particles of silicon dioxide varies significantly, reaching 11 ± 1 micromol OH/m^2 or 6.6 OH $groups/nm^2$.

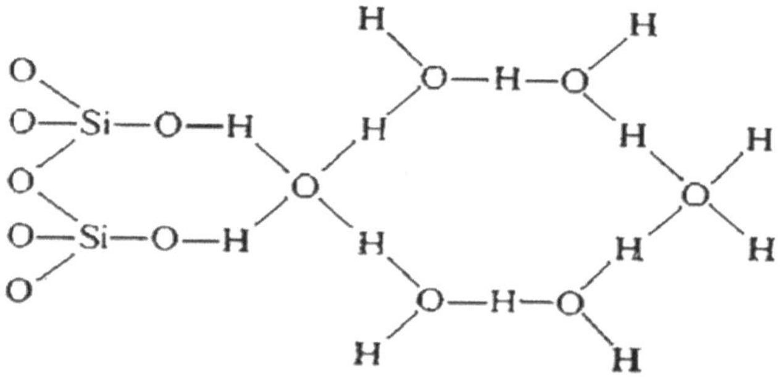

In the presence of water and water vapor, the surface of silicon dioxide with a large number of silanol groups, being hydrophilic, actively absorbs H_2O, forming a cluster according to the scheme [3]:

The surface of crystalline quartz differs significantly from the surface of amorphous silica in that it forms a film of monosilicic acid in water, in the structure of which the silicon atom coordinates four oxygen atoms, as in crystalline quartz [3]. Consequently, SiOH groups have a regular ordered arrangement on the crystalline surface, which determines the process of formation of any sorbed molecules. As for monosilicic acid, including that on the crystal surface of quartz, it quickly polymerizes, forming at first polysilicic acids with low molecular weights, and then polymeric varieties with a high molecular weight in the form of colloidal particles.

It is stated that H_2O molecules adsorbed on isolated SiOH groups react with neighboring siloxane surface sites, forming two new SiOH groups. The adjacent, vinyl SiOH groups also adsorb water, which then hydrates the surface of siloxane. It is believed that water molecules bound by a strong hydrogen bond are dissociated to a much greater degree and, thus, are able to break the siloxane bond and open the Si–O–Si bridge. SiOH groups are the sorption centers of organic and other molecules. Therefore, to achieve maximum adsorption, the surface of silica (silicon dioxide) should not contain adsorbed water, but should have the highest concentration of SiOH groups [3].

On the other hand, it is the presence of SiOH groups, the presence of an amorphous (up to 100) layer or a layer of monosilicic acid that determines the possibility and degree of dissolution of silicon dioxide in water and alkalis. Moreover, for quartz, this process has a number of features [10]. The surface of the particles in water gives off "mosaic" silica in the form of particles of colloidal size. The amount of emitted silica is noticeably larger than it corresponds to the monomolecular layer. In the case when quartz in water is intensively crushed into powder, the damaged surface layer can reach 35% with a specific surface area of 70 m^2/g, while solubility increases from 0.001% to 0.007% [3]. Moreover, when exposed to water, followed by removal of the outer part of the layer, the lower-lying deformed layer returns to its normal crystalline state.

More significant dissolution of quartz after intensive grinding, exceeding the amount of the amorphous phase is explained by G. S. Khodakov [11] by the fact that "... in the process of dispersion in quartz, one more additional bond of the silicon atom can be released, as a result of which the element with two hydroxyl groups preferably goes into solution compared to the molecule with three bonds with neighboring atoms. When a silicon atom with two hydroxyl groups goes into solution, two other bonds are destruction released on the surface of quartz, i.e., the process has a continuous chain nature." The breakage of bond proceeds according to the scheme [12]:

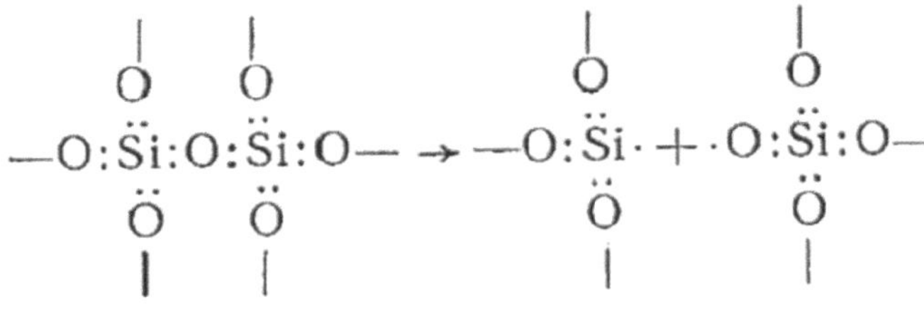

As a result, uncompensated charges, radicals and other active centers lead to anomalous phenomena of sorption and dissociation of adsorbed molecules.

The information presented in this section of the review on the features of the surface structure of quartz particles is of fundamental importance in the analysis of all processes occurring during the mechanochemical processing of this material.

Questions

1. What are the main crystalline modifications of silicon dioxide?
2. Describe the features of the structure of quartz, the variability of the angle of the siloxane bond Si–O–Si and the piezoelectric properties of quartz.
3. What is the role of hydroxyl (silanol) groups on the surface of quartz in the processes of sorption and the formation of polymeric formations?
4. What is the main difference between the surface of crystalline quartz particles and that of amorphous silica?
5. Describe the features of changes in the structure of the surface of quartz during grinding and the continuous chain nature of its dissolution after intensive grinding.

4.2 Mechanochemistry of Quartz, Features of Structural Changes during Dispersion of Quartz—Surface Radicals and Their Transformation

Quartz is one of the most common model systems for studying mechanically activated solids [11–14]. With many hours of quartz grinding in a vibratory mill, the long-range order is completely destroyed with the formation of an X-ray amorphous phase. Mechanical activation in a planetary mill leads to a gradual amorphization of particles with a local loss of structure stability. As follows from [14], this is expressed in the fact that despite the violation of the short-range order recorded by the degeneracy of the Si–O vibrations of quartz bonds, the material retains a two-dimensional order, according to the reflection on the electron diffraction pattern, 334 nm is one of the interplanar distances matrices. The violation of the short-range order was identified by the change in the IR spectra of activated quartz, in which the number of absorption bands gradually decreased with increasing processing time: first, the deformation vibrations of Si–O – bonds disappeared at frequencies of 515 and 790 cm^{-1}, and then their stretching vibrations at 1180 cm^{-1} (Fig. 4.3).

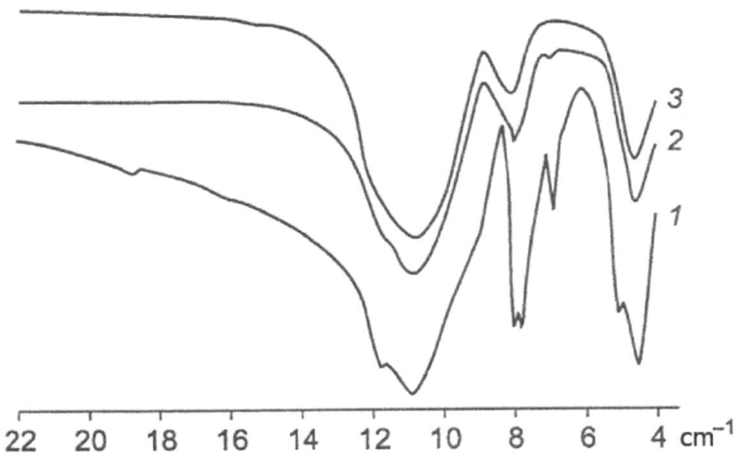

Figure 4.3 IR spectra of quartz: 1 – of the initial and mechanically activated quartz for: 2–30 min, 3–60 min [14].

After 60 min of activation, Si–O – bonds of quartz become threefold degenerate, and on the IR spectra, instead of seven, only three broad bands characteristic of amorphous SiO_2 remain, at frequencies of 460, 785, and 1070 cm^{-1} and shifted by 10–20 cm^{-1} towards lower frequencies compared to the spectrum of the original sample.

Moreover, not only amorphization of particles, chemisorption of water vapor from air, but also the formation of a polymer-like state was observed. At the same time, the density of samples decreases from 2.70 × 10^3 to 2.40 × 10^3 kg/m^3. In the process of mechanical action due to an increase in the amplitude of atomic vibrations, as well as during shear, the tense Si–O–Si bond also becomes polar, this contributing to appearance of a hydrogen bond with quartz with orientation of a water molecule, as shown in Fig. 4.4.

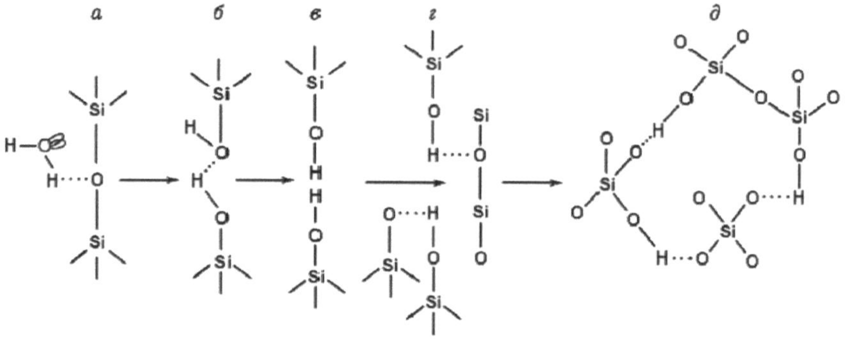

Figure 4.4 Scheme of the destruction of chemical bonds in quartz with participation of water vapor of air in the process of mechanical activation [14].

At the second stage, a reaction occurs between the water molecule and the silicon bridge, in which one of the water electrons is captured by silicon, and the proton is captured by the oxygen of the bridge tense bond. This transition leads to formation of two new bonds – one between silicon and water oxygen, and the other between the oxygen belonging to the bridge bond and hydrogen. As a result, the hydrogen bond is broken and the surface is saturated with OH groups. When the Si–O–Si bonds are broken, the interaction of quartz with water does not stop, hydrogen bonds are formed with neighboring groups, and the substance acquires a disordered polymer-like structure.

In addition to destruction of particles, amorphization of surface layers, saturation with hydroxyl groups in the process of grinding quartz and formation of a new surface, the atoms located on the boundary layer at the time of its formation have unsaturated valencies. Such a surface is energetically unstable and quickly relaxes [15, 16]. Nevertheless, some of the radical centers persist for a long time. So, when quartz is destroyed, so-called E′-centers are formed, corresponding to $\equiv Si^{\cdot}$-radicals and being formed in vacuum and inert atmosphere [17]. These centers are chemically active and interact with gases, for example:

$$\equiv Si^{\cdot} + CO_2 \rightarrow \equiv Si^{+}CO_2^{-}. \tag{4.1}$$

At T = 473 K, new compounds decay with the restoration of E′-centers.

When quartz was destroyed in a mill in an inert medium, the EPR signal was first observed by P. Yu. Butyagin and A. A. Berlin [18]. Later, the authors concluded that it is associated with formation of paramagnetic centers (PMCs), which are located in a thin surface layer. The results of studies [19–21] showed that the spectra recorded during mechanical destruction should be associated with PMCs of the O–Si type. This conclusion is supported by the data on the mechanisms of breaking of Si–O – bonds and formation of mechanical PMCs in silicon dioxide. It is known that mechanical destruction of quartz is accompanied by emission of electrons (mechanical emission). If we assume that Si^+ and O^- ions are separated on the resulting surfaces, we can consider mechanical emission as a kind of field emission due to strong electric fields in the cracks of the destroyed body. These fields pull out and accelerate electrons.

In [20], the EPR spectra of ground quartz in He at 300 K were traced. When investigating the nature and some chemical properties of paramagnetic centers formed during mechanical destruction of quartz, the EPR spectra arising from interaction of PMC with various gases (O_2, CO) were obtained and analyzed. It is shown that during mechanical destruction of silica, Si–O–Si bonds are destroyed with formation of two types of centers:

$$\equiv Si^{\cdot} \leftrightarrow \text{and} \leftrightarrow \equiv SiO^{\cdot}$$

The interaction of these radicals with molecules of various gases (O_2, CO, CO_2, N_2O, H_2) and the occurrence of the following free-radical reactions were stated:

$$\equiv Si^{\cdot} + O_2 \rightarrow \equiv SiOO^{\cdot}, \tag{4.2}$$

$$\equiv Si^{\cdot} + CO_2 \leftrightarrow (\equiv SiOCO)^{\cdot}, \tag{4.3}$$

$$\equiv Si^{\cdot} + N_2O \rightarrow (\equiv SiONN)^{\cdot} \rightarrow \equiv SiO^{\cdot} + N_2, \tag{4.4}$$

$$\equiv SiO^{\cdot} + CO \rightarrow (\equiv SiOCO)^{\cdot} \rightarrow \equiv Si^{\cdot} + CO_2, \tag{4.5}$$

$$(\equiv SiOCO)^{\cdot} + O_2 \rightarrow \equiv SiOO^{\cdot} + CO_2. \tag{4.6}$$

On the surface of crushed quartz, along with radical ones, there are also nonparamagnetic centers of chemisorption. The presence of these centers is particularly evident in the chemisorption of

CO_2 and N_2O. Under certain conditions, absorption of these gases is not accompanied by changes in the radical forms of the chemisorbed particles, i.e., it proceeds on non-paramagnetic centers. Pretreatment of the surface with CO_2, N_2O, CO, O_2 leads to the fact that the subsequent chemisorption of gases such as N_2O, CO and O_2 takes place only on radical centers of the surface. In this case, using the chemical properties of radicals, you can turn them from one type to another:

$$\equiv SiO^{\cdot} \rightarrow \equiv Si^{\cdot} \rightarrow \equiv SiOO^{\cdot},$$

i.e., free radical reactions can repeatedly proceed on one sample.

The presence of a defective structure in the surface layer of particles also ensures an increase in their reactivity [22–24]. In the works of V. A. Radzig [25, 26], studies of the interaction of activated quartz with various organic compounds were carried out. For example, the interaction of cyclopropane with free-radical centers, $(\equiv Si{-}O{-})_3\ Si{-}O^{\cdot}$, stabilized on the surface of SiO_2 at 300 and 150 K. The mechanism of interaction was investigated by the EPR method, and the products of this free-radical reaction were registered: a cyclopropylene radical. Experimental and theoretical studies of the structure and properties of intermediates were also carried out based on the results of the studies on the reactivity of SO_2 molecules with respect to the radical centers $(\equiv Si{-}O{-})_3Si^{\cdot}$, $(\equiv Si{-}O{-})_3Si{-}O^{\cdot}$, $(\equiv Si{-}O{-})_3Si{-}O{-}O^{\cdot}$, stabilized on the surface of SiO_2.

The experiments [25] with active quartz showed that at low activation temperatures, the formation of a siloxane bond between unbound silicon atoms becomes most energetically favorable. With an increase in temperature, the formation of high-energy centers on the surface — cyclic siloxane bridges and silanol groups — is preferable. These centers have a significantly greater adsorption ability and are more reactive with respect to organo-element compounds. When choosing the optimal pretreatment modes, the properties of the final product significantly change as a result of modification of the surface layers. In this case, active centers are not necessarily formed. Under mechanical action, strained surface Si–O–Si groups are able to form radicals upon contact with various organic and inorganic substances.

Questions

1. What are the main structural changes taking place during mechanical activation of quartz in a planetary mill?
2. Describe the formation of hydrogen bonds on the surface of quartz particles during grinding and formation of a disordered polymer-like surface structure.
3. Describe the formation of paramagnetic centers (PMCs) in a thin surface layer of particles during destruction of quartz and their stability.
4. What phenomenon is observed during quartz grinding, as a result of the appearance of an electric field in the cracks of the destroyed particles?
5. Desribe the interaction of activated quartz with various organic compounds and modification of surface layers of particles.

4.3 Modification and Radical Polymerization of the Surface of Quartz Particles during Mechanochemical Processing

Modification of oxide materials, including silicon dioxide of various structural forms (generally represented as silica), by chemical treatment is a well-developed method of obtaining materials with new properties as a result of chemical grafting of organic, inorganic and metal complex compounds [27, 28]. Such objects possess the chemical identity of the grafted compound and the properties of silica as a solid. Fixation of the active component on the surface of the carrier can be realized on account of physical and chemical interactions.

When chemically modifying silica, amorphous silica is most often used. Regardless of the type of structural modification, on the surface of the particles there are silanol and siloxane groups participating in the process of surface modification. At present, various methods have been developed that make it possible to fix virtually any class of chemical compounds on the surface of a matrix of silicon dioxide [29].

In any case, modification proceeds via heterogeneous functional hydroxyl groups, as well as through the destruction of deformed siloxane bonds with a stepwise reaction to them [30]. Depending on the set up task, hydrophilic or hydrophobic coatings are formed on the particles of silica. A wettable surface can be obtained by applying a layer, outside of which, groups of primary alcohol (surface etherification) are arbitrarily located. Moreover, such a surface has several advantages compared to a surface that carries only single terminal hydroxyl groups [3]. The reaction of alcohols with the surface of SiO_2 is carried out in an anhydrous medium at elevated temperatures. Primary, secondary and tertiary alcohols may be involved in these reactions. Chemical modification of silicas by the example of alcohols was first studied in 1950 by A. V. Kiselev [31]. Subsequently, numerous reactions by Ailer, Ebert, Wartmann and Doyle, Steeber, and Neimer were devoted to these reactions involving primary, secondary, and tertiary alcohols. The most thoroughly studied reactions are those of silica with methanol and butanol.

In the case when during interaction of organic compounds with the surface of silicon dioxide, hydrocarbon groups, in particular $Si(CH_3)_3$ are formed on it, the surface becomes hydrophobic. Such a surface is of great practical importance because of its water-repellent properties, in particular, in the flotation processes during the enrichment of mineral raw materials. To obtain alkoxy derivatives of silica, its treatment with n-butanol found the greatest distribution. Butoxyaeros (butasil) are widely used as fillers of polymer compositions, lubricant thickeners, components of adhesive compositions [32].

The activation energy of the esterification reaction of amorphous silica (silica gel) with butanol is 91.7 kJ/mol. The activation energy of the reverse reaction of ester bond hydrolysis is only 91.7 kJ/mol [3]. Therefore, the removal of water from the sphere of the reaction is a prerequisite for obtaining maximum surface coverage with a hydrophobic modifier. The grafting of the butoxyl group proceeds at a significant rate only at high temperatures. In vacuum, the grafted butoxy groups are stable up to temperatures 380–400°C, in the presence of oxygen their oxidation starts at 180°C. The maximum achieved surface coverage by the butoxy groups is 5 $\mu mol/m^2$, which corresponds to the value of the surface area A_m = 0.33 nm^2 per butoxy group [33]. The interaction of alcohols with silica refers to nucleophilic substitution reactions of the silicon atom. The presence of amino and carboxyl groups in the graftable compounds facilitates the substitution reaction. A detailed consideration of the mechanisms of interaction of alcohols with silica is given in the work of V. A. Tertykh [34].

The reactivity of powders can be significantly increased by treating the surface of particles with various inorganic and organometallic compounds, as well as subjecting them to mechanical impact [35, 36]. In most cases, the resulting activated product can be considered as chemically modified, since new covalent bonds are formed on the surface. So, for modification, silica is activated using different methods: plasma, γ- and UV-irradiation and mechanochemically (rubbing, crushing in the environment of the modifier). These methods are most widely used for modifying silica with polymers. So, during the grinding of silica in the medium of liquid saturated hydrocarbon, the hydrophobization of silica occurs, indicating the course of chemical modification [37, 38].

Modification of silica with organosilicon compounds is the main method in the synthesis of chemically modified silica (CMS). It allows you to get a variety of CMS with the desired properties, namely, with high hydrolytic and thermal stability, which are caused by a stable system of Si–O–Si–C bonds. Basically, when modifying silica, polymers are used with anchor groups that are capable of chemical interaction with silanol, siloxane, or activated in one way or another by surface silica groups. Strong polymer-silica bonds are formed when silyl anchoring groups are used. Such polymers are usually obtained by copolymerization of various vinyl monomers with unsaturated silanes. Modification of silica carriers with compounds that do not have anchor groups can be carried out under the action of various physical factors that cause the activation of the substance near the surface of the silica. Such activation leads to the possibility of the formation of covalent bonds with the carrier and simultaneous polymerization of compounds in the surface layer. It is known [39, 40] that polymerization of styrene, methyl styrene, butadiene, and chloroprene takes place in the process of grinding silica in a ball mill. The amount of non-extractable polymer was 0.2–2% by weight of the CMS. In [41], it was concluded that mechanically initiated polymerization occurs by a free-radical mechanism.

When creating sufficiently dense compounds on the surface of particles, the particle microencapsulation occurs. Microencapsulation is one of the branches of chemical technology that makes it possible to obtain microparticles of a chemical with new properties as a result of applying a film-forming substance to the surface [42, 43]. Using microencapsulation, they reduce the reactivity of substances, lengthen the stage life of unstable and rapidly decaying substances, mix immiscible and reacting with each other compounds, reduce the toxicity of products, provide a slow release of the active substance or its release at the right time, give the products new physical properties—decrease volatility, change in the density of products, etc. Microencapsulation of metals and various chemical substances, such as catalysts, stabilizers, plasticizers, oils, liquid and solid fuels, solvents, dyes, insecticides, pesticides, fertilizers, drugs, flavoring agents, food additives and fibers, as well as enzymes and microorganisms has been realized [43].

As a material of shells, any substances possessing film-forming properties under microencapsulation conditions are used. These include high-molecular compounds, low-molecular-weight fusible and soluble products of synthetic or natural origin. The choice of shell material depends on the purpose, properties, and method of release of the encapsulated substance, as well as on the chosen method of microencapsulation. One of the ways to obtain microencapsulated substances is mechanical dispersion of massive particles in the medium of encapsulating material.

In all of the above examples of modifying or microencapsulating particles, the surface undergoes significant changes, fundamentally changing the properties of the material as a whole. The scale of structural adjustment of the surface is nanometer size. Directed design of the microstructure of the surface in various ways is an effective way to create nanoformations and their complexes with different reactivity, allowing to regulate the functional properties of materials of a particular purpose [17], this being especially important for sorbents and catalysts.

Questions

1. Desribe the chemical modification of silica with alcohols - surface esterification.
2. How is the structure of silica modified with organic compounds formed?
3. Describe the formation of a hydrophobic and hydrophilic surface during interaction of organic compounds with the surface of silicon dioxide.
4. How does the hydrophobization of silica occurs by grinding in a liquid saturated hydrocarbon medium?
5. What is the reason for the choice of polymers with anchor groups when modifying silica?
6. What is microencapsulation during mechanical dispersion of massive particles in an organic film-forming material?

References

1. Avvakumov, E. G., Senna M., Kosova N. V. *Soft Mechanochemical Synthesis: A Basis for New Chemical Technologies*. Kluwer Academic Publishers, Boston, 2001. 200 p.
2. Balaz P. *Mechanochemistry in Nanoscience and Minerals Engineering*. Berlin; Heidelberg: Springer, 2008. 413 p.
3. Ailer R. *Chemistry of Silica*. Mir, Moscow, 1982. Part 1, 416 p.; Part 2, 712 p. (In Russian).
4. Schneider H., Majdic A., Vasudevan R. Kinetics of the quartz-cristobalite transformation in refractory-grade silica materials. *Mater. Sci. Forum.*, 1986. vol. 7. pp. 91–102.
5. Pryanishnikov V. P. *Silica System*. Stroyizddat, Leningrad, 1971. 240 p. (In Russian).
6. Smagin A. G., Yaroslavsky M. I. *Quartz Piezoelectricity and Quartz Resonators*. Energy, Moscow, 1970. 488 p. (In Russian).
7. Gorshkov V. S., Saveliev V. G., Fedorov N. F. *Physical Chemistry of Silicates and Other Refractory Compounds*. Higher. School, Moscow, 1988. 400 p. (In Russian).
8. Tarasevich Yu. I. *Structure and Chemistry of the Surface of Layered Silicates*. Kiev: Naukova Dumka, 1988. 248 p. (In Russian).

9. Martin I., et al. Improvement of crystalline silicon surface passivation by hydrogen plasma treatment. *Appl. Phys. Lett.*, 2004. vol. 84, no. 9. pp. 1474–1476.
10. Wang X., Chen Q., Hu H., et al. Solubility and dissolution kinetics of quartz in NH_3–H_2O system at 25°C. *Hydrometallurgy*, 2011. vol. 107, no. 1–2. pp. 22–28.
11. Khodakov G. S. *Grinding Physics*. Science, Moscow, 1972. 307 p. (In Russian).
12. Molchanov V. I., Selezneva O. G., Zhirnov E. N. *Activation of Minerals*. Nedra, Moscow, 1988. 208 p. (In Russian).
13. Tkâcovâ, K. *Mechanical Activation of Minerals*. Elsevier, Amsterdam, Oxford, New York, Tokyo, 1989. 155 p.
14. Chaykina M. V. *Mechanochemistry of Natural and Synthetic Apatites*. Novosibirsk: Publishing House of the Siberian Branch of the Russian Academy of Sciences, Geo Branch, 2002. 223 p. (In Russian).
15. Kostetsky, B. I. The structural-energetic concept in the theory of friction and wear (synergism and self-organization). *Wear*, 1992. vol. 159. no. 1. pp. 1–15.
16. Naito M., Kondo A., Yokoiama T. Applications of comminution techniques for the surface modification of powder materials. *ISIJ Int.*, 1993. vol. 33. no. 9. pp. 915–924.
17. Lygin V. I. Models of "hard" and "soft" surface. The design of the microstructure of the surface of silica. *Russ. Chem. J.*, 2002. vol. XLVI, no. 3. pp. 12–18 (In Russian).
18. Butyagin P. Yu., Berlin A. A. *Formation during Mechanical Destruction of Vitrified Polymers*. Science, Moscow, 1979. 270 p. (In Russian).
19. Radtsik V. A., Bystrikov A. V. Study of chemically active centers on the surface of quartz by the EPR method. *Kinet. Catal.*, 1978. vol. 19, no. 13. pp. 713–718 (In Russian).
20. Radzig V. A. Formation of free radicals due to interaction of groups with materials ($\equiv$ Si–O-$)_2$Si $\equiv$ O_2 H_2, CH_4, C_2H_6. *Chem. Phys.*, 1995. vol. 14, no. 2. pp. 416–427 (In Russian).
21. Ahmed-zadeh H. D., Baptismansky V. V., Zakrevsky V. A., Tomashevsky E. E. Paramagnetic centers formed during destruction of silicon dioxide. *Phys. Solid Body*, 1972. vol. 14, no. 2. pp. 422–426 (In Russian).
22. Schmalzried, H. Influence of structural defects on the reactivity of solids. *Reactiv. Solids. Proc. 8th Intern. Symp.* Göteborg, 1977. N.-Y. London, 1977. p. 237251.

23. Haber J. The role of surfaces in the reactivity of solids. *Pure Appl. Chem.*, 1984. vol. 56. no. 12. pp. 1663–1676.

24. Gomes, W. P., Dekeyser W. Factors influencing the reactivity of solids. *Treatise Solid State Chem.*, vol. 4. N.-Y. London, 1976. pp. 61–113.

25. Radzig V. A. Reaction intermediates on the surface of solids. *Chem. Phys.*, 1995. vol. 14, no. 8. pp. 125–135 (In Russian).

26. Radzig V. A. Interaction of cyclopropane with free–radical centers (≡Si–O-)$_3$Si–O, stabilized on the surface of SiO_2. *Chem. Phys.*, 1996. vol. 15, no. 6. pp. 77–83 (In Russian).

27. Minaeva Yu. V. Adaptive modification of the particle swarm method based on dynamic correction of the trajectory of movement of individuals in the population. *Bus. Inform.*, 2016. no. 4(38). pp. 52–59.

28. Sivanandini M., Dhami M. K., Dhami S. S., Pabla B. S., Surface modification of neutral colloidal silica nanoparticles. *Int. J. Res. Pharm. Chem.*, 2015. vol. 5, no. 1. pp. 167–171.

29. *Modified Silica in Sorption, Catalysis and Chromatography.*, ed., Lisichkina G. V., Chemistry, Moscow, 1986. 248 p. (In Russian).

30. Weinstein P. M., Ezhovsky Yu. K., Gavrilina I. P. Structure and reactivity of functional groups of the surface of amorphous silicon dioxide. *Chem. Phys.*, 1996. vol. 15, no. 2. pp. 104–110 (In Russian).

31. Kiselev A. V., Lygin V. I. *Infrared Spectra of Surface Compounds and Adsorbed Substances.* Science, Moscow, 1972. 440 p. (In Russian).

32. Neymark I. E. *Synthetic Mineral Adsorbents and Catalyst Carriers.* Kiev: Naukova Dumka, 1982. 216 p. (In Russian).

33. Strelko V. V., Kanibolotsky V. A. Classification of reactions involving the surface of dispersed silicas and the study on substitution processes of hydrogen bound with surface silicon atoms. *Colloid J.*, 1971. vol. 33, no. 5. pp. 750–754 (In Russian).

34. Tertykh V. A., Belyakova L. A. *Chemical Reactions Involving the Surface of Silica.* Kiev: Naukova Dumka, 1999. 278 p. (In Russian).

35. Wu W., Lu S., Wang J. Mechanochemical surface modification of particles by polymer grafting. *Chemistry in the Interests of Sustainable Development.*, 2005. no. 13. pp. 149–154.

36. Schiestel T., Brunner H., Tovar G. E. M. Controlled surface functionalization of silica nanospheres by covalent conjugation reactions and preparation of high density streptavidin nanoparticles. *J. Nanosci. Nanotechnol.*, 2004. vol. 4. pp. 504–511, (doi:10.1166/jnn.2004.079).

37. Pashchenko A. A., Voronkov M. G., et al. *Hydrophobization*. Kiev: Naukova Dumka, 1973. 237 p. (In Russian).
38. Wang Y. A., Li J. J., Chen H., Peng X. Stabilization of inorganic nanocrystals by organic dendrons. *J. Am. Chem. Soc.*, 2002. vol. 124, pp. 2293–2298, (doi:10.1021/ja016711u).
39. Blyskosh G. S. On some features of the surface structure of fillers modified by the method of radiation grafting of polymers. *High-molecular Compounds. Sulfur.*, 1969. vol. 11, no. 12. pp. 900–902 (In Russian).
40. Yin Y., Alivisatos A. P. Review article on colloidal nanocrystal synthesis and the organic–inorganic interface. *Nature*, 2005. vol. 437. pp. 664–670, (doi:10.1038/nature04165).
41. Crocker R., Schneider M., Hamann K. Polymer reactions on the surfaces of powders. *Successes Chem.*, 1974. vol. XLIII, no. 2. pp. 349–369 (In Russian).
42. Darbandi M., Thomann R., Nann T. Single quantum dots in silica spheres by microemulsion synthesis. *Chem. Mater.*, 2005. vol. 17 pp. 5720–5725, (doi:10.1021/cm051467h).
43. Solodovnik V. D. *Microencapsulation*. Chemistry, Moscow, 1980. 216 p. (In Russian).

Chapter 5

Activation and Modification of Quartz in Mechanical Reactors: Synthesis of Nanocomposition Quartz Particles Capsulated in Carbon-Containing Shells

5.1 Changes in the State, Structure, and Properties of Activated Quartz

Using mechanochemical treatment (MCT) as a way to transform the structure of the material and create quartz based composite powder systems with a nanostructured surface, studies were conducted to reveal the regularities of transformation of the structure of the dispersible quartz itself. Such information is necessary to state the optimal conditions of mechanical treatment and the choice of the most effective surface modifier, providing production of the material with a nanocomposite morphology and the required functional purpose.

The first characteristic of the material subjected to grinding under these or those conditions is dispersion of the powder. Its values were determined by a microscopic method using a POLAM R-211 polarization microscope. Powders with a particle size of up to 200 μm were subjected to grinding for a period of time from 5 to 60 min in a planetary centrifugal mill with an acceleration

Mechanochemical Synthesis of Composite Materials
Zulkhair A. Mansurov, Nina N. Mofa, Tlek A. Ketegenov, and Bakhtiyar S. Sadykov

ISBN 978-981-4800-88-4 (Hardcover), 978-1-003-12081-0 (eBook)
www.jennystanford.com

of 20 g and a predominance of impact. The size of particles and their distribution in fractions change depending on the processing conditions. As an example, Fig. 5.1 shows the results of particle size distribution for quartz ground for 5 min—more than 50% of particles in this powder of the size 25–45 μm. With an increase in processing time, the fractional composition changes. Figure 5.2 shows the size of particles constituting more than 50% of the powder, depending on the processing time on the grinding apparatus. A characteristic feature is the manifestation of periodicity in changing the particle size and homogeneity of the powder fraction under consideration.

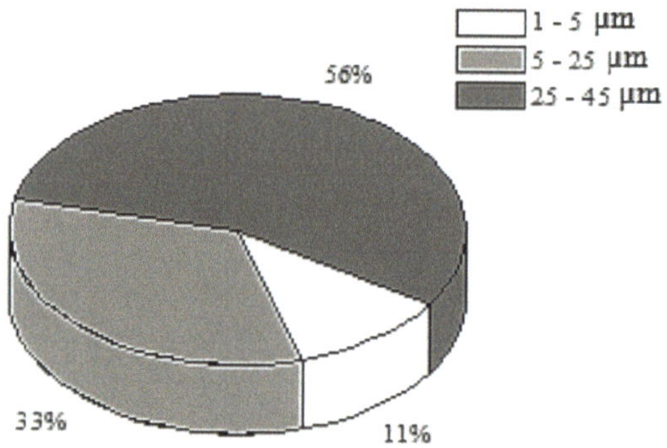

Figure 5.1 Particle size distribution for quartz treated in a mechanical reactor for 5 min.

For the material processed in a planetary mill with an acceleration of 20 g after 15, 30 and 40 min of processing, the particle size of the main fraction is 5–20 μm. With an increase in the time of mechanical action for more than 40 min, the particles become larger sticking together due to their agglomeration by sintering and the resulting powder gets more polydisperse. Agglomeration introduces a significant error in the determination of the true size of particles by a microscopic method. A characteristic feature of the quartz powder after 20 min of processing is its homogeneity (the minimum variation of particle size in the range of 10–20 μm).

Another structural characteristic closely related to the dispersion of the powder material is the bulk density. Its decrease for a quartz powder after 10–15 min of mechanical action and an increase after more than 40 min of treatment (Fig. 5.3) correlates well with the decrease in the size of particles in the first case and their growth in the second one due to agglomeration while maintaining a certain share of the highly dispersed fraction (in case of polydispersity of the powder) to impact the material. Of particular interest is the processing time interval of 20–35 min, when the bulk density increases with high dispersion (small particle size). This is possible when the changes take place not only the size, but also in the configuration of particles (they become flatter), and that, according to [1], provides a higher degree of packing. Measurements of the specific surface of quartz powder showed that its values also change with time not monotonously.

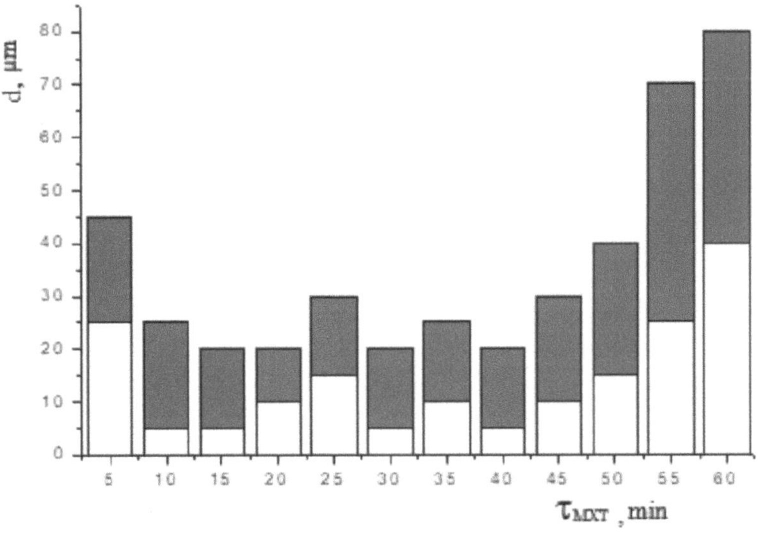

Figure 5.2 Dependence of the size of quartz particles (*d*), constituting more than 50% of the powder volume, on the time of treatment (τ_{MCT}) in a planetary centrifugal mill.

According to the data presented in Fig. 5.3, it should be noted that the period of processing for 20–35 min is the time providing specific features in the transformation of the structure, and, as a result, in changing the specific surface of the samples. An increase in the specific surface with a longer treatment time (40 min or more) with a simultaneous increase in the bulk density, particle size, and their polydispersity generally indicate quite complex structural changes in the particles during grinding in mechanical reactors.

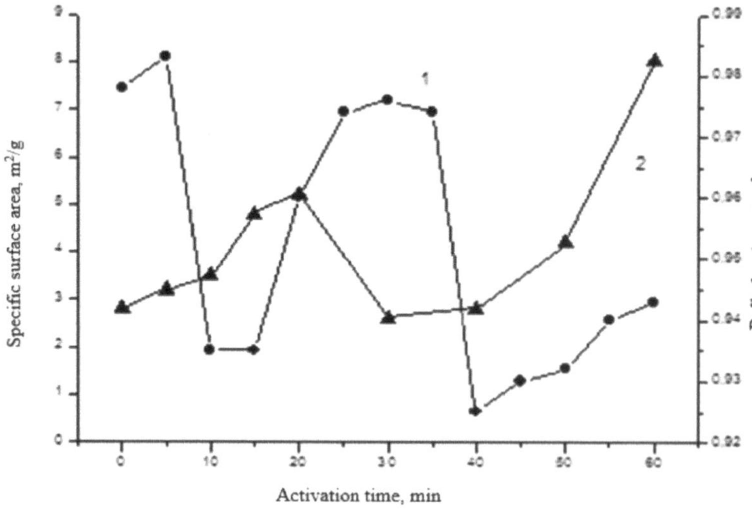

Figure 5.3 Dependence of the change in bulk density (1) and specific surface (2) of quartz powder from the time of mechanical treatment in a planetary centrifugal mill.

Thus, the grinding process has a clearly defined periodic nature, which is manifested in changes in the particle size, surface state, and structure in the volume of particles and, very likely, their shapes. According to the data presented in Figs. 5.1–5.3, the most significant structural changes in all parameters of the quartz being ground take place during 15–20 min of mechanical impact.

Since, as a result of intensive dispersion in a planetary centrifugal mill, there takes place not only grinding of particles but also their saturation with defects determining the different level of the energy state of the material and, consequently, its chemical activity, an assessment of the defectiveness of quartz particles with the help of X-ray structural analysis is given. Defectiveness of quartz after mechanical treatment was estimated by the broadening of X-ray lines at half-height of the peak. According to the results of measuring the broadening of quartz lines for the interplanar distance of 1.98 Å, where there is no influence of the preferred direction, the crystallite sizes were calculated with an accuracy of ± (8–10%). The measurement results (Fig. 5.4) showed that both the linewidth (FWHM) and the crystallite size (L) of quartz do not change linearly with the processing time in a mechanical reactor. The periodicity in increasing and decreasing these characteristics, i.e., in the defectiveness of the structure of particles is traced. The observed periodicity indicates the process of material dispersion and the formation of a defective structure (accumulation and annihilation of defects) in particles [2]. Broadening of the lines is a total reflection of the increase in dispersion, accumulation of defects and amorphization of the surface.

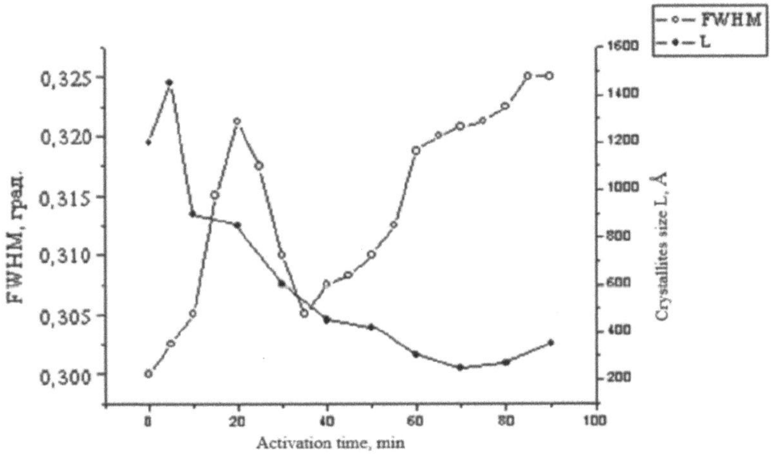

Figure 5.4 Dependence of the change in the broadening of X-ray lines (FWHM) and crystallite size (L) of quartz powder particles on the time of treatment.

The obtained regularities make it possible to reveal a number of extreme points reflecting the peculiarities of the structural rearrangements in quartz. So, when the time of processing quartz is 15–20 min, a significant increase in the broadening of X-ray lines and inhibition of crystallite size changes are observed, which may be associated with an increase in particle dispersion, accumulation of surface defects and, conversely, regeneration of bulk defects or their release to the surface. The processing time interval of 30–40 min, judging by the intensity of the decrease in both the width of the X-ray line and the crystallite size, is characterized by the most significant volume changes in the structure of the material being dispersed. At the same time, recombination of surface defects and agglomeration of particles are possible, which is reflected in a decrease in the specific surface of the powder particles after 40 min of treatment. Judging by the presented results of measurement of these characteristics, a longer treatment of quartz (from 50 min or more) leads to a decrease in the size of crystallites and a change in the structure and state of the surface layers of particles under conditions of reduced dispersion due to powder agglomeration.

Thus, as was shown earlier [3, 4] and confirmed by the results of the research [2], one of the important factors of the mechanical treatment process is the periodicity of structural changes in the volume and on the surface of ground quartz particles, where broken bonds play a significant role [5, 6].

The degree of imperfection of the structure associated with the change in the number of broken bonds can be traced, in particular, by the change in the relative amount of silicon in quartz with the activation time. This dependence is also periodic in freshly activated quartz (Fig. 5.5). The relative amount of silicon is higher than that according to the stoichiometry for SiO_2 (i.e., above 46.7%). This fact indicatesthe breaking of bonds along the surface of the particle and the decrease in the amount of oxygen in the amorphized surface and near-surface layers, where defect "preservation" is possible, unlike the surface that is active to the gaseous processing medium. After a month of aging of activated quartz, the relative amount of silicon decreases, the periodicity disappears, and the effect of processing time is practically leveled out [7].

The change in the structure and state of the material after mechanical treatment can be investigated by various physico-chemical methods [8, 9]. One of the most structurally sensitive for determining the state of the material methods is a dilatometric analysis. Thermal expansion, like other thermophysical characteristics of a substance, reacts subtly to all changes, both in the phonon and electronic subsystems of the crystal lattice of a solid. The defects accumulated by the material under mechanical action affect the spectrum of phonon oscillations and render an influence on the electron density in the defective parts of the crystal lattice. Both facts counted for thermophysical properties, including the intensity of thermal expansion of the material.

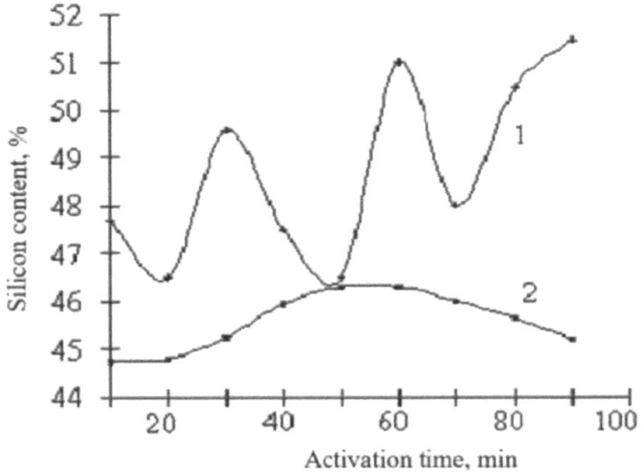

Figure 5.5 Dependence of silicon content in quartz on the time of mechanical treatment of the sample. (1) freshly activated quartz, (2) aged for a month.

The dilatometric curves of quartz have a specific form [10] and consist of three sections with different linear thermal expansion coefficient (LTEC) in the range of heating temperatures up to 1000°C. The first section of the curve describes the expansion of α-quartz, the third one-that of β-quartz, and the second one describes LTEC, much higher than the other two, corresponds to the state of preparation for transition $\alpha \rightarrow \beta$, which is characterized by weakening of bonds predetermining the rearrangement of the trigonal lattice into a hexagonal one. Figure 5.6 shows the dilatometric curves of quartz in the initial and activated states at different exposure times in a planetary centrifugal mill.

After grinding of quartz (Fig. 5.6), there occurs a decrease in the intensity and degree of thermal expansion. The most significant change in the course of the curve form takes place after 20 min of activation. The decrease in the intensity and level of thermal expansion after mechanical treatment can be associated only with a significant compression of the crystal lattice and a decrease in the amplitude of phonon oscillations at its nodes.

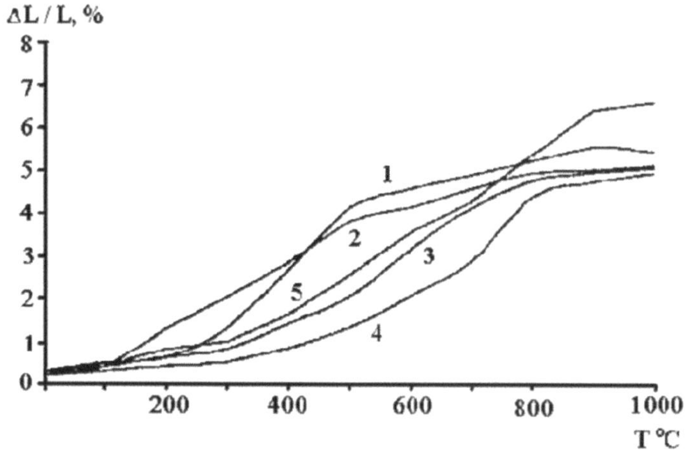

Figure 5.6 Dilatometric curves of quartz in the initial state (1) and activated in a planetary centrifugal mill for 10 (2), 15 (3), 20 (4) and 30 min (5).

Since compression in a planetary centrifugal mill prevails over other types of deformation, quartz processed in such a mill shows a decrease in the effect of thermal expansion. After annealing, which removes the deformation distortions of the lattice, thermal expansion of the samples approaches to that of the initial nonactivated material.

In order to quantify the change in the state of the material after mechanical treatment, the coefficient of thermal expansion (α) was calculated for each of the three periods of quartz state during heating. These are the α-quartz region, the region corresponding to the transition state $\alpha \leftrightarrow \beta$ (400–573°C), and the β-quartz region. The dependence of LTEC (α) on the activation time for three periods of quartz heating is shown in Fig. 5.7. The most significant changes in the state of the material, which are manifested both in the region of relatively low temperatures (α-quartz) and during heating above 500°C, are associated with mechanical treatment of up to 18 min. LTEC of quartz as for α-, and β-state from the processing time within 20–35 min does not change. Longer grinding leads to some changes in the thermal expansion characteristics of quartz. Significant changes in the state of the material occur at 10–16 min of mechanical action, as a result, in the β-region, after 12 min of activation, LTEC decreases, and after 14 min—on the contrary, its sharp increase is observed. For the β-region, the minimum and maximum of LTEC fall on 10, 15, and 12, 16 min of processing, respectively.

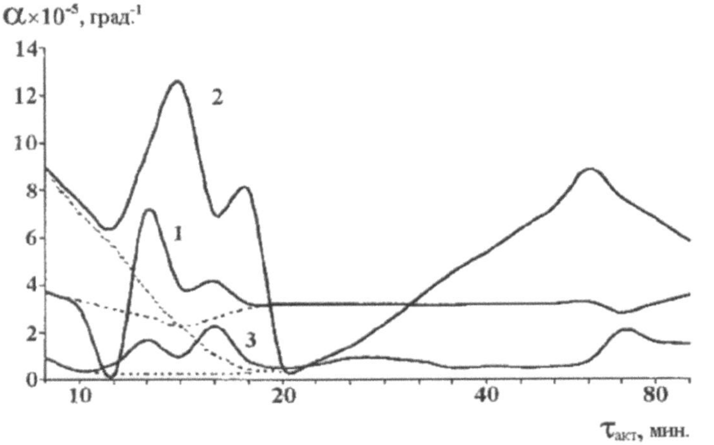

Figure 5.7 Dependence of the coefficient of linear thermal expansion of a quartz sample on the time of preliminary mechanical activation. Measurements were carried out for temperature regions of α-quartz (1), before transition $\alpha \rightarrow \beta$ (2) and β-quartz (3).

The strongest impact of preliminary mechanical treatment of quartz manifests itself at the stage of pre-state of $\alpha \leftrightarrow \beta$ transition (400–573°C). In addition to a profound change in the structure of the material after 10–15 min of grinding, there is a sharp decrease in LTEC after 20 min of processing. A further increase in the time of mechanical action contributes to an increase in the coefficient of thermal radiation of the sample during heating in this temperature range. For quartz, processed in a planetary centrifugal mill, there is no effect of increasing LTEC in the temperature range after 12–16 min of pre-activation. In this case, the minimum on the dependence curve $\alpha = f(\tau_{\text{act}})$ after 20–25 min of activation is no less clearly expressed.

Thus, preliminary mechanical treatment is most pronounced at temperatures that determine the polymorphic transformation of a substance. This fact is very important when choosing the optimal modes for the subsequent synthesis on pre-activated materials, taking into account the most effective time of its mechanical processing. The dependence of LTEC on the time of material activation indicates quite well the change in its state as a result of mechanical activation and the implementation of these changes in the process of subsequent heating.

The above-considered regularities in the change of thermophysical properties after mechanical treatment of quartz in an activator mill are characteristic of quartz, which has a phase transition at 575°C. Dilatometric studies, for example, for iron oxides (Fe_2O_3) after mechanical activation, showed that LTEC changes only upwards with activation time, and after annealing decreases to the initial value (Fig. 5.8).

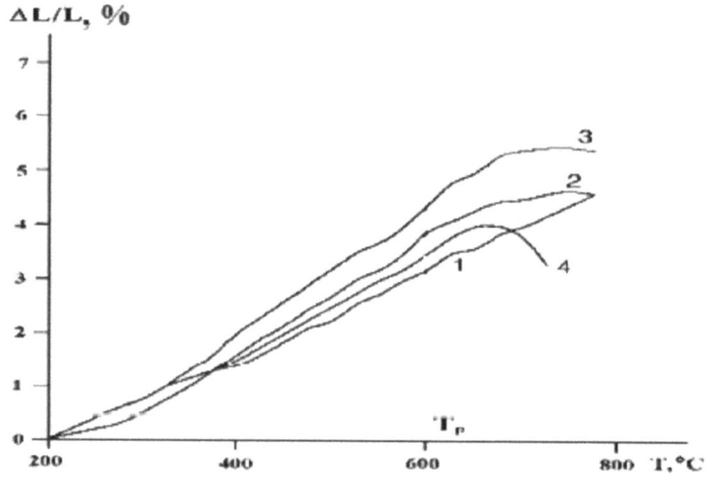

Figure 5.8 Dilatometric curves of iron oxide in the initial state (1) and after 20 (2), 30 (3) min of activation and after annealing of the activated sample (4).

Similar changes occur during the dispersion of alumina (Al_2O_3). According to [9], processing of these materials in mills of dynamic action occurs with an intense decrease in the size of particles, without significant amorphization of their surface.

From the presented results, it follows that the structure of quartz and especially the surface layers of particles after mechanochemical treatment undergoes significant changes. The material is highly electrified. An assessment of the state of quartz associated with structural changes can be carried out by measuring its electrical resistivity. Measurements of the electrical resistance of ground quartz were carried out on a screened powder fraction of less than 60 μm compressed in ampoules to a density (3.7–4.0) × 10^{-3} g/mm^3. The measurement results showed a decrease in the resistivity of quartz after grinding by more than an order of magnitude [11]. The longer the grinding, the lower the resistance of the quartz powder (Fig. 5.9a).

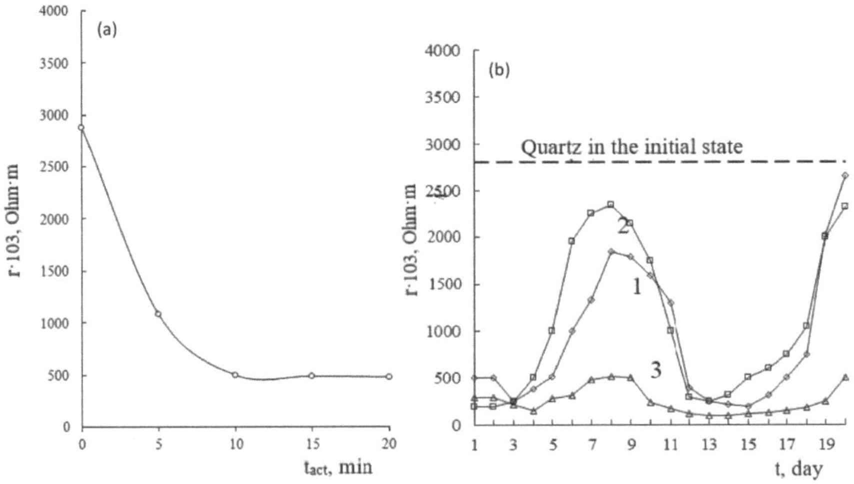

Figure 5.9 The change in electrical resistivity of quartz powder depending on the time of mechanical processing (a) and subsequent weathering (b) of the samples after activation of 5 (1), 10 (2) and 20 (3) min.

The resistivity of the mechanically activated quartz powder continues to decrease during the first 2–3 days, and then its increase, followed by a decrease after 10–12 days and an increase in values to the level of quartz, after 18 or more days of weathering (Fig. 5.9b) is observed. The results indicate the periodicity in the development of relaxation postprocesses in the material, associated with the redistribution of the defective structure, the removal of elastic stresses, possibly the recombination of radicals in the surface layer and a decrease in the surface charge.

Thus, from the presented results, it follows that the process of activation of quartz in the mills of dynamic action is of a periodic character. The preferred accumulation of defects during grinding first occurs in the volume of the particle, then is replaced by an enhanced restructuring of the structure on the surface. Deep changes in the structure of the surface layers are manifested in the change in the thermal and electrical properties of the particles. Accumulation of these changes contributes to agglomeration of fine particles during long-term grinding of powders, i.e., formation of structural complexes. The processing time in the mill with an acceleration of 20 g for quartz for 20 min, judging by all the characteristics of the activated powder, provides the most significant changes in the structure, both in volume and on the surface of the resulting particles. This creates a prerequisite for active chemical interaction under these conditions when co-processing quartz with other materials and creating new structures and compounds on the surface of quartz particles.

Questions

1. What is the frequency of the grinding process and changes in the surface state and structure in the volume of particles?
2. Give the assessment of the degree of imperfection of quartz particles as a result of mechanical processing.
3. What is the dilatometric analysis, as a method of state change in the phonon and electronic subsystems of the crystal lattice as a result of mechanochemical processing of a solid?
4. Describe the features of the change in thermal expansion of quartz after mechanical processing associated with polymorphic transformations of the substance.
5. What changes in the electrical properties of particles associated with the features of the structure of the surface layers occur after mechanical treatment?

5.2 Structure, State, and Properties of Quartz Powder after Mechanical Treatment with Modifiers

Investigations on the change in the structure and state of quartz particles during processing in a mechanical reactor showed the presence of extreme points in the change of various characteristics of the activated material. Processing for 20 min provides the most significant changes in the structure and especially the surface layers of quartz particles. This fact is especially important when choosing the optimal time regime for grinding quartz with various additives that provide surface modification of the particles. In this regard, this section presents mainly experimental material on the mechanochemical modifying treatment of quartz for 20 min in a planetary centrifugal mill. For a comparative analysis in a number of cases, the results obtained at 10 and 30 min of treatment are presented.

As is known [9], alcohols, being surface active substances, actively contribute to grinding. Therefore, various alcohols (one-, two-, and three-atom) were used as modifiers. In this regard, the action of monohydric alcohols is especially effective. The dispersion of the quartz powder, processed in the presence of (5 wt.%) ethanol (C_2H_5OH) and butanol (C_4H_9OH), increases significantly, and the dispersion in particle sizes decreases (Fig. 5.10).

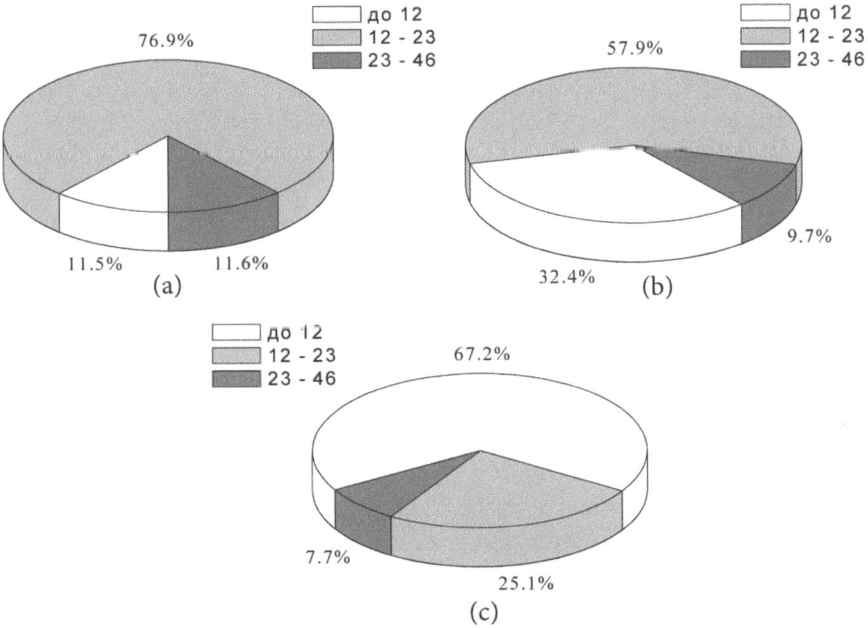

Figure 5.10 Particle size distribution after 10 min of grinding for quartz (a) and quartz treated with ethanol (b) and butanol (c).

At the same time, a distinctive feature of particles treated in the presence of butanol is a slight increase in the particle size of the main fraction, which constitutes more than 50% of the powder volume (Fig. 5.11). A similar picture takes place when ethylene glycol ($C_2H_4(OH)_2$—dihydric alcohol) and glycerol ($C_3H_5(OH)_3$—trihydric alcohol) are used as additives.

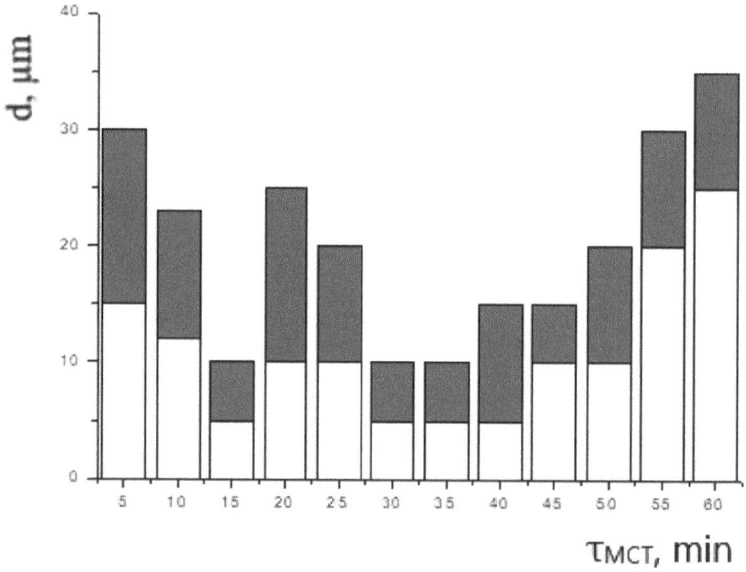

Figure 5.11 The dependence of the particle size of the main fraction, constituting more than 50% of the volume of the powder obtained in the presence of butanol (5 wt.%) on the processing time.

Some coarsening of the particles and an increase in the polydispersity of the powder after 20–25 min of grinding with these modifiers can serve as indirect evidence of the change in their sizes due to modification (growth) of the surface layers. An increase in the grinding time of more than 40 min leads to coagulation of small particles. The fractional composition of the modified powder after long treatment is more monodisperse, i.e., the scatter in the particle size becomes much smaller than for unmodified quartz (Fig. 5.11).

The specific surface and specific pore volume of particles with a modified surface are more sensitive characteristics of changes in the structure and micro-composition of the surface layers of particles treated in the presence of organic additives. Table 5.1 shows the results of measurements of these characteristics for quartz, treated with alcohols, and also, for comparison, with carbon, which, above all, provides mechanical coating of particles dispersed in its presence. For the comparative assessment, the most effective time of mechanochemical processing (τ_{MCT}), and the same amount of the modifier introduced were chosen.

Table 5.1 Specific surface area (S_{SSA}), specific pore volume (V_{SPV}), apparent (ρ_a), pycnometric (ρ_p) density and degree of modified volume (ζ) of quartz powder after 20 min of mechanical treatment with various modifiers

Material	τ_{MCT}, min	S_{SSA}, m^2/g	V_{SPV}, cm^3/g	Density ρ, g/cm^3 (ρ_a)	(ρ_p)	ζ, %
SiO_2	0	2.64	0.01			
	10	3.56	0.04			
	20	5.20	0.11	2.55	2.65	
SiO_2+5%C_2H_5OH	20	5.24	0.18	2.03	2.06	17.1
$SiO_2$5% C_4H_9OH	20	6.50	0.32	1.67	1.69	32.4
SiO_2+5% C_2H_4 $(OH)_2$	20	10.50	0.24	1.59	1.70	25.8
SiO_2+5%C_3H_5 $(OH)_3$	20	15.20	0.29	1.46	1.69	18.5
SiO_2+5% C*	20	77.60	0.72	0.88	0.91	72.7

*C, activated coal.

For quartz, a significant increase in the specific surface after 20 min of treatment is associated with an increase in the proportion of fine particles, and an increase in the specific pore volume is associated with the loosening of the surface layer. An increase in the specific surface and at the same time the specific pore volume of quartz particles processed in the presence of various alcohol additives can only be a consequence of a change in the structure of the surface layers of the particles, i.e., their modification. The largest specific pore volume of particles modified with butanol indicates the formation of a loose surface of particles, which ensures a sufficiently high sorption of benzene vapor. However, its value is still significantly less than when using activated carbon in a mixture with quartz.

The dispersion of the powder, the structure and surface state of the particles significantly affect the bulk density of the powder. Table 5.1 shows the pycnometric density of the powder in comparison with the data on apparent density. Based on these measurements, the degree of modified volume of quartz powder can be calculated:

$$\zeta = [1 - (\rho_a - \rho_{mod\cdot})/(\rho_{SiO2} - \rho_{mod\cdot})]\ 100\%,$$

where ρ is the density of the material, g/cm^3; m weight of the material, g; ρ_{SiO2} true (pycnometric) density of a quartz particle, g/cm^3; ρ_a apparent density of the material after processing, g/cm^3; ρ_p pycnometric density of the material after processing, g/cm^3; and ρ_{mod} = (ρ_p − ρ_a) the density of the modified part of the material, g/cm^3.

The decrease in the pycnometric density and the difference between the pycnometric and apparent density indicate an increase in the porosity of the analyzed powder. The calculation of the degree of the modified volume of the surface layers of quartz showed that, among the selected alcohols, the most significant changes in the surface of quartz particles (up to 32.4%) occur in the presence of butanol. An increase in the degree of the modified volume of over 70% when treating quartz with 5% activated carbon can take place only in case when not only enveloping the quartz with carbon, but also their active interaction with changes in the structure of the surface layer of the particles occurs.

Acrylic acid ($H_2C=CH-COOH$) and styrene ($CH_2-CH-C_6H_5$) are known [12, 13], as one of the most active modifiers of inorganic materials. Already with a weak mechanical action (abrasion in the mortar), acrylic acid and styrene polymerize. Polystyrene with intense mechanical abrasion is susceptible to degradation and repolymerization [12, 14]. Therefore, acrylic acid and polystyrene were used as modifiers with a pronounced ability to polymerize and degrade. Table 5.2 shows the results on measurements of density, specific surface, specific pore volume and the degree of modification of the powder volume by these organic compounds, as well as in a mixture with sodium chloride and activated carbon, which, as is known [15], also actively affect the process of destruction and polymerization.

Large values of the degree of modification of the volume of quartz powder treated with acrylic acid and polystyrene are a consequence of a change not only in the structure, but also in the morphology of the particles, and it is possible that there is both a polymerized layer on the particle surface and a polymer unbound to the particle, which is also part of the powder.

Table 5.2 Specific surface area (S_{SSA}), specific pore volume (V_{SPV}), apparent (ρ_a) and pycnometric (ρ_p) density, degree of modified volume (ζ) of quartz powder after 20 min of treatment

Material	S_{SSA}, m²/g	V_{SPV}, cm³/g	Density ρ, g/cm³ (ρ_a)	Density ρ, g/cm³ (ρ_p)	ζ, %
SiO_2	5.20	0.11	2.55	2.65	—
SiO_2+5% AA*	56.6	0.23	0.82	0.94	74.0
SiO_2+5% AA+5%C	52.7	0.43	0.87	0.97	70.0
SiO_2+5%AA+1%NaCl	45.8	0.39	0.97	1.05	65.8
SiO_2+5%PS**	21.0	0.30	0.65	0.67	72.7
SiO_2+ 10% PS	46.7	0.36	0.58	0.62	73.0
SiO_2+10% PS +5%C	42.3	0.37	0.62	0.69	79.0
SiO_2+5% PS+1% NaCl	54.3	0.47	0.73	0.77	74.0

*AA, acrylic acid, **PS, polystyrene.

Analysis of the dispersity of the modified particles showed (Fig. 5.12) that their size increased significantly, respectively, to 90 and 120 μm, compared with quartz ground without a modifier, when the largest particles did not exceed 50 μm. In the presence of sodium chloride, the polymerization process increases and the amount of coarse fraction (up to 90–120 μm) of particles in the powder increases [16]. Thus, the grinding of quartz in the presence of various organic additives significantly affects grinding process, the degree of dispersion of the material and the transformation of the structure of the surface layers of the particles.

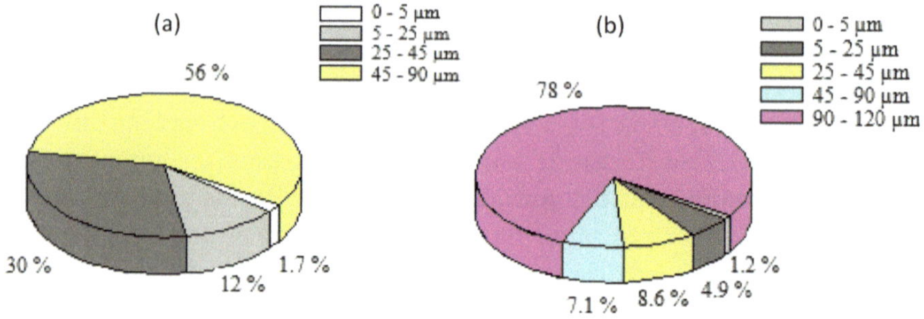

Figure 5.12 Particle size distribution for quartz treated with acrylic acid (a) and in the presence of a mixture of acrylic acid with NaCl (b).

The results of X-ray analysis show that the presence of an organic modifier affects the change in the defective structure in the particle volume, i.e., in a quartz core. Organic formations on the surface of a particle are not registered by the X-ray diffraction method, in this case they are "transparent" for it. The presence of modifiers, especially acrylic acid, after 20 min of processing clearly contributes to a decrease in crystallite size (Fig. 5.13), i.e., an increase in the volume defectiveness of samples and, consequently, a change in its energy state [17].

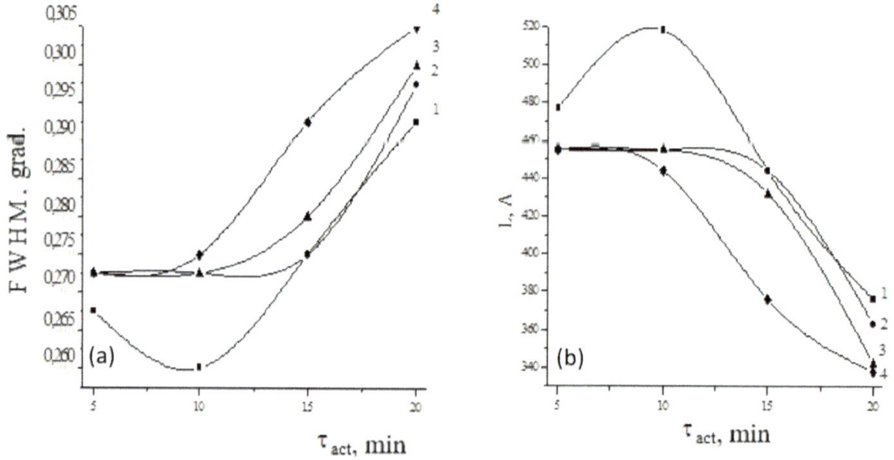

Figure 5.13 Dependence of the change in X-ray line width (FWHM) and crystallite size (L) on the processing time of quartz activated without (1) and in the presence of modifiers: butanol (2), acrylic acid (3) and polystyrene (4).

Quartz, crushed in the presence of polystyrene, shows the greatest broadening of the x-ray line, a change in the size of the crystallites. Therefore, changing the modifying organic additive can significantly affect the change in the internal structure of the dispersible inorganic material.

The fact that the synthesized particles are complex micro-compositional objects is verified by the results of dilatometric studies, indicating simultaneously the change in the energy state of the material. After treatment of quartz with alcohols, acrylic acid and polystyrene, the LTEC calculated in the temperature range (T1p and T2p) from 200 to 573°C increases noticeably (Table 5.3).

Table 5.3 Linear thermal expansion coefficient of quartz after mechanochemical treatment with various modifying additives

	$\alpha \times 10^{-5}$						
τ_{mct}, min	—	C_2H_5OH	C_4H_9OH	$C_2H_4(OH)_2$	$C_3H_5(OH)_3$	*AC	**PS
10	3.49	3.55	4.76	4.25	4.5	5.92	6.07
20	1.75	3.72	5.26	4.78	5.10	6.17	6.29

*AA, acrylic acid; **PS, polystyrene.

Polymer compounds synthesized by mechanical treatment and modifying quartz, are destroyed in the process of heating, which clearly follows from the course of the dilatometric curves (Fig. 5.14). However, the features of the dilatometric curves for samples modified with acrylic acid and polystyrene, namely, the presence of a site in the range up to 150–200°C, suggests that not just organic polymers are formed, but more complex compounds that significantly reduce the LTEC of the material. The presence of organic additives and their burnout would intensify the thermal expansion of the sample. The observed course of the curves indicates the modification of the structure of quartz particles, the presence of new compounds on their surface.

A certain idea of the surface of the particles after treating quartz with organic additives can also be obtained from the results of measuring the electrical resistance of the modified material. According to the measurement results (Table 5.4), the electrical resistivity of quartz modified with organic compounds immediately after treatment for 10 min is significantly higher than for quartz treated without modifiers.

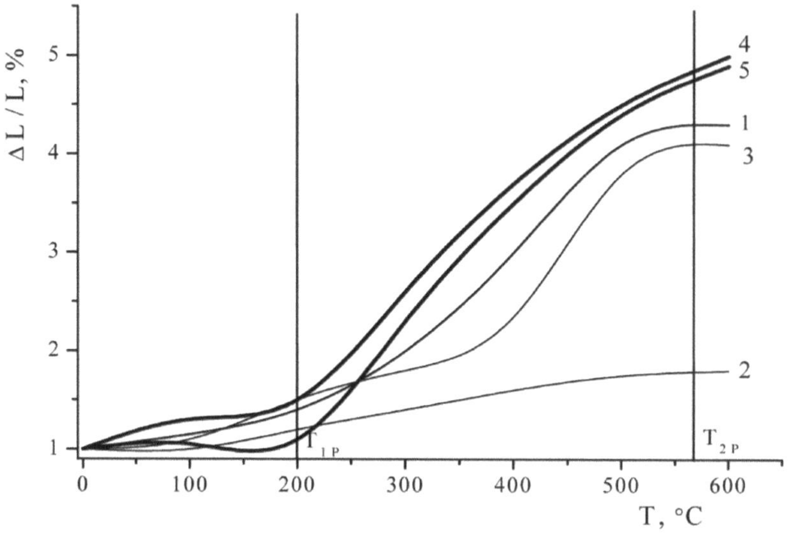

Figure 5.14 Dilatometric curves of quartz, in the initial state (1), activated without modifiers (2) and in the presence of 5 wt.% butanol (3), acrylic acid (4) and polystyrene (5).

Table 5.4 Electrical resistivity of quartz powder after mechanochemical treatment with various modifying additives

τ_{mct}, min	$\rho \times 10^6$, Ohm·m Modifying additives							
	—	C_4H_9OH	AA*	PS**	AC***	NaCl	AA*+ 1%NaCl	PS**+ 1% NaCl
10	0.565	5.520	1.915	2.525	6.045	0.180	0.730	2.100
20	0.627	6.700	1.080	2.930	4.750	0.150	0.540	1.855

*AA, acrylic acid; **PS, polystyrene, ***AC, activated coal.

With an increase in the time of the mechanochemical treatment of quartz with modifiers, a decrease in the electrical resistance of the material is usually observed, especially if sodium chloride is present in the system. However, the subsequent state changes associated with stress relaxation and restructuring of the structure after grinding, occur differently in systems with different modifiers [11, 18]. When using organic modifiers, the electrical resistance of the powder continues to decrease or stabilizes during the first two or three days after obtainig the material. Then the growth of resistance begins and when modifying quartz with alcohols, its values eventually approach the values of unmodified quartz. When using acrylic acid and polystyrene as modifiers, the values of electrical resistance of the samples even after a month of aging of the material remain significantly lower than for unmodified quartz, i.e., there takes place stabilization of the state.

As a result of complex analysis using different methods, which make it possible to evaluate structural adjustment at the mechanochemical treatment of quartz in the presence of organic additives at various levels, it is shown that the degree of change in the surface layer with participation of organic substances depends on the type of modifier. In the presence of alcohols, the dispersion of particles is enhanced, and a number of features are noted in the change in the structural characteristics of quartz particles in the presence of butanol. The observed changes may be due to the formation of a loose surface layer of the modified particle.

The use of more structurally complex organic modifiers (acrylic acid, polystyrene) for all analyzed characteristics results in formation of particles with a pronounced micro- or nanocomposite structure. Features of the structure of such powder materials with a complex morphology of particles are displayed in the change in their heat and electro-physical properties. The observed analogy in changing the electrical properties of quartz modified with acrylic acid, polystyrene is a consequence of the fact that carbon resulting from the decomposition (mechanocracking) of organic compounds plays a decisive role in formation of the surface layer of particles. A similar decomposition to carbon and hydrogen during mechanical processing was stated when processing natural gases in mechanical reactors in [19], which was especially pronounced in such cases when a small amount of quartz was used in a mechanical reactor.

Questions

1. Describe the features of dispersion of quartz particles in the mill in the presence of monatomic, diatomic and triatomic alcohols.
2. How do the changes in the structure of surface layers of quartz particles and formation of a modified surface happen?
3. Estimate the degree of the volume of quartz powder modified with organic compounds during mechanical processing.
4. How does an organic modifier effect the change of the defective structure on the surface and in the volume of the particle during mechanical treatment?
5. Describe the features of the change in thermal and electrophysical properties of modified quartz particles during mechanical processing.
6. How does the modification occur as a result of the decomposition (mechanocracking) of organic compounds and the stabilization of the state of ultrafine particles?

5.3 Morphological and Structural Features of Activated and Modified Quartz

5.3.1 Electron Microscopy of Modified Quartz

The most informative method for studying the structure and morphology of highly dispersed particles is electron microscopic analysis. The possibilities of this method in studying the transformation of surface layers of modified quartz particles are very extensive and make it possible to visually follow the entire path of creating composite systems with participation of organic and inorganic materials. When grinding quartz in a mill without modifying organic additives, there proceeds partial amorphization of the surface layer of the particle and the introduction of rubbed ultrafine iron rubbed from the walls of the grinding vessel and balls into it, this substantially changing the electrical properties of the powder material. The main part of the volume of such particles remains crystalline, which is confirmed by the results of electron microdiffraction [20]. When an organic additive is introduced into dispersible quartz, the particle structure undergoes significant changes. The surface layer can be a multilayer formation with a different density structure (Fig. 5.15b).

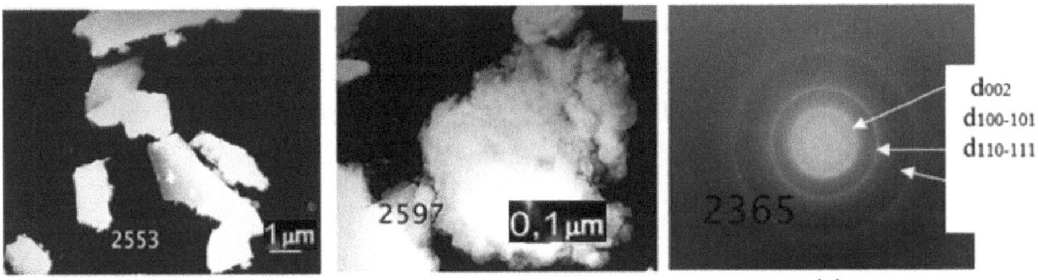

(a) (b) (c)

Figure 5.15 Electron-microscopic images (a, b) and electron microdiffraction of quartz (c) in the original (a) and modified with butanol (b, c) state.

Electron microdiffraction of quartz particles modified in the presence of butanol (Fig. 5.15c) indicates the formation of carbonaceous structures in the surface layer of a quartz particle with a thickness from 10 to 40 nm [11, 18–21]. Using the results of electron microdiffraction, you can trace the whole process of phased transformation of quartz particles (Fig. 5.16a) into a microcomposition material containing particles with partially amorphized (Fig. 5.16b) and a modified surface, up to the formation of dense carbonaceous structures on the surface of quartz (Fig. 5.16c), processed, in particular, with butyl alcohol.

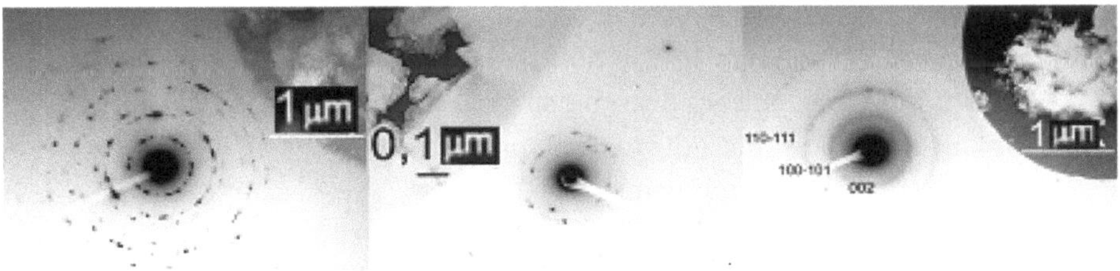

(a) (b) (c)

Figure 5.16 Electron diffraction patterns of quartz in the original (a), activated (b) and modified (c) state.

The morphology and carbonization of the surface layer of particles treated in the presence of butanol are similar to those observed when treating quartz directly with carbon. The more complex the carbon-containing modifier used in its structure (monatomic and polyatomic alcohols, acids, and polymers), the greater the diversity observed in the morphology of the modified surface layer of quartz particles during mechanochemical processing. These changes are primarily associated with participation of carbon as a modifying additive. So, when using polyatomic alcohols, for example, glycerol, a feature of the structural composition is the formation of spherical particles and their dendriform-like aggregates with a high specific surface (Fig. 5.17).

(a) (b) (c)

Figure 5.17 Electron microscopic images of quartz particles modified in the presence of glycerol for 10 (a), 20 (b) and 30 min (c).

Modification of quartz particles with acrylic acid during mechanical treatment leads to formation of a dense, fairly uniform organic film on the surface and the presence of highly dispersed crystallites in the surface polymer layer. Formation of a homopolymer unbound to the surface of quartz is also observed. In this case, the microdiffraction pattern indicates the formation of quartz-organic composites (Fig. 5.18).

When introducing chlorides, in particular sodium chloride, in addition to acrylic acid used as a modifier quartz particles become more rounded "spherical" in shape, and the modified layer of quartz particles is characterized by different density and heterogeneous micro-composition.

Electron microscopic images clearly illustrate that a quartz particle modified by mechanochemical processing is a multilayer formation of various structural combinations and forms with nanoscale organometallic compounds on the surface [2]. The microdiffraction pattern in Fig. 5.18a refers to a three-dimensionally ordered (up to graphite structural packing [22]) carbonaceous matter on the surface of a particle. The diffraction pattern in Fig. 5.18b shows the presence of a metal-silicon-carbon compound (point reflections) and a weakly crystallized silicon-carbon substance (reflexes are shown by arrows).

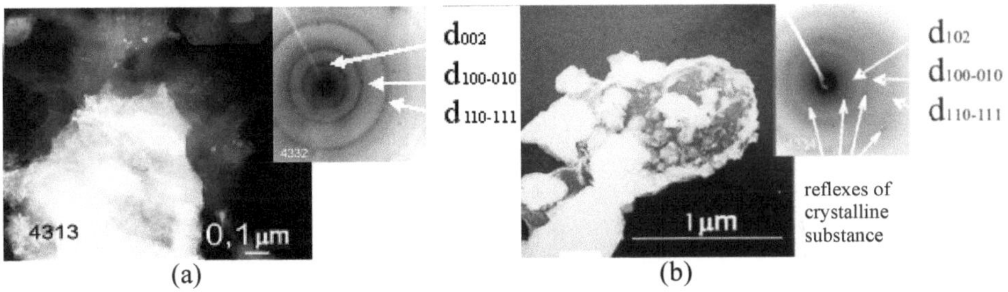

Figure 5.18 Electron microscopic images and electron microdiffraction of quartz particles modified in the presence of acrylic acid (a) and acrylic acid with sodium chloride (b).

The variety of structural forms of the modified surface of quartz is most pronounced when using polystyrene as a modifier. When joint processing of quartz with polystyrene, the destruction of the quartz particle and the destruction of polystyrene molecules into polyine and aromatic components proceed simultaneously. In this case, the structure of the active surface of quartz changes with participation of polycyclic aromatic molecules with the simultaneous participation of polyinc compounds, this corresponding to the known concepts of soot formation depending on the raw material used [22]. Depending on the amount of modifier used (from 3% to 10%), the time of mechanical impact, as well as variations in the force conditions of exposure, structured or delicate films are formed cross-linked with the surface or rolled up in the form of tubes of various configurations, the size of which reaches 50–70 nm (Figs. 5.19, 5.20).

Complication of the structure of particles and formation of different conglomerates proceed differently at different stages of the dispersion of mixtures (quartz+polymer). First, a mixture is formed with the preferred formation of a homopolymer, then a growth of neoplasms of a rather complex structure on the surface of the particle and an organic, dense film encapsulating it is observed.

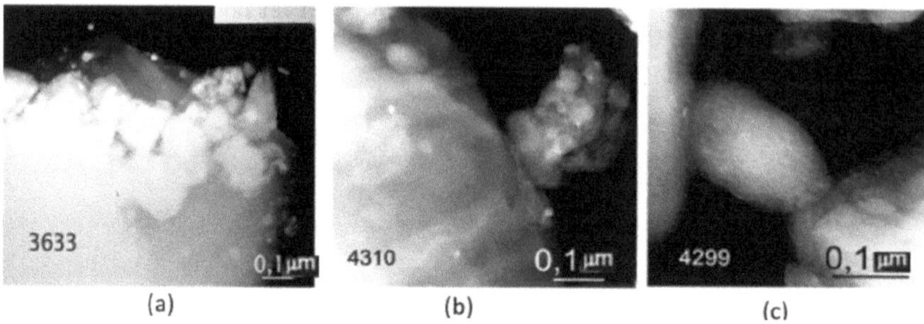

Figure 5.19 Electron-microscopic images of particles of quartz, modified by mechanical processing with 5 wt.% polystyrene for 10 (a) and 15 (b, c) min.

An increase in the processing time ultimately leads to coagulation of dispersed particles with a modified surface layer and formation of dendriform-like formations, the growth of which occurs according to the principle of similarity of fractal structures. On the surface of such particles, flat and rolled (tubular) film, as well as openwork nanostructured formations grafted to the surface are formed. So, the morphology of the system changes in an unexpected way and the main element involved in the construction of nanoscale objects during the mechanochemical treatment of quartz is carbon [2, 11]. As shown by the results of electron microdiffraction, a distinctive feature of carbon structures on the surface of quartz subjected mechanochemical treatment in the presence of polystyrene is the presence of texturing, that is, the preferential orientation of carbon particles inside the film formations (Fig. 5.20e).

Thus, in the process of mechanochemical treatment of quartz with organic compounds, a complex multistage process of the formation of new carbon-containing structures on the particle surface takes place, which, according to [23], can be considered as a result of inoculation to radical centers ($\equiv Si^{\cdot}$ and $\equiv SiO^{\cdot}$) of degraded organic compounds arising on the surface of the split groups. The degree and form of "carbonization" of the quartz surface is determined by the type of the modifier used.

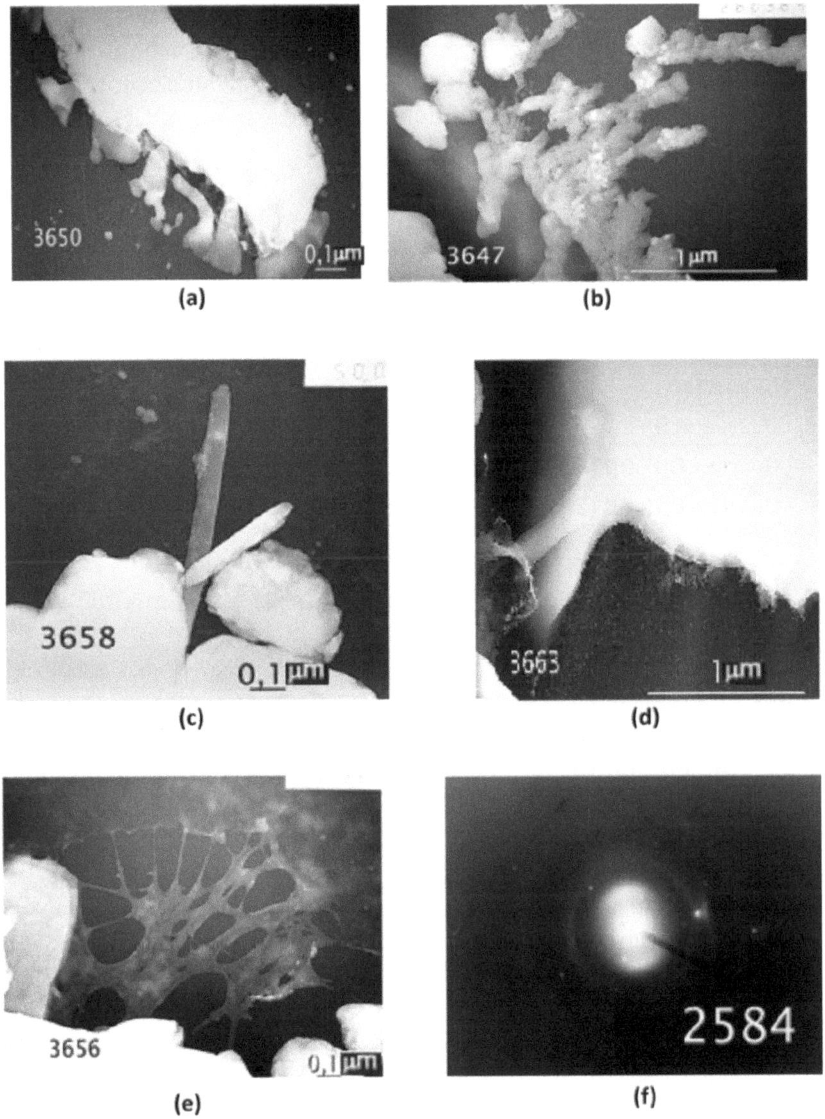

Figure 5.20 Electron-microscopic images (a–e) and electron micro-diffraction (f) of quartz particles modified by mechanical treatment with polystyrene for 20 (a, b) and 30 (c–e) min at a ratio of Mn/Mn = ½ (a–d) and ¼ (e).

The amount of carbon bound in the modified quartz was determined by the adsorption-weight method (Table 5.5). With the same weight amount of the modifier introduced and the same treatment time, the maximum amount of carbon—2.6%—was found in samples with polystyrene, which showed the greatest variety of structural forms on the surface of the modified particle [2, 24].

Table 5.5 Content of carbon and iron in the modified surface layer of quartz particles after mechanochemical treatment with various modifiers

Material	τ_{act}, min	C, %	Fe, %
Quartz (initial)		0.08	0.05
Quartz (act)	10	0.08	2.32
Quartz (act)	20	0.08	3.34
Quartz + C_2H_5OH (5 %)	20	1.06	11.62
Quartz + C_4H_9OH (5 %)	20	1.76	14.86
Quartz + $C_2H_4(OH)_2$ (5 %)	20	1.45	6.78
Quartz + $C_3H_5(OH)_3$ (5 %)	20	1.53	4.94
Quartz + AA* (5 %)	20	2.03	5.62
Quartz + PS** (5 %)	20	2.60	7.08

*AA, acrylic acid; **PS, polystyrene.

In addition to the data on the carbon content in the volume of modified quartz, the table shows the results of X-ray analysis of the iron content in quartz after its mechanochemical treatment. Iron is the second element introduced into quartz during its grinding, and actively participating in the structure formation of the surface layer of modified silica powder particles. The amount of iron in the treated quartz varies with the type of modifier used. The highest content of iron was found in samples ground in the presence of alcohols, and especially when using butanol.

Thus, electron microscopic images of quartz particles modified by mechanochemical treatment with various organic additives clearly demonstrate the formation of nanoscale structures on the surface of particles. The modified surface layers can have both loose and dense structures and in some cases are completely new formations that can be attributed to tubular forms. Silicon, oxygen, carbon, and iron are involved in the formation of such structures. Information for a more specific description of the structure and composition of the substance of surface nanolayers of modified quartz particles can be obtained using spectral analysis methods.

5.3.2 IR Spectroscopy of Activated and Modified Quartz

Spectroscopic methods of analysis provide information on the components of the surface structure of both dispersed and extended objects. In addition, it is possible to imagine the relationship between the elements that make up the various groups from which the structure of the material is formed. According to the calculated and experimental data [25, 26], the IR spectra of quartz, as well as other modifications of SiO_2, consist of seven main optical absorption bands, which are combined into doublets and triplets and occupy four frequency ranges: 1250–1000 cm^{-1}, 835–700 cm^{-1}, 550–400 cm^{-1}. The band lying at 690 cm^{-1} is specific only for the quartz lattice, which distinguishes it from other modifications of silica. Specific peaks of the bands are shifted depending on the purity of the material, the degree of lattice disorder and its defectiveness. The IR spectra can be used to judge the mineral deposit and its structural features [27].

Figure 5.21 shows the IR spectra of quartz in the initial state, in which all the above research results were obtained. It is a mineral from the Tekturmyk (T) deposit. The pattern of infrared absorption spectra of Si–O–Si groups characteristic of quartz consists of the region of frequencies of stretching vibrations (1200–850 cm^{-1}) and deformation vibrations (600–400 cm^{-1}). In the studied mineral, there is adsorbed water, the absorption bands of which are of low-intensity and occur at 3427 cm^{-1}.

The deformation vibrations of Si–OH groups are represented by the average intensity band of 1428 cm^{-1}. These frequencies also include deformational vibrations of Si–OC groups. The spectra of the Si–O–Si groups of the quartz under study are represented by the main absorption band at 1085.92 cm^{-1}, the other two stretching vibration bands at 1173.75 and 965.39 cm^{-1} are weakly expressed. The spectra of deformation vibrations at 798–783 cm^{-1} and 513 cm^{-1} are also of low-intensity. Deformation vibrations at 468 cm^{-1} are represented well. There are also a number of spectra of low intensity on the right slope of the main Si–O–Si band, in the frequency range 1000–800 cm^{-1}, which can be assigned to Si–OC, Si–C and C–H groups [26, 27].

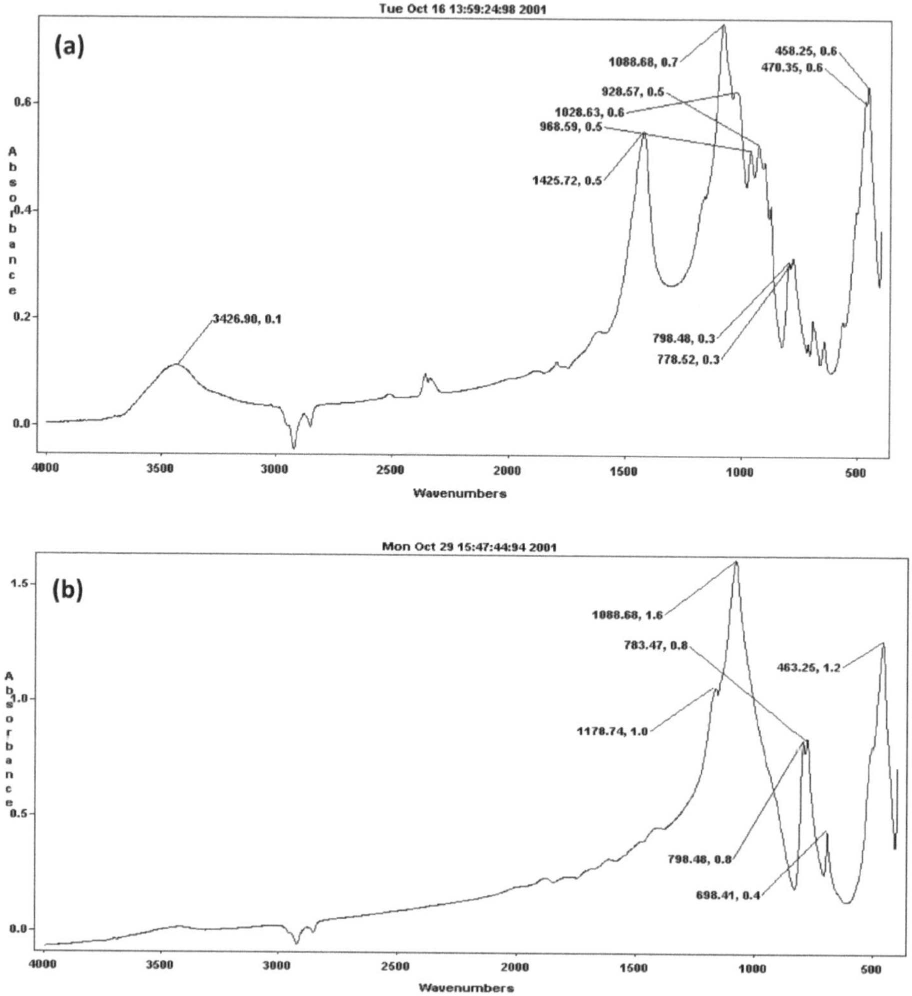

Figure 5.21 IR spectra of quartz (T) in the original (a) state and after mechanical activation for 20 min (b).

After mechanical activation for 20 min, the IR spectra of quartz differ much from the spectrum of the original sample [28].

All absorption bands for the hydroxyl groups ν_{OH}, δ_{Si-OH}, δ_{H-OH} disappear. In the region of the absorption bands of Si–O–Si stretching vibrations, the intensity remains high, but the shoulders disappear on the low-frequency slope. In the region of Si–OH (1420 cm^{-1}), the disappearance of the band may be due to the occurrence of a known reaction with the formation of siloxane bonds [25]:

$$\equiv Si{-}OH + OH{-}Si\equiv \ \leftrightarrow \ \equiv Si{-}O{-}Si\equiv + H_2O,$$

which is confirmed by an increase in the intensity of all the Si–O–Si bands, both stretching vibrations (1088.68 cm^{-1}) and deformation vibrations (798–783 cm^{-1} and 463 cm^{-1}). It should be noted that for a given quartz, all high-frequency lines are doublet, which is a consequence of a certain degree of lattice disorder, the presence of defects and impurities in quartz.

For a comparative analysis in Fig. 5.22, the presented IR spectra for vein quartz (A) of the Aktas deposit are of higher purity and crystallization, which have more clearly defined absorption bands of both valence and deformation vibrations. Hydroxyl groups with hydrogen bonds are present in the sample in a small amount as a result of water chemisorption, which is indicated by the absorption bands at 3622 and 3496 cm^{-1}, but there is no band of deformation vibrations of Si–OH at 1432 cm^{-1}. Thus, the obtained IR spectrum characterizes the really high purity and crystallization of the quartz sample.

After treatment in the reactor for 5 and 10 min, changes in the spectrum are insignificant: the intensity of low-frequency oscillations decreases and is amplified for high-frequency ones at 1160 cm^{-1}. At the site of the deformation peak at 470.7 cm^{-1}, a doublet is formed with frequencies of 475 and 466 cm^{-1} (Fig. 5.22). After 20 min of processing, changes in the IR spectrum become more significant. The intensity of the whole spectrum increases, but at the same time the contribution of low-frequency stretching vibrations becomes relatively less. As in the case shown in Fig. 5.21, the surface of the particles dehydrates and all spectra shift to high frequencies.

These data indicate the fact that during the process of quartz grinding, there takes place accumulation and redistribution of defects, resulting in increased lattice disorder, and the number of groups with weakened bonds increases but this is less pronounced than for quartz (T). When grinding quartz in the presence of monohydric alcohols, in particular butanol, it is possible to achieve high dispersion without amorphizing the particles, which is used in the preparation of samples for X-ray phase analysis. In this case, it is natural to expect that the surface of the particles absorbs the alcohol. The infrared spectra of such samples (Fig. 5.23) show the presence of adsorbed water in the frequency range 3425.36 cm^{-1} (ν_{OH}) and associated hydroxyl groups at 1428.9 cm^{-1} ($\delta_{Si\text{-}OH}$), the same frequency range accounts for Si–OC groups.

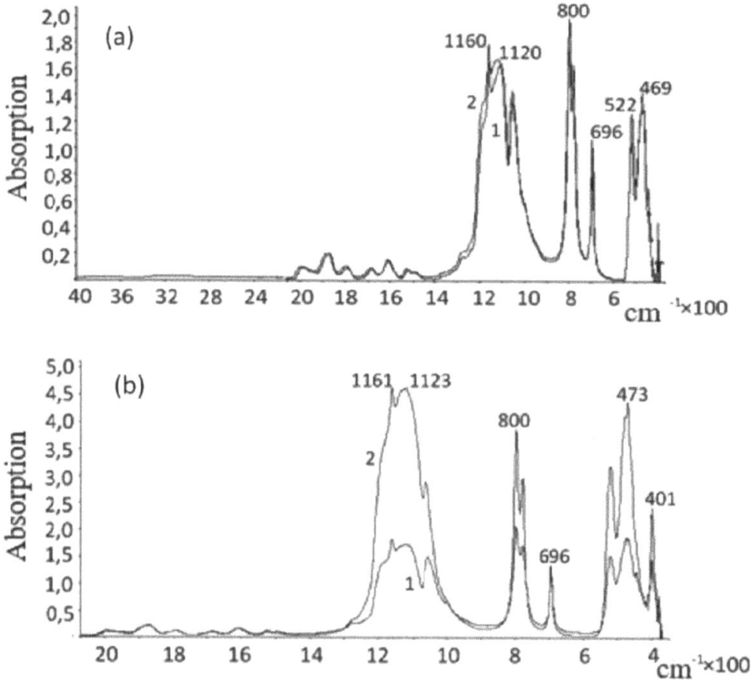

Figure 5.22 IR spectra of quartz (A) in the initial state (a, curve 1) and after activation for 5 (a, curve 2), 10 and 20 min (b, curves 1 and 2, respectively).

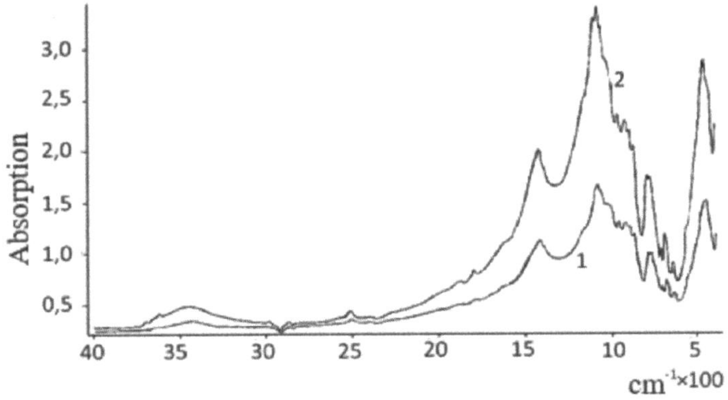

Figure 5.23 IR spectra of quartz, crushed: (1) with butanol and (2) with silicic acid.

The broadening of all bands of the quartz spectrum and the enhancement of a number of bands on the right-hand slope of the main spectrum of Si–O–Si in the range of 968.59–928.57 cm^{-1} are also observed. The spectrum of quartz milled with butanol is almost identical to the spectrum of a quartz sample milled with silicic acid. The main difference between these spectra (Fig. 5.23) is the greater intensity of all the Si–O–Si lines of both valent and deformation vibrations of samples modified with silicic acid. Hence, it follows unambiguously that a layer of silicic (or polysilicic) acid is formed on the surface of the particles during grinding with alcohol.

The use of polyatomic alcohols as a modifier, in particular ethylene glycol, leads to more significant changes, both the basic spectrum of the quartz core, and the spectrum of groups that modify the surface of the base particles (Fig. 5.24).

The spectra of the mixture obtained by grinding quartz with 5% ethylene glycol lead to the degeneration of the spectrum of deformation vibrations of Si–O–Si and the high-frequency spectrum of stretching vibrations (1165.18 cm^{-1}). The spectra of hydroxyl groups have a high intensity, especially at frequencies of 3393.19 cm^{-1}. In addition, there are valence vibrations of CH_2 and CH_3 groups at frequencies of 2947.29 and 2884.89 cm^{-1}. With an increase in the quartz grinding time in a mechanical reactor in the presence of ethylene glycol, hydroxyl groups are destroyed and the intensity of the corresponding absorption bands decreases.

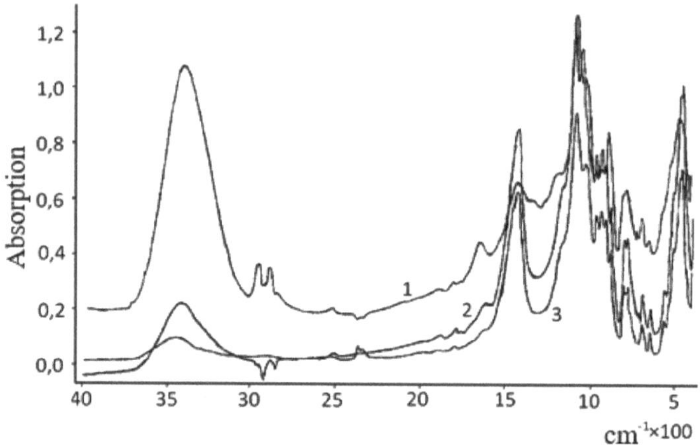

Figure 5.24 IR spectra of quartz ground with ethylene glycol (1) in a mechanical reactor for 5 min (1) and 20 min. (2) and with polystyrene for 20 min (3).

The quartz spectrum is expressed more clearly, on the right side of which lines from C–O-groups are preserved. A similar pattern of the spectra, but of lower intensity, is ovbserved when using polystyrene as a modifier. A distinctive feature of this spectrum is the presence of low-intensity absorption bands in the region of 2500–2300 and 2000–1700 cm^{-1}.

When modifying quartz with acrylic acid, also in the amount of 5%, against the background of the intense spectrum of the quartz base (matrix), the structure of the modified surface appears weakly (Fig. 5.25). More specifically, we can talk about the features in the transformation of the spectrum of Si–O–Si bonds due to accumulation and redistribution of structural defects depending on the processing time. As in case when using other modifiers, after short-term mechanical grinding, there takes place a decrease in the intensity and degeneration of a number of absorption bands, and after an increase in the processing time to 20 min, the resolvability of all the spectra is enhanced. But in all cases for the modified samples in the frequency range 1500–2100 cm^{-1} a set of low-intensity spectra is observed.

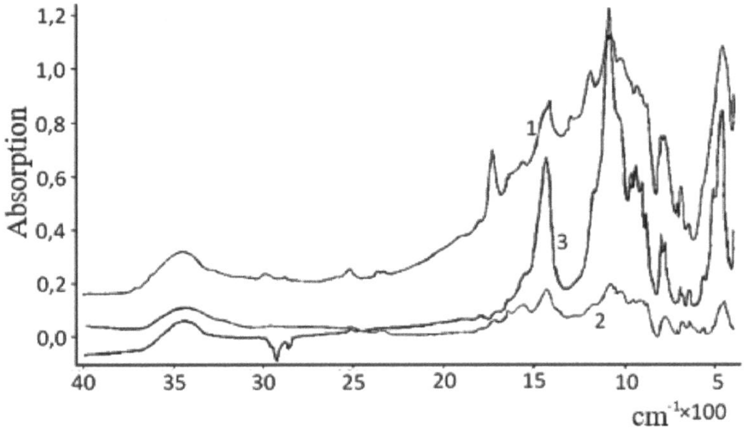

Figure 5.25 IR spectra of quartz milled in a mechanical reactor with acrylic acid for 5 (1), 10 (2) and 20 min (3).

For a more detailed analysis of the change in the structure of the modified volume of quartz, studies were also carried out on a more pure and crystallized material (quartz A). Figure 5.26 shows the spectra obtained on quartz (A) modified with butanol and polystyrene in an amount of 10% for 10 min. Changes in the spectrum of the quartz matrix mainly relate to the intensity of all absorption bands. The presence of polystyrene leads to a decrease in intensity, and the presence of butanol, on the contrary, to a significant increase of it.

Transformation of the spectrum type concerns only the frequency range of stretching vibrations (999–1186 cm^{-1}). Here there is a degeneration of the bands 1162 and 1062 cm^{-1} and the mixing of the middle band into the region of lower frequencies (up to 1109 cm^{-1}) for particles treated with polystyrene. After treatment with butanol, all the bands in this group are resolved well and the oscillations of the middle frequencies are amplified, which are shifted, on the contrary, upwards to 1159 cm^{-1}. Of particular interest is the frequency range above 1400 cm^{-1}, which consists of seven absorption bands (Fig. 5.26b), which correspond to the presence of carbonate and carboxyl groups of different isotopic composition on the Si–O–Si surface [29, 30]. In addition, there are bands of weak intensity above 2400 cm^{-1}, associated with hydroxyl groups of different nature of bond with the surface of quartz.

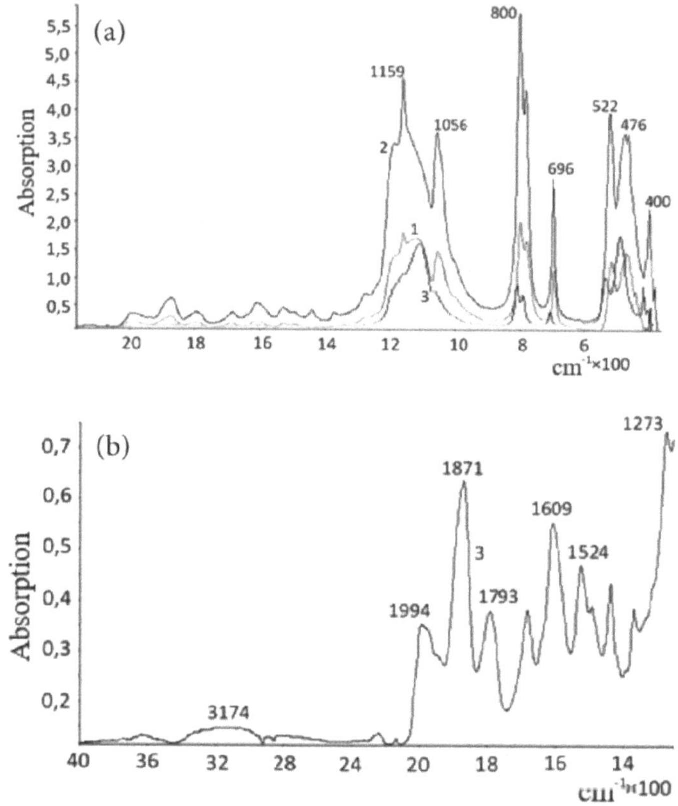

Figure 5.26 IR spectra of quartz (A) after activation for 10 min without a modifier (a curve 1) and in the presence of 10% butanol (a curve 2) or 10% polystyrene (a, b curve 3).

The study of quartz (A), modified with acrylic acid (5 wt.%) also showed that in the frequency range of 1400–2100 cm^{-1} an absorption spectrum is observed (Fig. 5.27a,b), but somewhat different from that obtained for specimens modified with butanol (Fig. 5.27b). If the frequencies 1800–2000 cm^{-1} can be associated with carbonate groups, then the area 1400–1750 cm^{-1} belongs to the new compounds.

The observed spectra belong to iron acrylate, which is synthesized in the process of mechanochemical processing of quartz with acrylic acid. This is confirmed by the IR spectra obtained directly from iron acrylate synthesized by the usual chemical method [31].

Iron for the compounds synthesized during mechanochemical processing, including acrylic acid salts, enters the mixture as a result of rubbing abrasive particles of quartz from the surface of grinding vessels and balls. The amount of iron in the treated quartz varies with the type of modifier used. Its elemental content varies from 4% to 14%. In the bound state in the form of various oxides and salts, iron was registered by X-ray phase analysis and Mössbauer spectroscopy.

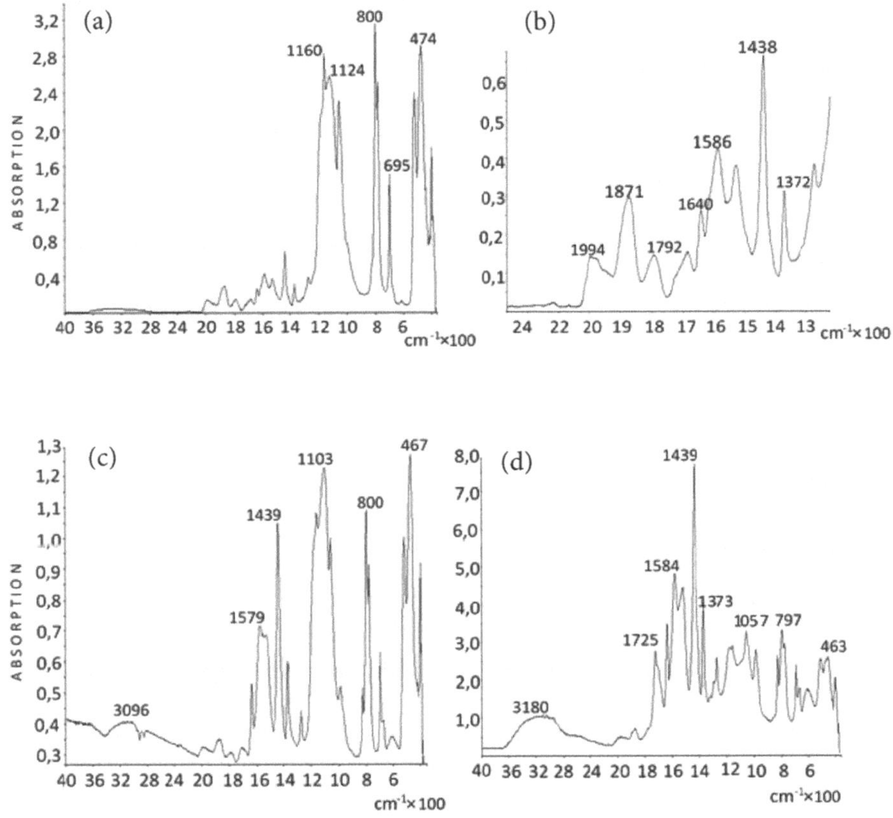

Figure 5.27 IR spectra of quartz (A) modified with acrylic acid during mechanochemical treatment for 10 min in a mill with 5% (a, b), 15 (c) and 40% (g) acrylic acid.

Thus, the results of infrared spectroscopy of activated and modified quartz powder showed that during mechanochemical processing, changes occur in the main spectra (frequency shift and band deformation) due to stress and weakening of bonds in the Si–O–Si siloxane groups. The change in the intensity of the absorption bands and the degeneracy of a number of spectra, starting with deformation vibrations, are associated with breaking of siloxane bonds, formation of silane bonds and formation of a destructible modifier on the surface of particles of groups of elements and radicals. Such groups include Si–OC and Si–C, the spectra of which are of low-intensity against the background of the spectrum of the quartz matrix, since they are included in the structure of the surface layer of a particle ~ 50 nm thick.

Of particular interest are the data indicating synthesis of compounds with iron in the volume of a modified quartz powder. This is most clearly shown when using acrylic acid as a modifier. Synthesis of iron acrylates during mechanochemical treatment on the surface of quartz particles shows the possibilities of such synthesis when creating micro (or nano-) composite materials. By variation of processing modes, selection of modifiers with active participation of metal in these chemical processes, in this case-of iron, grinding balls and the surface of a mechanical reactor, powder materials of different composition, structure and quality can be obtained. It is important to answer the question in which form (elemental or bound) the iron on the surface of the modified quartz particles is. In solving this problem, the most informative methods are EPR and Mössbauer spectroscopy combined with X-ray phase analysis.

5.3.3 EPR, Mössbauer Spectroscopy and X-Ray Phase Analysis of Quartz Modified by Mechanochemical Processing

The electron paramagnetic resonance (EPR) method makes it possible to detect the presence of iron in its various forms, both in the form of coordination-unsaturated Fe^{3+} ions, and in the metallic state, which are registered by different resonance and nonresonant absorption spectra. An example of such studies are the results of work [32] on the study of the features of the state of iron in zeolites, which are used as catalysts. Iron was injected into zeolite samples by impregnating them with $FeCl_3$ solution with subsequent calcination in vacuum, in oxygen and in air. In this case, the EPR and FMR spectra were refistered. It is believed that ferromagnetic resonance in such a spectrum is associated with the appearance of highly ordered polynuclear iron-containing structures in the zeolite nanoparticles.

The EPR spectrum taken for quartz (T) in the initial state (Fig. 5.28a) shows a relatively low content of ferric iron in two states. The spectrum has a wide line with ΔH = 90 mT and with g = 2.14 and a narrower line with ΔH = 39 mT and g = 3.9. The wide lines correspond to Fe^{3+} ions in a slightly distorted octahedron field, while the line with g = 3.9 is associated with Fe^{3+} ions in a strong crystalline field of low symmetry due to rhombic distortions of the coordination node.

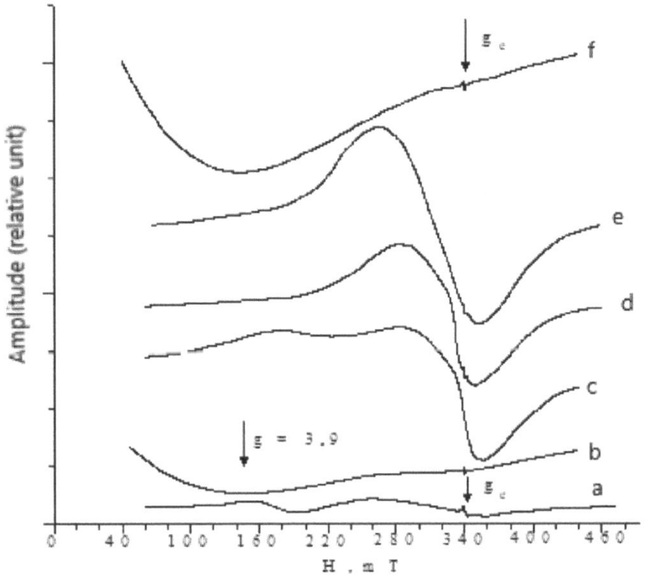

Figure 5.28 EPR spectra of quartz in the initial state (a), activated for 20 min (b), SiO_2 + 5% $FeCl_3$ systems at τ_{act} = 5 min (c), SiO_2 systems + 5% $FeCl_3$ at t_{act} =10 min (d) SiO_2 + 5% $FeCl_3$ systems at t_{act} = 20 min. (e) SiO_2 + butanol systems at t_{act} = 20 min (f).

After mechanical activation, the EPR spectrum of a quartz sample changes dramatically [7, 33, 34]. Instead of resonance lines, a line appears (Fig. 5.28b), extending from the region of weak magnetic fields to maximum field values. This line has a minimum at H = 120 mT for the activation time τ_{act} = 5 min. As the activation time increases, the absorption increases in the region of weak fields, and the minimum absorption of microwave power shifts to a stronger field. In the region of g = 2, a weak resonance signal is observed against the background of this line, which may be due to the presence of a phase of iron silicate origin in the sample. The totality of the data considered suggests that the line in Fig. 5.28b is most likely associated with ultrathin iron films stripped of quartz particles from the body and the balls of the mill.

With the introduction of various modifier additives into activated quartz, the EPR spectrum parameters change. Addition of ferric chloride to quartz during its mechanical activation was due to the well-known fact [3] that salts are well destructed by mechanical action and the surface of quartz should be enriched with iron. Addition of ferric chloride significantly changed the appearance of the spectra. After mechanoactivation of the system SiO_2 + 5% $FeCl_3$ at τ_{act} = 5 min. an anisotropic spectrum of ferric iron is observed (Fig. 5.28c). Such a line shape of the resonant signal may be due to the presence of axial symmetry in the anisotropy of the g-factor. The line width of this spectrum can be conditionally taken as the line component in the g = 2 region, which is equal to 77.5 mT. A narrower line is superimposed on this wide line, with an estimated linewidth of 13.5 mT. With an increase in activation time up to 10 min, the line anisotropy decreased markedly, the amplitude of the line increased, and the line itself narrowed somewhat (Fig. 5.28g), but the bends on the line still indicate superposition of at least two resonant lines. With a 15 min activation of the system under study, a further increase in the amplitude of the spectrum is observed, and the width of its line also increases. After 20 min of activation, both the amplitude of the spectrum and its width (ΔH = 88 mT) continue to increase. The line becomes smoother (Fig. 5.28d), although it retains some asymmetry.

With the introduction of 10% $FeCl_3$ into the quartz powder and after its activation for 5 min, the spectrum anisotropy became noticeably less. But even for this series of samples, spectra with kinks on the resonance line are characteristic, this indicating the superposition of several lines with similar values of g-factor s. As in case of the adding 5% $FeCl_3$, the smoothest spectrum is registered for τ_{act} = 20 min. It should also be noted that the behavior of the linewidth depending on the activation time repeats the previous case: the narrowing of the line at τ_{act} = 10 min. and its increase in subsequent activation times, the increase in the amplitude of the spectrum with the activation time. A similar picture takes place with an increase in the amount of added iron chloride supplement to 15%. The exception is τ_{act} = 20 min, when the signal amplitude falls. With an increase in the amount of $FeCl_3$ introduced, the level of the amplitude of the spectrum increases, especially after 15 min of treatment. Activation within 20 min leads to a decrease in the intensity of the spectrum. The EPR line in this case narrowed, this swowing complex processes occurring in the system during activation and leading to an increase in exchange interactions.

There are several reasons leading to a decrease in the amplitude of the signal at τ_{act} = 20 min. and the content of 15% $FeCl_3$. First, trivalent iron in the process of mechanical activation could change the valence. Secondly, clusters of trivalent iron could be formed, followed by their reduction to metallic iron. The magnetic resonance spectrum of such nanoparticles of metallic iron, moreover, probably containing impurities, can have a large line width, which makes it difficult to fix it. In addition, iron can react with silicon and impurities, forming non-magnetic compounds.

All the results on the change in the parameters of the EPR spectrum of quartz activated with ferric chloride are presented in Table 5.6. In order to assess stability of the structures formed as a result of mechanoactivation and their possible transformation over time, repeated measurements of the same samples were carried out after 39 days of caking in air. In this case, changes in the parameters of the spectra were detected, which are mainly reduced to a decrease in the amplitude of the spectrum. The width of the spectrum changes ambiguously and only with the introduction of

15% ferric chloride in activated quartz, a steady increase in the width of the spectrum occurs with a decrease in its amplitude.

Table 5.6 Parameters of the EPR spectra of a mechanically activated quartz + $FeCl_3$ system

$FeCl_3$, %	Time of the MCT, min	Initial measurements		Repeated measurements after 39 days	
		Line width (mT)	Amplitude (relative unit)	Line width (mT)	Amplitude (relative unit)
5	5	77.5	0.3	72	0.20
	10	68.0	0.9	68	1.0
	15	76.0	3.9	79	3.30
	20	88.0	6.0	86	2.7
10	5	70.5	0.43	68.0	0.3
	10	66.5	1.23	68.0	1.1
	15	77.5	3.6	84.0	4.1
	20	85.5	7.0	76.0	3.3
15	5	65.0	0.3	68.0	0.3
	10	70.5	2.0	72.0	2.0
	15	92.0	7.2	98.5	6.8
	20	81.5	5.2	88.0	5.4

Thus, according to the presented results, the presence of at least two types of magnetic centers is typical for all the systems studied, as indicated by kinks on the resonance curves. After τ_{act} = 20 min, the lines become smoother. Apparently, when activated for 20 min, the highest homogeneity of the systems under study is achieved. It can be expected that the particle sizes of quartz reached approximately the same values. With shorter activation times, the sample is more heterogeneous. In this case, it is possible to implement a larger number of different structures and forms of compounds. It should be expected that such a quartz powder exhibiting ferromagnetic resonance should have sufficiently high ferromagnetic properties. According to the EPR results, such a state can be described by a change in the intensity of the spectrum in the region of resonant absorption of a trivalent iron ion (Fig. 5.29). The ferromagnetic properties of quartz increase with the processing time most intensely when

15% ferric chloride is contained in the mixture. With prolonged weathering (aging), the magnetic characteristics of powders activated from 5 to 20 min decrease regardless of the amount of the added additive (Table 5.6). Thus, the characteristics of the activated material can be judged by the characteristics of the EPR spectrum.

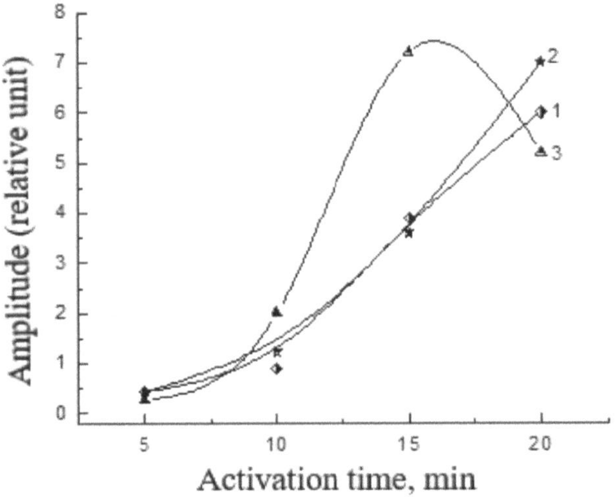

Figure 5.29 Dependence of the intensity of the EPR signal of a trivalent iron ion on the activation time of quartz activated with $FeCl_3$ in the amount of: 5 (1), 10 (2) and 15% (3).

When activating quartz with the addition of alcohols: ethanol, butanol, and ethylene glycol, the form of the spectrum is in general similar to the spectrum of activated initial quartz. The same is the absence of clearly pronounced resonance lines in the spectra and strong absorption of microwave power in weak magnetic fields. However, absorption of microwave power in weak fields in this case very much depends on the type of organic additive. According to the results of X-ray spectral analysis (Table 5.5), as noted earlier, these samples are enriched with iron. It is known that in thin layers of such iron, domain walls absorb microwave power in zero fields. If the amplitude of microwave power absorption at a certain value of the magnetic field is taken as the value of its intensity, then these amplitudes will characterize the relative content of ferromagnetic iron in activated quartz. The obtained data on the content of ferrophase in activated quartz, depending on the type of additive and activation time, are presented in Fig. 5.30, from which it follows that the absorption intensity in the field of weak fields, and, consequently, the ferrophase content increases with activation time (except for the data for butanol at τ_{act} = 15 min.) in the range of additives: ethylene glycol® ethanol® butanol.

In order to find out how much iron gets from the balls and from the reactor walls to quartz when it is activated and evaluate the role of additives in this process, we sequentially measured, using the method of X-ray fluorescent, X-ray spectral and X-ray phase analysis, the concentration of iron in the original mineral and after its activation, as well as in samples treated with butanol for 5, 10, 15 and 20 min. and with ethanol and ethylene glycol after 20 min of activation [2]. It is stated that the initial sample of quartz contains 0.05% iron. When activated for 10 and 20 min, the content of iron increases to 2.3% and 3.3%, respectively. But, if butanol is added to quartz when it is activated, the content of iron in it after τ_{act} = 20 min increases 4.4 times, which is 14.8%.

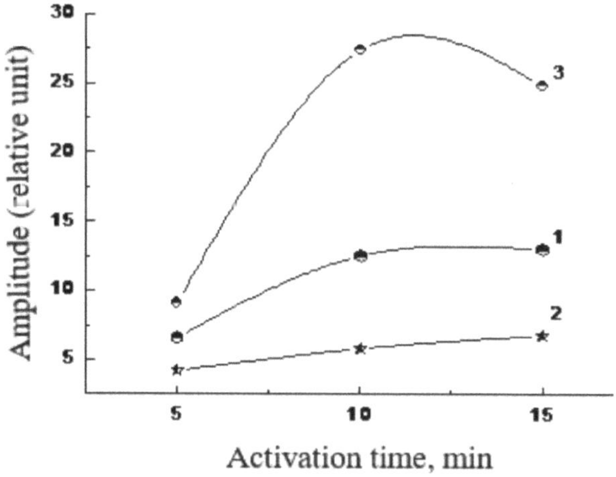

Figure 5.30 The dependence of the absorption of microwave power in weak magnetic fields of quartz with various additives: (1) ethylene glycol, (2) ethanol, (3) butanol.

Introduction of alcohols into quartz makes it possible to obtain a higher degree of dispersion of the sample in a shorter activation time, which increases the abrasive properties of the material and increases the number of nano-sized iron particles rubbed from the walls of the mechanical reactor and the surface of the balls. When using butanol, this effect is most pronounced. Grinding of quartz and iron "rubbing" in the presence of ethanol and ethylene glycol occurs to a lesser extent and amounts to 11.62% and 6.78%, respectively, after 20 min of grinding, although the degree of grinding with ethanol is much higher than with other additives.

Thus, with the introduction of organic additives into quartz powder processed in a mechanical reactor, ferromagnetism manifests itself due to an increase in the content of fine iron (Fe nanoparticles are 10 to 50 nm in size). At the same time, the dependence of the ferromagnetism of quartz on the amount of iron in it is not unambiguous, this indicating complex structural transformations of the surface layers of the particles. These structural rearrangements are associated with the destruction of organic additives, followed by sorption or chemical interaction of the resulting radicals with active centers on the surface of the silica quartz until the formation of monomeric (or polymer) layers encapsulating the particle [15], as shown by electron microscopic studies. The surface layers can be analogous to [36] systems—"claspol" (metal clusters in a polymer matrix). Electromagnetic properties of such systems depend on the specific features of the emerging structures.

It was of interest to consider the system with the simultaneous introduction of ferric chloride and alcohols, in particular: SiO_2 + 5% $FeCl_3$ + 5% C_2H_5OH, into activated quartz. After τ_{act} = 5 min a line from Fe^{3+} ions with ΔH = 50 mT and $g \approx 2$ was detected in the spectrum of such a sample, and absorption of microwave power in weak magnetic fields was also observed. Activation of the system for 10 min leads to a slight line broadening (ΔH = 62.5 mT) and an increase in its amplitude by 2.5 times. Absorption in weak fields increased 1.7 times. After τ_{act} = 15 min the line width increased slightly (ΔH = 65 mT), and its amplitude dropped by 2 times. It is very difficult to interpret this behavior uniquely in the change of the EPR spectra depending on the time of activation of this system. However, if we take into account the fact that the absorption in weak fields also increased by 2 times, it can be assumed that part of Fe^{3+} ions in the activation process could form clusters with their subsequent reduction to fine iron. It is also possible that in the process of activation the valence of iron changes or the formation of new iron-containing non-magnetic compounds occurs. Thus, when two types of additives are introduced into quartz at the same time, a total manifestation of their action is observed in the spectrum: absorption appears in weak fields and a spectrum due to ferric iron, which ultimately enhances the manifestation of the magnetic characteristics of the material. However, the magnetization of quartz treated both in the presence of iron salts and with the participation of alcohols, decreases after treatment by 25–30% after 20–30 days of aging.

The most intense nonresonant absorption in the region of weak fields, i.e., the manifestation of ferromagnetism, was obtained on quartz powders modified by mechanochemical treatment in the presence of acrylic acid and polystyrene (Fig. 5.31). The selected organic additives are characterized by both increased mechanical destruction and synthesis of new polymers with participation of inorganic materials in the process of their grinding.

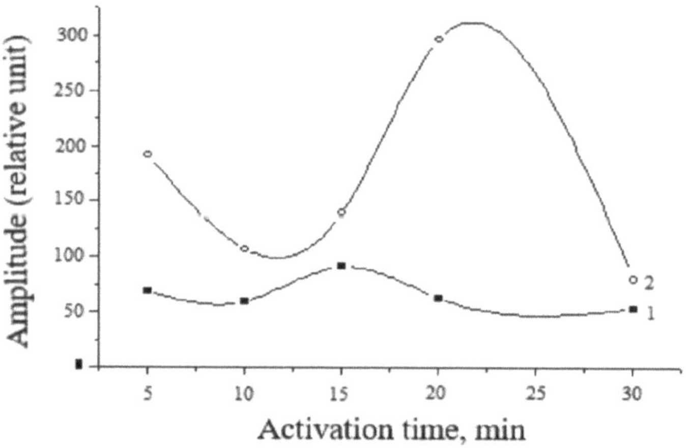

Figure 5.31 The dependence of the absorption of microwave power in weak magnetic fields on the time of activation of quartz with various additives: (1) polystyrene, (2) acrylic acid.

The nonlinear dependence of the amplitude of the EPR spectra and magnetic permeability on the time of treatment indicate complex structural changes in the material: formation of metallic iron clusters and trivalent iron complexes on the surface of quartz particles, and also formation of iron carbide. Formation of iron carbide is a consequence of the destruction of acrylic acid and polystyrene up to carbon. The surface layers of the particles are saturated with carbon and its compounds, and mechanochemical "carbonization" of quartz occurs.

Thus, in the process of mechanochemical treatment in the presence of ferric chloride and organic additives, a significant structural rearrangement of the surface of dispersible quartz particles occurs. First of all, it is enriched with iron as a result of the mechanochemical decomposition of $FeCl_3$ and "rubbing" of ultrafine metallic iron from the walls of the mechanical reactor and grinding balls. The presence of iron embedded in the surface of quartz, of course, is the determining factor in the ferromagnetism of quartz after its mechanical treatment. However, the degree of magnetization is due to a specific combination of surface structures. The EPR spectra of quartz after mechanochemical treatment show a variety of iron states depending on the specific processing modes. Iron in the trivalent state forms complexes of exchange interaction, which is especially pronounced after the optimal processing time—20 min. But such structural formations with iron are metastable and the ferromagnetic characteristics of the particles associated with them significantly decrease with time.

The more stable "ferromagnetism" of quartz is associated with metallic iron nanoparticles and the orderliness of its structural formations with organic compounds covering the surface of quartz during quartz processing in the presence of various modifiers. The particles of quartz powder modified in our experiments with acrylic acid and polystyrene are distinguished by the highest level and stability of ferromagnetic resonance indices in the EPR spectrum. Electron microdiffraction of such particles captures the "carbonized" surface of the particles, up to the release of pure carbon. Consequently, ferromagnetism after the mechanochemical treatment of quartz should be associated with the formation on its surface of nanoscale compounds, including iron and carbon.

Direct evidence of the presence of nanodispersed iron particles in the metallic state and in the form of various compounds on the surface of quartz was obtained by Mössbauer spectroscopy [35–38]. As the initial samples, the mineral of the Tekturmys deposit was investigated; quartz (T). An elemental analysis performed by the X-ray fluorescence method showed that in natural quartz (T) the mass fraction of iron impurities is about 0.8%. This amount of iron is sufficient to determine its chemical form using Mössbauer spectroscopy. Figure 5.32a shows the Mössbauer spectrum of quartz in the initial state.

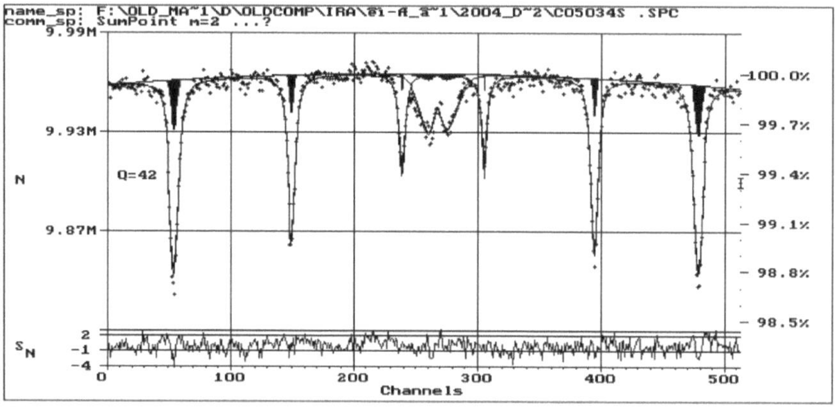

(a)

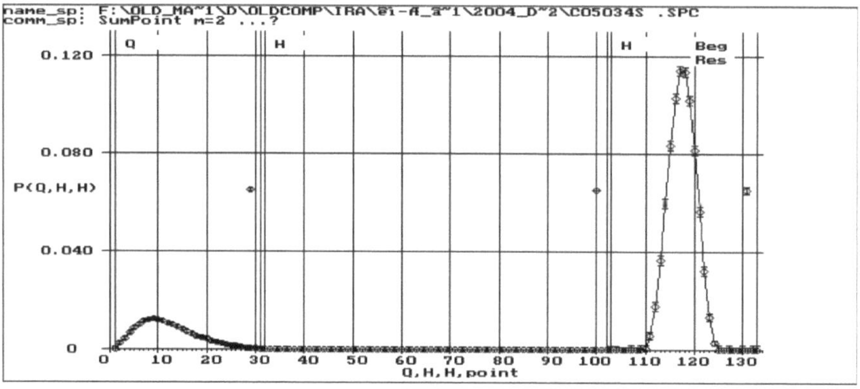

(b)

Figure 5.32 Mössbauer spectra using ^{57}Fe source for the systems (a) and the distribution function of hyperfine fields (b) of quartz (T) in the initial state.

The spectrum was processed by restoring the distribution functions of the quadrupole displacement $p(\varepsilon)$ and the hyperfine magnetic fields $p(H_n)$. The results of processing are presented in Fig. 5.32b, in which it can be seen that splitting of the hyperfine fields $p(\varepsilon)$ and $p(H_n)$ has a two-model form. This indicates the presence of two subspectra from iron atoms in magnetically ordered states. As can be seen from the processing results, the spectrum is described quite well by a sextet with Mössbauer parameters characteristic of hematite Fe_2O_3 and a quadrupole doublet from iron found in the crystal lattice of pyrite FeS_2. Estimation of the relative redistribution of iron in phases shows that about 81% of it is in the oxide state (Fe_2O_3) and 19% in the form of pyrite (FeS_2).

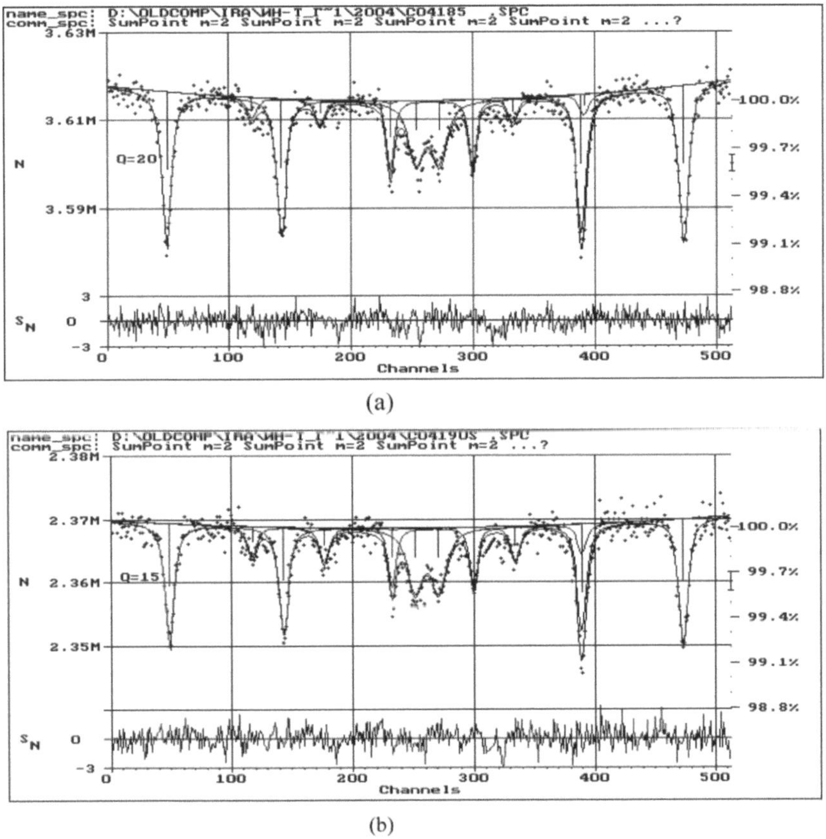

(a)

(b)

Figure 5.33 Mössbauer spectra using ^{57}Fe source for the systems quartz (T), activated during 10 (a) and 20 min of processing (b).

The spectra of samples subjected to mechanical activation are a superposition of three subspectra—two Zeeman sextets and one quadrupole doublet. Figure 5.33 shows the Mössbauer spectra of quartz, subjected to activation for 10 (a) and 20 min (b).

The predominant phase in them is iron oxide (Fe_2O_3). The parameters of hyperfine interactions (I_s, Q_s, H_s) do not change their values when the activation time changes. The second component of the spectrum is a quadrupole doublet, which also preserves the parameters of the hyperfine interaction. The parameters of the quadrupole doublet $-I_s \approx 0.29$ (2) mm/s and the quadrupole shift $Q_s \approx 0.37(2)$ mm/s indicate that this partial spectrum belongs to iron atoms associated with sulfur atoms. The noticeable broadening (up to 0.6 mm/s) of the resonance lines of the quadrupole doublet may be due to local inhomogeneity for Fe atoms in the pyrite matrix. It was not possible to detect significant changes in the relative content of Fe_2O_3 and FeS_2, depending on the time of activation of the samples.

Beside Fe_2O_3 and FeS_2 in activated sample, α-Fe is present in a lesser amount the relative content of which increases with the activation time, that is apparent from Table 5.7. It contains the Mössbauer parameters of the spectra (isomeric shift I_s, hyperfine magnetic field H_s, quadrupole splitting Q_s, width of the resonance lines G of the quadrupole doublet, the percentage of the expected phases and their chemical formula) for the studied samples before and after activation.

Table 5.7 Mössbauer parameters of activated quartz (T) spectra

Material	τ_{act}, min	Mössbauer parameters					Phases
		I_s, (mm/s)	H_s, (kOe)	Q_s, (mm/s)	S, (%)	G, (mm/s)	
SiO_2		0.373±0.001	515.7±0,1	—	80.78		Fe_2O_3
		0.301±0.013		0.109±0.001	19.22	0.59±0.02	FeS_2
				0.301±0.013			
SiO_2	10	0.373±0.002	515.7±0.2	—	69.5		Fe_2O_3
		0.0263±0.013	330.6±1.1	0.101±0.002	10.5		α-Fe
		0.296±0.015	—	0.026±0.013	20.0	0.61±0.04	FeS_2
				0.37±0.02			
SiO_2	20	0.371±0.003	516.1±0.3	0.105±0.003	62.4		Fe_2O_3
		0.02±0.01	330.6±0.9	0.040±0.011	14.9		α-Fe
		0.28±0.02	—	0.38±0.02	22.6	0.61±0.05	FeS_2

After treatment of quartz with different modifiers, changes in the Mössbauer spectra are observed both in the phase composition of iron compounds on the particle surface and in the parameters of their spectra. Table 5.8 shows the characteristics of the spectra of modified quartz. A distinctive feature of the spectra of quartz, modified with butanol and acrylic acid, is the absence of hematite and a small amount of pyrite. Samples treated with acrylic acid for 10 min contain a small amount of magnetite. After longer treatment (up to 20 min), the oxide forms of iron in the sample are absent.

Table 5.8 Mössbauer parameters of the spectra of modified SiO_2 (T) at different processing times

		Mössbauer parameters					
Material	τ_{act}, min	I_s, (mm/s)	H_s, (kOe)	Q_s, (мм/с)	S, (%)	G, (mm/s)	Phases
SiO_2+5% butanol	10	0.005±0.001	332.98±01	-0.009±0.001	95.2		α-Fe
		0.009±0.001		0,000	4.85	0.59±0.02	FeS_2
	20	0.004±0.001	33325±01	-0.005±0.001	95.1		α-Fe
		0.050±0.001		0.332±0.02	4.94	0.61±0.042	FeS_2
SiO_2+5% AA*	10	-0.219±0.003		0.455±0.003	6.33		Fe_3O_4
		0.004±0.001	334/96±09	-0.008±0.001	91.1		α-Fe
		-0.075±0.02		0.38±0.02	2.53	0.61±0.05	FeS_2
	20	0.002±0.001	334.96	-0.004	93.1		α-Fe
		0.016±0.001		0.271	6.92		FeS_2
SiO_2+5% PPS**	10	0.365±0.003	516.8±0.3	-0.104±0.003	57.6		Fe_2O_3
		0.018±0.008	330.6±0.7	-0.020±0.008	18.4		α-Fe
		0.29±0.02		0.38±0.02	24.0	0.62±0.04	FeS_2
	20	0.369±0.002	515.8±0.2	-0.103±0.002	55.6		Fe_2O_3
		0.015±0.005	330.6±0.4	-0.020±0.005	22.9		α-Fe
		0.29±0.01		0.39±0.01	21.5	0.57±0.03	FeS_2

*AA, acrylic acid; **PS, polystyrene.

From this, it follows that in the process of machining quartz with the indicated modifiers, there is an intensive reduction of iron from its compounds to the metallic state. When using polystyrene as a modifier, almost half of the iron-containing phases are represented by hematite (a weak ferromagnet compared to metallic iron and magnetite). In this case, the spectra have a number of specific features (Fig. 5.34), peculiar only to the material modified with polystyrene.

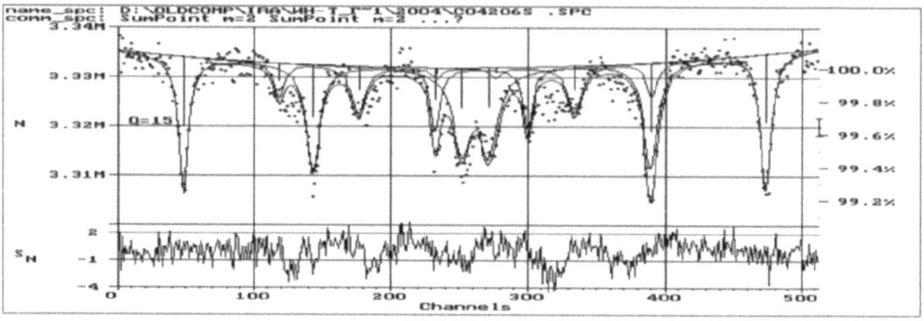

(a)

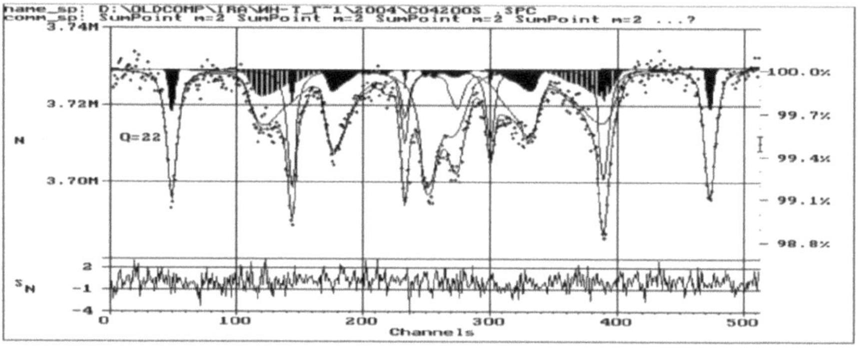

(b)

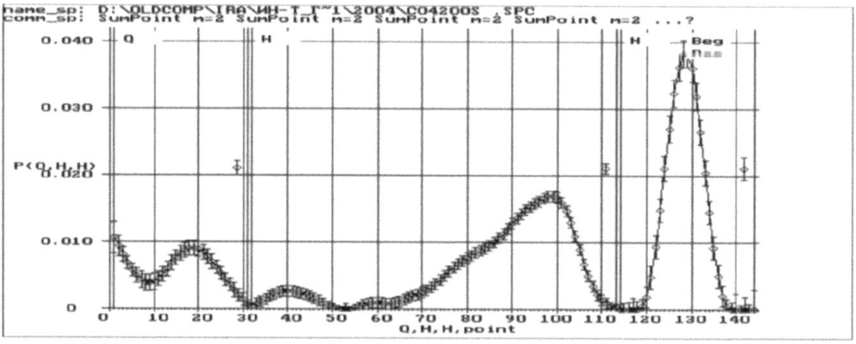

(c)

Figure 5.34 Mössbauer spectra using ^{57}Fe source for the systems quartz (T), modified with 5% polystyrene for 10 (a) and 20 min (b). Distribution functions of hyperfine magnetic fields $p(H_n)$ and quadrupole displacements $p(\varepsilon)$ in quartz after 20 min of processing.

The results of spectrum processing by restoring the distribution functions of the superfine magnetic fields $p(H_n)$ and quadrupole displacements $p(\varepsilon)$, performed using the DISTRI programs, show that distribution $p(H_n)$ is quite complex. Distortion of the α-Fe subspectra by impurity atoms is observed, which leads to formation of a large number of nonequivalent positions of resonant atoms. Another likely cause of distortion of the spectrum is amorphization of α-Fe during the activation process, and the products of destruction of polystyrene fix the amorphous state of iron. The third fact of undoubted interest is the formation of Fe_3C iron carbides in the modifying treatment process, the spectrum of which falls on 60 Q, H, H (Fig. 5.34c).

Thus, the results of Mössbauer spectroscopy clearly demonstrated the presence in quartz dispersed, both without and in the presence of modifiers, of iron in a metallic state and in the form of oxides and compounds with sulfur. Metallic iron gets introduced into quartz not only in the process of "rubbing" with a solid quartz particle from the balls and the surface of the vessel, but also as a result of its reduction from oxides and pyrite, present as impurities in the quartz mineral.

Reduction can occur due to the products of the mechanical destruction of the organic modifiers used, i.e., carbon and hydrogen. Carbon formed during the destruction of polystyrene actively interacts with iron to form carbides under conditions of weak heating of the material during its grinding. Thus, we have an example of a non-thermal development of the reaction in formation of iron compounds with carbon. Non-interacting iron particles have signs of a defective structure and the presence of impurity atoms in the lattice.

Some details in the description of the features of the structure of iron particles present in modified quartz are obtained by X-ray phase analysis of the synthesized material [20]. Figure 5.35 shows the radiographs of quartz in the initial state and treated with butanol, polystyrene and acrylic acid. For modified samples, low-intensity iron lines are registered. A quantitative analysis of the X-ray patterns obtained showed that the content of metallic iron in quartz depending on the time of treatment (from 5 to 30 min) with butanol varies from 4–14.7%, with acrylic acid—from 2.7 to 5.62%, and with polystyrene—from 7.08 to 12.63%. In addition, in the latter two cases, reflections from solid solutions of carbon in iron and a change in the background depending on the processing time, as well as the type and amount of the modifier used, were registered.

The background value is given as the difference between the values observed in the sample after processing and in its initial state. The value of background in the sample with 10% acrylic acid without treatment is 14 mm, with the background of 11 mm on the quartz standard. For samples with 5% acrylic acid after treatment for 10 and 20 min, the background is approximately the same when the difference in iron content, as a phase, is 2.8 times. It follows that part of iron is present in the form of an X-ray amorphous phase.

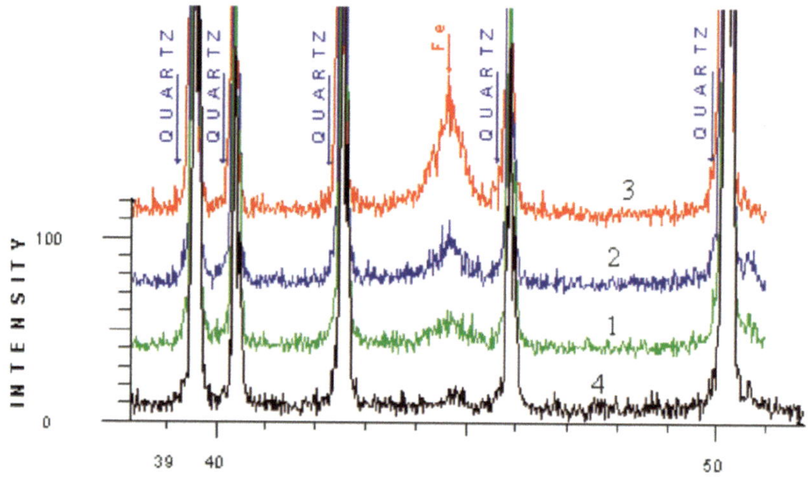

Figure 5.35 Radiographs of quartz after 10 min of activation (1) and after modifying treatment with butanol (2), polystyrene (3) and acrylic acid (4).

Table 5.9 shows basic XRD data for samples treated with acrylic acid and polystyrene. These radiographs show the background, the value of which depends on the total content of iron in the sample. In addition, a halo in the region of angles of 14.5°θ, belonging to the X-ray amorphous phase, was determined.

For samples modified with 5% polystyrene, the iron content, the magnitude of the background, and the amount of the X-ray amorphous phase are approximately doubled when compared with the data for samples obtained after 10 and 20 min of processing. After 30 min of processing, the results are slightly higher than after 20 min due to the loss of abrasive ability of quartz particles and the formation of a stable film surface of a new phase on a quartz particle. In these samples, a new phase was registered, represented by a line of 2.08 Å, and belonging to the solid solution of carbon in iron. Most of this phase is formed after 30 min of processing.

Table 5.9 The phase composition and parameters of radiographs of modified quartz samples at different times of treatment

Material	τ_{act}, min	SiO_2, %	Ferrite, mm	Fe,%	Background	Halo
SiO_2+5%AA*	10	86		1.4	45/34	5
	20	89		5.62	48/37	
SiO_2+10%AA*	10	78		0.6	75/61	12
	20	76		1.6	76/62	10
SiO_2+5%PS**	10	92	29	3.8	48/37	19
	20	81	41	6.1	65/54	20
	30	73	50	6.9	78/61	20

*AA, acrylic acid; **PS, polystyrene.

As shown by the results of Mössbauer spectroscopy in the process of mechanochemical processing, impurities containing iron in natural quartz reduced to metallic iron. Since the amount of impurities in the mineral used is small, XRD does not fix them. When $FeCl_3$ is introduced into the treated quartz during the process of mechanical treatment, decomposition of salt does take place. X-ray phase analysis in the sample recorded the presence of Fe and Fe_3O_4. This process is intensified in the case of the simultaneous introduction of a mixture of ($FeCl_3$ and alcohol) into the quartz powder. In this case, in the sample, 3.75% Fe and ~1% solid solution of carbon in iron are formed in 10 min, this indicating decomposition of alcohol to C and H with subsequent interaction with the surface of quartz (carbonization), reduction of iron from various compounds and possible dissolution of carbon in it. These results are in complete agreement with the EPR spectroscopy results of nonresonant absorption in the region of weak fields.

Thus, a comprehensive analysis of activated and modified samples of quartz powder showed that in the process of mechanochemical processing the surface of quartz particles is saturated with nanodispersed particles of iron or its oxides, as well as supersaturated solid solutions of carbon in iron. All of these formations are ferromagnetics. The presence of ferromagnetic compounds in the "carbonized" surface layer of a quartz particle allows us to suggest that a synthesized nanocomposite object obtained on the basis of a quartz core should exhibit magnetic properties.

Questions

1. Describe the morphological differences of the surface layer of quartz particles, during mechanochemical processing with carbon-containing modifiers: monohydric and polyhydric alcohols, acids and polymers.
2. What multilayer composite formations with nano-sized organometallic compounds are formed when modifying quartz particles with polymer compounds during mechanochemical treatment?
3. What is the role of "rubbed" iron in the structure formation of the surface layer of modified silica powder particles and its dependence on the type of modifier used?
4. Describe the changes in the IR spectra during quartz grinding, associated with accumulation and redistribution of defects, formation of groups with weakened bonds.
5. Describe the changes in the IR spectra associated with breaking of siloxane bonds, the formation of silane bonds and formation, on the surface of quartz particles, of groups of elements and radicals of the modifier being destructed.
6. How does the synthesis of iron acrylates happen during mechanochemical treatment on the surface of quartz particles and creation of micro (or nano-) composite materials?
7. What changes in the parameters of the EPR spectrum are observed depending on the conditions of mechanochemical treatment of quartz and modifier additives?
8. What types of magnetic centers are formed during the mechano-chemical treatment of quartz with modifiers that form monomeric (or polymeric) layers that encapsulate a particle?

9. Describe the features of the Mössbauer spectra of quartz subjected to mechanochemical activation.
10. What changes occur during the mechanochemical treatment in the degree of defectiveness of the material, the appearance of strained bonds and the formation of a layer of polysilicic acid on the surface of an activated quartz particle, according to the results of XRD, XRD, IR spectroscopy and electron microscopic analysis?

5.4 Features of Ferromagnetism of Quartz Powder Induced as a Result of Mechanochemical Processing

Production of "ferromagnetic" quartz due to directional modification of the surface layer has great prospects in the use of such material for different areas of practical application. When introduced in one or another way into the structure of a non-magnetic particle of a ferromagnetic element, in particular iron, one more important problem needs to be solved—stabilization of the structure and state of the material, ensuring its ferromagnetism.

To describe the change in the magnetic properties of quartz after mechanochemical processing, a physical characteristic was chosen — magnetic permeability (μ), which in relative units is defined as $\mu = 1 + \chi$, where χ is the magnetic susceptibility of the material. According to the change in magnetic permeability, one can judge the physicochemical structure, types of chemical bonding, and other structural forms in inorganic materials. The results of the studies showed [39–41] that both the structure and the magnetization effect depend on the processing modes and the modifiers introduced.

Measurements of the magnetic permeability of the samples were carried out by the developed diagnostic method for powder materials [42]. As the processing time increases from 5 to 20 min, the magnetic permeability (μ_{rel}) of quartz increases from 2 to 3 relative units (Fig. 5.36). Longer processing practically does not affect the value of the indicator μ.

Processing of quartz in a mechanical reactor in the presence of alcohol additives, surface modifiers, contributes to an increase in the magnetic characteristics of quartz. Were used one-, two- and triatomic alcohols, which were introduced in the amount of

5% by weight of the powder. In the case of ethanol, the value of μ_{rel} quartz after 5 min of processing was 5 units, and with an increase in exposure time increased, so that after 20 min it reached 10.5. When using butanol as a modifier, the magnetization of quartz during processing from 5 to 15 min increased from 3 to 5 units.

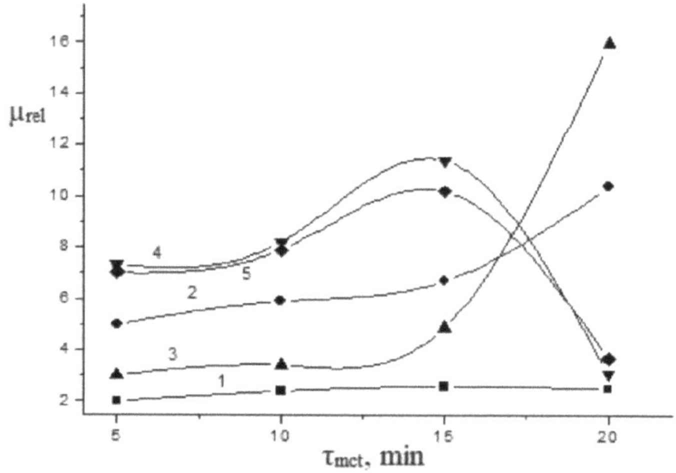

Figure 5.36 Dependence of magnetic permeability of samples on the time of mechanochemical treatment of quartz (1) and quartz with modifiers: ethanol (2), butanol (3), ethylene glycol (4) and glycerin (5).

However, after 20 min of processing, μ_{rel} of quartz increased to 17 (Fig. 5.36, curve 2). The use of ethylene glycol contributed to an increase in magnetic permeability of quartz in the first stages of mechanical treatment (from 5 to 15 min), to 7.5 and 11.5, respectively. However, after 20 min of processing, μ_{rel} decreased to 3.5 (Fig. 5.36, curve 3). When using glycerin as a modifier, the change in μ_{rel} of quartz is of a similar nature.

Thus, the degree of magnetization of quartz from depending on the time of mechanochemical treatment within 20 min when using various alcohols as modifying additives is a non-linear function. In the first stages of grinding, two- and triatomic alcohols have the most favorable influence on the magnetization effect. But with a longer force effect, the greatest magnetization of quartz is observed when using butanol.

According to the results of structural studies of modified quartz samples in the formation of the structure of the surface layer, carbon from organic additives—modifiers degraded in the process of mechanical action plays a significant role. In order to assess the interaction of carbon modifier and iron, which either enters the quartz powder from the surface of the balls and grinding vessels, or is recovered from various compounds in the process of mechanochemical processing, with the manifestation of magnetization of the modified silica, activated carbon was used as a modifier.

The results of measurements of the magnetic permeability of quartz modified with activated carbon (5 wt.%) for 15–30 min showed fairly high values of μ_{rel} = 12–17 (Fig. 5.37).

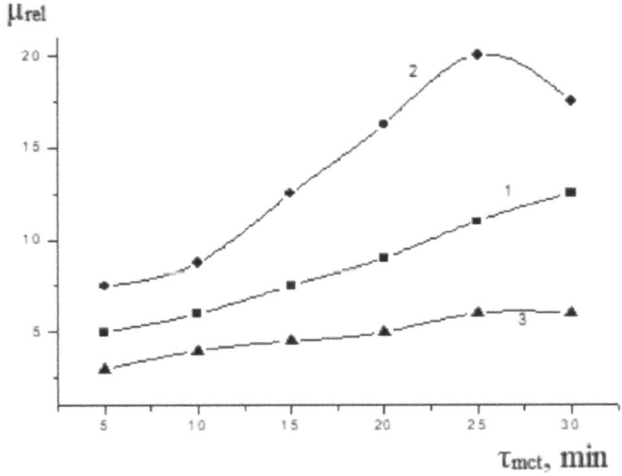

Figure 5.37 Dependence of magnetic permeability on the time of mechanochemical processing of quartz from 5 (1), 10 (2) and 50 mass % (3) of activated carbon.

After a longer treatment, magnetic permeability of the material drops sharply to 6–7, which may be due to formation of non-magnetic iron-carbon compounds (iron carbide) on the surface of quartz particles. A similar picture is observed when introducing more carbon from 10 to 50% into quartz, i.e., its magnetic permeability drops sharply and μ_{rel} is only 3–6 units.

Table 5.10 presents the values of iron content (according to the results of X-ray analysis) in the samples modified with coal, on which, first of all, the parameters of magnetic characteristics should depend. From a comparison of the presented results on the values of iron content in the modified quartz powder and magnetic permeability of the samples, it follows that the presence of ferromagnetic metal particles in quartz undoubtedly plays an important role in manifestation of its ferromagnetism, but formation of a specific structure of surface layers of the particles is more important.

Table 5.10 The content iron according to X-ray diffraction in SiO_2 after mechanochemical treatment in the presence of activated carbon

Material	Treatment time, min	Iron content, %
SiO_2+5% AC*	10	1.35
	20	1.78
	30	2.02
	40	2.23
SiO_2+10%AC	10	2.21
	20	2.58
	30	3.33
	40	5.02
SiO_2+50%AC	10	4.93
	20	5.23
	30	5.62
	40	5.67

AC*, activated coal.

An indirect confirmation of this is the measurement of magnetic permeability of quartz modified with polystyrene (Fig. 5.38).

The dependence of the values of μ_{rel} of the obtained material on the processing time and the amount of the modifier introduced is described by a curve with extreme points. The maximum values of magnetic permeability are reached after 30 min of processing and with a content of 20% modifier.

The results of X-ray diffraction analysis showed (Table 5.9) that quartz after mechanochemical treatment with polystyrene contains ~2–3% ferrite (solid solution of carbon in α-iron). Formation of ferrite is a consequence of polystyrene destruction in the process of mechanochemical processing before formation of free active carbon followed by its dissolution in the crystal lattice with a subsequent quenching of this state as a result of rapid heat removal from the source of interaction. In addition, a small amount of FeO, Fe_3O_4, Fe_2O_3 was found in the treated quartz. The results of X-ray analysis showed (Table 5.11) a significant amount of iron in the mixture depending on the processing time and the modifiers introduced.

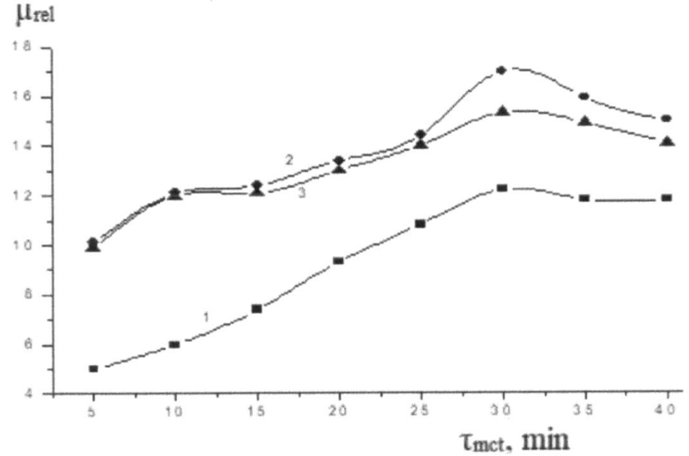

Figure 5.38 Dependence of magnetic permeability on the processing time of quartz with 5 (1), 10 (2), and 20% (3) polystyrene.

Table 5.11 Iron content (by XRD) in quartz modified with various amounts of polystyrene

Treating time, min	The iron content (%) in quartz with a different amount of introduced modifier-polystyrene		
	5%	10%	20%
5	7.08	8.72	12.65
10	6.53	11.41	14.05
20	12.63	13.81	15.00
30	14.2	14.7	16.7
40	14.7	15.1	17.3

Measurement of magnetic permeability of quartz modified with polystyrene indicated the fact that the use of this compound to impart magnetic properties to quartz powder is promising. Formation of a solid solution of carbon in iron (ferrite) favorably affects the manifestation of magnetic properties in quartz powder, since it is a ferromagnet. The use of acrylic acid as a modifier proved to be even more effective for magnetizing quartz. When using the same amount (5 wt.%) of polystyrene and acrylic acid, the effect of magnetization in the second case is higher (Fig. 5.39).

With an increase in the amount of acrylic acid added to quartz to 15%, the magnetic permeability of the material sharply decreased. This is due to the fact that, according to the results of IR spectroscopy, the main amount of iron was spent on the formation of non-magnetic iron acrylate. According to the XRF results, the amount of metallic iron in such samples is less than 2% regardless of the processing time, while in samples modified with acrylic acid in the amount of 5% within 20 min, it reaches 3.9%. Thus, the optimal amount of acrylic acid for modifying quartz in order to obtain a magnetic powder should not exceed 5 wt.%.

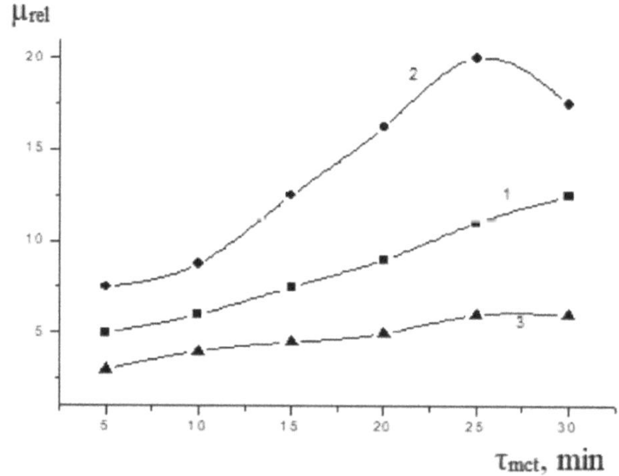

Figure 5.39 Dependence of magnetic permeability on the processing time of quartz with 5 mass. % polystyrene (1), 5 (2), and 15 wt.% acrylic acid (3).

To increase magnetic permeability of the material as a result of mechanochemical treatment, a mixture of iron oxides, namely

Fe_2O_3/Fe_3O_4 (at a ratio of 3/1) was introduced into quartz. According to the literature data [42] and the results of our own experiments presented in the previous sections, the iron oxide Fe_2O_3 processed in the mill of a dynamic action can be converted to Fe_3O_4 and Fe, i.e., to magnetic form. But after of 20 min of treatment, the effect of decomposition of Fe_2O_3 to Fe_3O_4 is negligible (according to the results of XRF it makes up about 20%). Measurements of magnetic permeability for mixtures of quartz and iron oxides showed its relatively low values (Table 5.12).

Table 5.12 Magnetic permeability of a mixture of quartz and iron oxides processed in the presence of modifiers

	Treatment time, min		
Material	**0**	**10**	**20**
$SiO_2+Fe_xO_y$*(10%)	12	12.5	13
[$SiO_2+Fe_xO_y$] (1:1)	14.5	14.5	15.0
[$SiO_2+Fe_xO_y$](1:1)+PS**(2,5%)		17.5	20.0
[$SiO_2+Fe_xO_y$] (1:1)+PS **(5%)		19.0	17.5
[$SiO_2+Fe_xO_y$] (1:1)+PS **(10%)		20.5	13.5
[$SiO_2+Fe_xO_y$] (1:1)+PS **(20%)		20.5	13.5
[$SiO_2+Fe_xO_y$](1:1)+AA***(2,5%)		15.3	16.7
[$SiO_2+Fe_xO_y$] (1:1)+AA***(5%)		16.0	18.0

*Fe_xO_y–Fe_2O_3/Fe_3O_4 in the ratio of 3/1; **PS, polystyrene; ***AA, acrylic acid.

So, for quartz treated in the presence of a 10% mixture of iron oxides, magnetic permeability does not exceed 13, and for a mixture of quartz with iron oxides in the amount of 50/50 (i.e., at a ratio of 1: 1), the μ_{rel} values only increase to 15. Higher values were obtained when iron and polystyrene oxides were simultaneously added to quartz (Table 5.12). In this case, magnetic permeability of the powder increased to 20.5 with a treatment time of 10 min. Longer mechanochemical treatment of such a mixture reduces the effect of magnetization. The use of acrylic acid as a modifier in such a mixture is also less effective. The obtained values of magnetic permeability of the modified powder do not exceed 18 units.

Thus, it can be concluded that the presence of iron-containing components is not the only cause of manifestation of ferromagnetism in the modified quartz powder. An important role in manifestation of magnetism in quartz material is played by the conditions and processing modes, the type and number of modifying additives. When favorable conditions are reached, a certain ordered micro (or nano-) composite structure is formed on the surface of a quartz particle, in which ferromagnetic and paramagnetic compounds are present, this rendering a positive effect on manifestation of magnetic properties in mechanically modified quartz [43].

Figures 5.35–5.39 and Table 5.12 show the results of μ_{rel} measurements for freshly prepared samples, or rather, within an hour after mechanochemical treatment. Repeated measurement of magnetic permeability of the same samples after 10 days of weathering ("aging") showed that for quartz, processed without modifiers, it practically does not change (Fig. 5.40a).

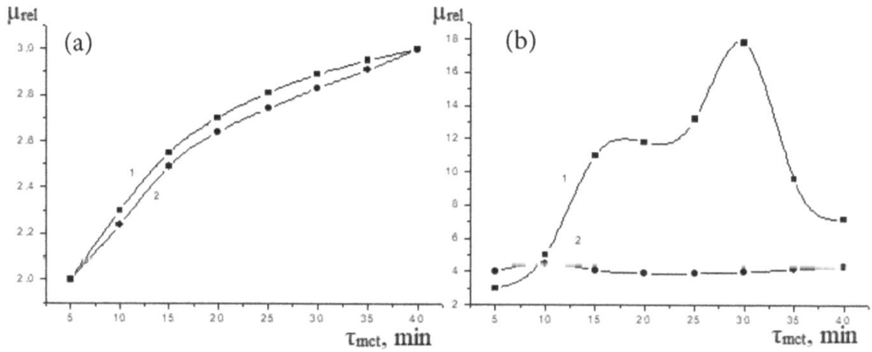

Figure 5.40 Dependence of magnetic permeability of the samples on time: immediately after mechanochemical treatment (1) and after 10 days of weatering (2) for quartz (a) and silica, modified with activated carbon (b).

Samples of quartz, treated with activated carbon for 10–30 min, after 10 days show a significant decrease in μ_{rel} (Fig. 5.40b). However, after longer treatment (40 min), the values of magnetic permeability change slightly with time. Similar results, i.e., the effect of aging, was also found on the samples modified by alcohols. This suggests that in such cases, metastable compounds with active centers are formed on the surface of quartz, which eventually regroup or relax, interacting with the external environment. However, the remaining ferromagnetic compounds included in the surface structure of a quartz particle have a stable magnetic potential. Quartz, modified by polystyrene and acrylic acid (Fig. 5.41), showed stable values of magnetic permeability after treatment for up to 25 min. After a longer treatment, the modified quartz is "ages".

The ambiguous dependence of μ_{rel} on the time of aging for quartz treated with various carbon-containing additives indicates that its state is determined by structural formations on the surface of the particles. In the process of sufficiently long processing of quartz, the formation of more stable structures and the state of the material takes place with the participation of certain modifying additives, which determine the stability of the complex of new electrophysical and magnetic properties of the modified quartz powder.

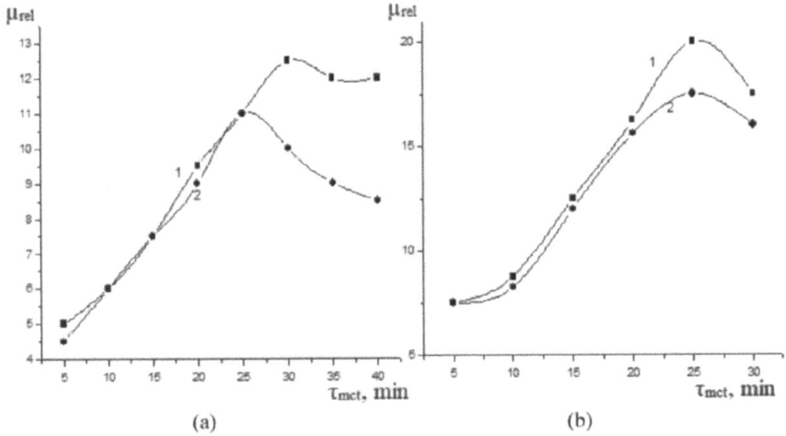

Figure 5.41 Dependence of the magnetic permeability of samples on time: immediately after mechanochemical treatment (1) and after 10 days of weatering (2) for quartz modified with polystyrene (a) and acrylic acid (b).

Table 5.13 shows for comparison the indices of the specific resistance and magnetic permeability of quartz modified by various organic additives in a freshly prepared state and after aging for a month.

Table 5.13 The change in electrical resistivity and magnetic permeability of quartz powder depending on the tine of aging after grinding for 20 min with various modifiers

	Electrical resistivity, $\rho \times 10^6$, Ohm · m		Magnetic permeability, μ_{rel}	
	Aging time, τ_{ag}, days		Aging time, τ_{ag}, days	
Material	0	30	0	30
SiO_2	1.1	6.0	2.0	2.0
SiO_2 + C_4H_9OH (5%)	6.0	15.0	4.0	2.8
SiO_2 +PS* (5%)	2.5	3.5	11.0	10.5
SiO_2 +AA** (5%)	1.2	2.5	20.0	16.5

*PS, polysterene; **AA, Acrylic acid.

Polystyrene and acrylic acid are effective modifiers from the point of view of obtaining not only higher indicators of properties, but also their stabilization with time. It was also found that quartz, modified by more complex organic substances (stearin and oil), has the ability to magnetization. The attempt to modify quartz with oil has its advantages over other additives in solving the problem of using modified materials as adsorbents for purifying water from contamination by its oil products.

There are known methods for synthesis of sorbents, when sorbate molecules are introduced into the volume of particles, thus ensuring the possibility of preserving the "imprints" of its molecules, which later will play the role of specific sorption centers [44]. Measurements of magnetic permeability of samples of quartz powder with such modifying additives yielded positive results (Fig. 5.42) in terms of the prospects for their subsequent practical application.

The magnetic permeability of quartz, modified with stearin, is higher than when using oil. In addition, the values of magnetic permeability were constant over time, i.e., repeated measurements after 10 and 30 days showed the invariance of their values (about 10–12 units). Stability of magnetic properties of a modified quartz material is one of the main requirements for synthesis of a magnetic sorbent.

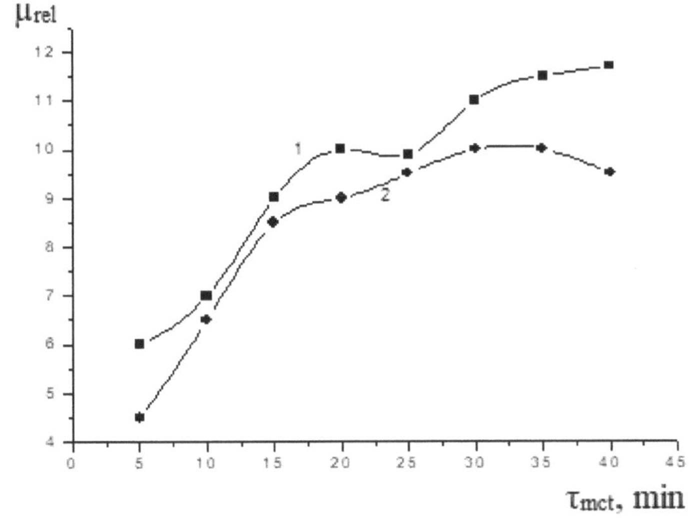

Figure 5.42 Dependence of magnetic permeability of samples on the time of mechanochemical processing with complex modifying additives: 5% stearin (1) and 5% oil (2).

Despite the fact that magnetic permeability for quartz mechanochemically treated in the presence of stearin and oil is small, an important fact is formation of ferromagnetic compounds that are stable in time. For the synthesis of magnetic sorbents based on mechanochemically modified natural mineral raw materials, it is necessary to state the dependencies of formation of surface compounds in the amorphized layer, which are encapsulated in a shell consisting of organoelement compounds playing the role of a protective membrane. Thus, despite the metastability of the state of the material obtained as a result of mechanochemical treatment of quartz, the structure of the surface layer stabilizes the acquired new physicochemical properties of the powder system.

The above-considered results on the investigation of ferromagnetism of quartz induced by mechanochemical treatment indicate that this effect is associated with a change in the state and structure of the surface layer of a quartz particle. Carbon and iron are the constituent elements involved in the restructuring of the quartz matrix. Iron enters the synthesized material as a result of "rubbing" it from the surface of the mechanical reactor in the process of grinding quartz, as well as during reduction of various iron compounds present in the treated mixture to metal. Carbon is introduced either directly into the mixture, or is obtained as a result of the destruction of organic modifiers. Of the modifiers studied, polystyrene was the most effective. The carbon atoms obtained as a result of destruction of polystyrene, are characterized by increased activity. Atomic carbon, diffusing into iron, forms supersaturated solid solutions (ferrite) and chemical compounds—iron carbide.

Encapsulated in a nanometer-scale carbon-containing shell, a quartz particle and ferromagnetic compounds included in the surface structure are a complex micro (nano-) composite system with a set of special physicochemical properties. This is confirmed by all the results of structural studies, including electron microscopy, IR, EPR and Mössbauer spectroscopy. The altered surface structure, enriched with carbon and carbon-containing compounds, implies a change in the chemical properties, namely the sorption capacity, of the synthesized material.

Thus, as a result of mechanochemical activation and modification of quartz by organic compounds, composite mixtures with a nanostructured surface of particles can be obtained. Modification of quartz with formation of composite nanostructured surface layers with participation of carbon, iron and silicon provides the manifestation of the ferromagnetic properties of the powder material synthesized by mechanochemical processing. The choice of optimal modifiers and processing modes is a way to stabilize "ferromagnetism" of modified quartz particles.

Questions

1. Describe the dependence of magnetic permeability of samples on the time of mechanochemical treatment of quartz and modifying additives.
2. How does the magnetization of the modified quartz depend on the interaction of carbon modifier and iron in the process of mechanochemical treatment?
3. What is the dependence of the amount of rubbed iron in the mixture on the processing time and the modifiers introduced?

4. How does the manifestation of magnetism in quartz material depend on the conditions and modes of processing, the type and number of modifying additives?
5. Describe the formation of an ordered micro- (or nano-) composite structure, with participation of ferromagnetic and paramagnetic compounds, and manifestation of magnetic properties of mechanochemically modified quartz?
6. Describe the metastability of the state of the material obtained as a result of the mechanochemical treatment of quartz, and stabilization of new physicochemical properties of the composite powder system, encapsulation of particles into an organometallic shell.
7. What conditions for selection of optimal modifiers and modes of mechanochemical processing ensure the stabilization the "ferromagnetism" of quartz particles?

References

1. Berryman J. G. Random close packing of hard spheres and disks. *Phys. Rev.*, 1983. vol. 27, no. 2. pp. 1053–1061.
2. Mofa N. N., Shabanova T. A., Mansurov Z. A. Mechanochemical synthesis of nanocomposite powder materials based on quartz. *Vestnik Kaz. NU. Chem. Ser.*, 2005. no. 3, no. 39. pp. 27–36 (In Russian).
3. Molchanov V. I., Selezneva O. G., Zhirnov E. N. *Activation of Minerals*. Nedra, Moscow, 1988. 208 p. (In Russian).
4. Chaykina M. V. *Mechanochemistry of Natural and Synthetic Apatites*. Novosibirsk: Publishing House of the Siberian Branch of the Russian Academy of Sciences, Geo Branch, 2002. 223 p. (In Russian).
5. Radzig V. A. Formation of free radicals in the interaction of groups with materials $(\equiv Si{-}O{-})_2Si{\equiv}O_2$ H_2, CH_4, C_2H_6. *Chem. Phys.*, 1995. vol. 14, no. 2. pp. 416–427 (In Russian).
6. Ahmed-zadeh H. D., Baptismansky V. V., Zakrevsky V. A., Tomashevsky E. E. Paramagnetic centers formed during destruction of silicon dioxide. *Phys. Solid*, 1972. vol. 14, no. 2. pp. 422–426 (In Russian).
7. Mofa N. N. Physico-chemical aspects of the synthesis of nanostructured systems for special purposes. *Materials of the III International Symposium "Physics and Chemistry of Carbon Materials/Nanoengineering"*. Almaty, 2004. pp. 83–86 (In Russian).
8. Molchanov V. I., Yusupov T. S. *Physical and Chemical Properties of Finely Dispersed Minerals*. Nedra, Moscow, 1981. 170 p. (In Russian).

9. Avvakumov E. G. *Mechanical Methods of Activation of Chemical Processes*. Science, Novosibirsk, Sib. Branch, 1986. 300 p. (In Russian).
10. Mofa N. N. Mechanochemical processing is a progressive technological process of creating new composite materials. (Review). *Sat. Chemistry and Chemical Technology. Modern Problems*. Almaty: Kazak University, 2004. pp. 163–198 (In Russian).
11. Mofa N. N., Kasymbekova D. A., Shabanova T. A., Kalugin S. A., Mansurova R. M. Mechanochemical synthesis of dispersed carbon-containing composites based on quartzite and acrylic acid. *Materials of the III International Symposium "Physics and Chemistry of Carbon Materials/Nanoengineering"*. Almaty, 2004. pp. 134–137 (In Russian).
12. Simionescu K., Oprea K. *Mechanochemistry of High-Molecular Compounds*. Mir, Moscow, 1970. 354 p. (In Russian).
13. Kochnev A. M., Galibeev S. S. Modification of the structure and properties of polymers. *Izv. VUZ. Chem. Chem. Technol.*, 2003. vol. 46, no. 4. pp. 3–10 (In Russian).
14. Zanin A. M., Kiryukhin D. P., Barelko V. V., et al. On the pulsating nature of the distribution of low-temperature chemical transformations of irradiated solid systems initiated by brittle fracture. *Rep. Acad. Sci. USSR.*, 1983. vol. 268, no. 35. pp. 1146–1149 (In Russian).
15. Berlin A. A., Wolfson S. A., Enikolopyan N. S. *Kinetics of Polymerization Processes*. Chemistry, Moscow, 1978. 240 p. (In Russian).
16. Mofa N. N., Kasymbekova D. A., Shabanova T. A., Kalugin S. A., Mansurova R. M. Mechanochemical synthesis of dispersed carbon-containing composites based on quartzite and acrylic acid. Materials of the III International Symposium "*Physics and Chemistry of Carbon Materials/Nanoengineering*". Almaty, 2004. pp. 134–137 (In Russian).
17. Mofa N. N. Features of the structure and staging of combustion systems based on quartz, modified by mechanochemical processing. *Combustion Plasma Chem.*, 2003. vol. 1. pp. 89–97 (In Russian).
18. Mansurov Z. A., Mofa N. N., Ketegenov T. A., Chervyakova O. V. Carbon containing microcomposition particles with quartz nucleus obtained by mechanochemical synthesis. *Proc. Inter. Symp.* "Carbon 2003". July, 2003. Ovideo, Spain. CD.
19. Gamolin O. E. Mechanochemical transformations of gaseous hydrocarbons: author. *Dis. Cand. Chemical Sciences*. Tomsk, 2005. 22 p. (In Russian).

20. Mofa N. N., Shabanova T. A., Ketegenov T. A., Chervyakova O. V., Mansurov Z. A. Electron-microscopic investigations of quartz particles modified by mechanochemical processing. *Eur. Chem. Tech. J.*, 2003. no. 5. pp. 297–303.

21. Mansurov Z. A., Mofa N. N., Shabanova T. A. Synthesis of powderlike materials with particles encapsulated in nanostructural carbon containing films. *Proc. Inter. Symp* "Carbon 2004". July, 2004. Providence, USA. CD.

22. Krestinin A. V., Kislov M. B., Raevsky A. V., et al. On the mechanism of formation of soot particles. *Kinet. Catal.*, 2000. vol. 41, no. 1. pp. 102–110 (In Russian).

23. Crocker, R., Schneider, M., Hamann, K., Polymer reactions on powder surfaces. *Usp. Chem.*, 1974. vol. XLIII, no. 2. pp. 349–369 (In Russian).

24. Mofa N. N., Chervyakova O. V., Ketegenov T. A., Mansurov Z. A. Synthesis of new materials by mechanochemical encapsulation of quartz particles in metal-complex carbon-containing shells. *Proceedings of II International. Symposium "Physics and Chemistry of Carbon"*. Almaty, 2002. pp. 119–122 (In Russian).

25. Vlasov A. G., Florinskaya V. A., Venediktov A. A., Dutova K. P., Morozov V. N., Smirnov E. V. *Infrared Spectra of Inorganic Glasses and Crystals.* L.: Chemistry, 1972. 304 p. (In Russian).

26. Kiselev A. V., Lygin V. I. *Infrared Spectra of Surface Compounds and Adsorbed Substances.* Science, Moscow, 1972. 459 p. (In Russian).

27. Eytel V. *Physical Chemistry of Silicates.* IL, Moscow, 1962. 204 p. (In Russian).

28. Chervyakova O. V., Mofa N. N., Mansurov Z. A. IR spectroscopic studies of changes in the structure of the surface of quartz as a result of mechanochemical processing. *Materials of International Scientific Conference: Problems of Turbulence, Heat and Mass Transfer and Combustion*, dedicated to the 70th anniversary of Prof. S. I. Isataeva. Almaty, 2002. pp. 126–127 (In Russian).

29. Grigoriev T. F., Vorsina I. A., Barinova A. P., Boldyrev V. V. Mechanochemical synthesis of dispersed layered composites based on kaolinite and higher carboxylic acids. *Rep. Acad. Sci.*, 1995. vol. 341, no. 1. pp. 66–68 (In Russian).

30. Radtsik V. A. Registration by IR spectroscopy in the overtone region of the groups (≡Si–O)2Si=O and $(\equiv \mathrm{Si{-}O})_2\mathrm{Si}<\mathrm{O}_2>\mathrm{C{=}O}$ on the silica surface. *Kinet. Catal.*, 2001. vol. 42, no. 1. pp. 53–61 (In Russian).

31. Tietze L., Aicher T. *Preparative Organic Chemistry.* Mir, Moscow, 1999. 704 p. (In Russian).

32. Volodin A. M, Sobolev V. I., Zhidomirov G. M. In situ ESR study of the state of iron ions in FeZSM-5 zeolites. *Kinet. Catal.*, 1998. vol. 39, no. 6. pp. 844–857 (In Russian).
33. Ryabikin Yu., Mofa N., Zashkvara O., Ketegenov T., Chervyakova O., Mansurov Z. EPR in mechanoactivated quartz-containing systems. *International conference "Modern Development of Magnetic Resonance"*. Kazan, 2004. pp. 153–154.
34. Ryabikin Yu. A., Zashkvara O. V., Mofa N. N., Chervyakova O. V., Mansurov Z. A. Magnetic properties of silicon dioxide modified in a mechanical reactor. *5th International Conference "Nuclear and Radiation Physics"*. Almaty, Kazakhstan, 2005. pp. 142–145 (In Russian).
35. Ketegenov T. A., Urakaev F. Kh., Tyumentseva O. A., Mofa N. N., Kozhakhmetov S. K., Kozhakhmetov S. M. Features of the formation of nanostructured composite materials based on quartz in mechanical reactors. *Problems of Modern Materials Science (Proceedings of the VIII Session of the Scientific Council on New Materials of the IAAS.* May 29, 2003 Kiev). Gomel, 2004. pp. 130–133 (In Russian).
36. Ketegenov T. A., Tyumentseva O. A., Mofa N. N., Urakaev F. Kh. Modeling the reaction of the material of milling bodies with the substance under treatment for quartz treated with steel furniture as an example. *Proceedings of International Symposium on Fracture Modeling and Assessment of Structural Integrily (FMASI).* Korea, June, 2003. pp. 127–133.
37. Ketegenov T. A., Urakaev F. Kh., Tyumentseva O. A., Mofa N. N., Mansurov Z. A. Synthesis of iron silicates on the surface of quartz particles in the process of their mechanical processing. *Rep. Natl. Acad. Sci. Republic Kazakhstan.*, 2003. no. 2. pp. 66–72 (In Russian).
38. Vereschak M. F., Manakova I. A., Dosmaganbetov T. D., Zhaporova A. B., Ketegenov T. A., Mofa N. N., Mansurov Z. A. *Mossbauer Researches of Quartz Subjected to Mechanical Activation//ICAME.* France, 2005. pp. 245–246.
39. Mofa N. N., Ketegenov T. A., Mansurov Z. A., Soh D. W. Nanocomposite magnetic powder materials using mechano-chemical synthesis. *Trans. Electronic Mater.*, 2004. vol. 5, no. 1. pp. 24–33.
40. Soh D. W., Lim B., Soh H., Mofa N. N., Ketegenov T. A., Mansurov Z. A. Magnetism of nanocomposite quartz powder using of MCR method. *Proceedings of International Symposium on Maritime and Communication Sciences,* Myong-ji Universiy Young. Korea, 2004. pp. 113–116.

41. Mofa N. N., Ketegenov T. A., Ryabikin Yu. A., Chervyakova O. V., Ksandopulo G. I. Magnetism of iron-containing particles in the quartz matrix after mechanochemical processing. *Inorg. Mater.*, 2002. vol. 38, no. 2. pp. 1–6 (In Russian).

42. Kulebakin V. G. *The Use of Mechanochemistry in Hydrometallurgical Processes*. Science Sib. Sep., Novosibirsk, 1988. 137 p. (In Russian).

43. Mofa N. N. Mechanochemical synthesis of magnetic oxide materials with a multilayer particle structure. *Materials of International. Scientific-Practical Conference "Chemistry: Science, Education, Industry. Opportunities and Prospects for Development."* Pavlodar, 2001. pp. 101–109 (In Russian).

44. Melikhov I. V., Berdonosova D. G., Sigeikin G. I. The mechanism of sorption and prediction of the behavior of sorbents in physicochemical systems. *Uspekhi Khimii.*, 2002. vol. 2, no. 71. pp. 159–178 (In Russian).

45. Hansen M., Anderko K. *Double Alloy Structures*. Metallurgy, Moscow, 1962. vol. 1. 895 p. (In Russian).

Chapter 6

Theoretical Preconditions for Creation of Mechanochemical Synthesis of Composite Nanostructured Systems Based on Quartz

6.1 The Main Processes in Mechanochemistry of the Quartz Particle Modification

The main processes of changing the structure of quartz being crushed in a mechanical reactor are primarily associated with accumulation of defects and destruction, i.e., with a change in the state and breakage of bonds between the constituent elements of a crystalline object. Siloxane bonds ≡Si–O–Si≡ break with formation of reaction centers ≡Si$^{\cdot}$ and ≡SiO$^{\cdot}$. Thus, on the newly formed surface after the split of the particles there are unsaturated valences. Radical centers actively interact with various gases, this being studied in detail in [1–4]. In the presence of oxygen, oxygen-containing electron-defect centers form at radical centers:

≡Si–O^{+} – Si≡,
≡Si–O^{+} – H,
≡Si–O–O$^{\cdot}$

Mechanochemical Synthesis of Composite Materials
Zulkhair A. Mansurov, Nina N. Mofa, Tlek A. Ketegenov, and Bakhtiyar S. Sadykov

ISBN 978-981-4800-88-4 (Hardcover), 978-1-003-12081-0 (eBook)
www.jennystanford.com

In this work, the most intense are siloxane bonds, which only break in contact with active gas-phase reagents and form new bonds and compounds with them. The chemistry of these processes in general form obeys the laws described in the works of V. A. Radzika [5, 6] and R. Ayler [7], who considered changes in the structure on the surface of silica with varying conditions of exposure to different gases, as well as alcohols (the process of esterification). The occurrence of strained siloxane bonds after mechanochemical treatment of quartz, both without and in the presence of modifiers, is quite clearly demonstrated by the results of IR spectroscopy on the shift of absorption bands in the frequency range of 1000–1250 cm^{-1}.

In general, the change in the structure of the surface layers of a quartz particle under mechanical action can be represented by a set of active centers and groups resulting from deformation and destruction. These are, first of all, silicylene $\equiv Si^{\bullet}$, $=Si^{\bullet\bullet}$, silanone $\equiv SiO^{\bullet}$ and peroxide centers $\equiv SiOO^{\bullet}$. In addition, there are silanol centers $\equiv SiOH$ and deformed (stressed) siloxane groups $[\equiv Si{-}O{-}Si\equiv]_{def.}$, which are the main basis for the formation of surface structures. As shown in the works of V.A. Radtsik [5, 6], in the process of surface transformation, not separate reaction centers, but structural fragments with a defective element, i.e., in fact, clusters containing broken bonds are silylene $(\equiv Si{-}O)_3Si^{\bullet}$, $(\equiv Si{-}O)_2Si^{\bullet\bullet}$ and silanone $(\equiv Si{-}O)_3Si{-}O$, $(\equiv Si{-}O)_2Si{=}O$, the latter of which is of the greatest interest due to the relatively high stability in time, and due to the double bond, is distinguished by its activity with respect to various elements. When quartz is destroyed in the atmosphere of air, the process of formation of silanone groups on the surface of a quartz particle, according to [8, 9], can be represented by the following reactions:

Oxidation to dioxysilane form

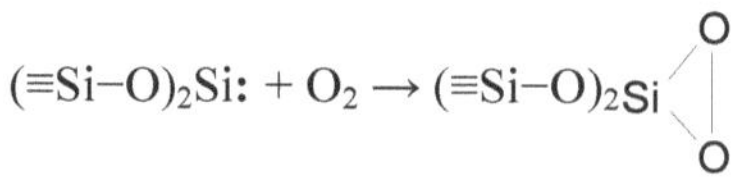

(6.1)

Reduction to silanone from an intermediate carbonate group

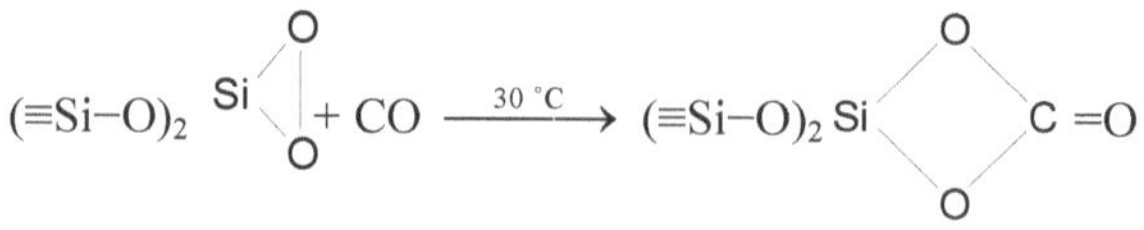

(6.2)

$$(\equiv \text{Si–O})_2\,\text{Si}\!\left\langle\begin{matrix}\text{O}\\ \text{O}\end{matrix}\right\rangle\!\text{C}=\text{O} \underset{30^{\circ}C}{\overset{280^{\circ}C}{\rightleftarrows}} (\equiv \text{Si–O})_2\text{Si}=\text{O} + \text{CO}_2 \tag{6.3}$$

The applicability of this scheme is possible because of heating (up to 200–300°C) of the mixture being treated in the mill with its sufficiently long grinding. The resulting silanone groups actively interact, primarily, with hydrogen and water vapor from the atmosphere of the mill with the formation of silanol groups:

$$(\equiv \text{Si–O})_2\text{Si=O} + \text{H}_2 \leftrightarrow (\equiv \text{Si–O})_2\text{Si(H)(OH)} \tag{6.4}$$

$$(\equiv \text{Si–O})_2\text{Si=O} + \text{H}_2\text{O} \leftrightarrow (\equiv \text{Si–O})_2\text{Si(OH)(OH)} \tag{6.5}$$

Thus, it is a process of not only accumulation of defects on the particle surface, i.e., its loosening and amorphization, but also formation of silicic acid, which quickly polymerizes according to the standard Ashley and Ineson mechanism [10]. This process is due to the presence of both radical and ionized groups on the surface of the particle, which is associated with the electrification of the deformable particle surface.

$$\equiv \text{SiO}^- + \text{HO Si}\equiv \leftrightarrow \equiv \text{Si–O–Si}\equiv + \text{OH}^- \tag{6.6}$$

According to [10], the polymerization of monomer silicic acid is subjected to an autocatalytic effect, as concentration of $SiO^{\bullet}$ increases with the formation of the polymer. This confirms that during the grinding of quartz in the atmosphere of air, there takes place not only amorphization but also polymerization of the surface layer of a quartz particle. As a result, anomalous solubility of the surface occurs [10–12] and the absorption lines belong to OH hydroxyl groups on the IR spectra.

Both silanol groups and monosilicic acid are capable of forming Si–O–M bonds with polybasic metal cations, in particular with iron. Polysilicic acid, being a more polyfunctional compound, forms a chelate-like bond with a metal atom:

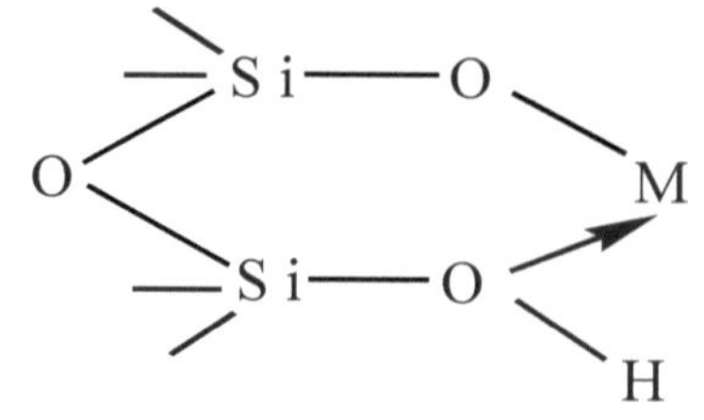

The formation of silicates of metals is also possible if the surface silica is in a weakly polymerized state. These reactions are particularly active with solutions of salts, or metal hydroxide, for example iron [7], under the conditions of low acidity of silicic acid (pH 2–3). Thus, interaction of silicic acid with metal ions, especially at low pH values, can lead to formation of an association of polymer compounds on the surface of a quartz particle, including both the silica (silanol) component and metal cations.

Thus, the transformation of the surface of a quartz particle in the process of its grinding in an activator mill can be represented by the following scheme:

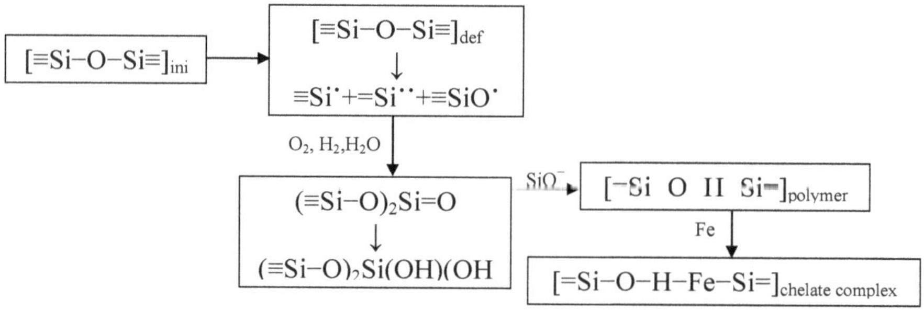

With the introduction of various alcohol additives into the processed quartz mixture, the treatment atmosphere is saturated with vapors of organic compounds used. In addition to the processes discussed above, sorption of the dissociation groups of the constituent organic compounds (R–OH) and binding of polysilicic acid with organic hydroxylated compounds, according to [13], proceed via formation of stable hydrogen bonds. In general, this process of grafting alcohol groups to the surface of a deformed quartz particle runs as follows:

Grafting by silanol groups:

$$\equiv Si–OH + HO–R \rightarrow \equiv Si–O–R + H_2O \quad (6.7)$$

grafting on siloxane bonds:

$$[=Si–O–Si=] + R–OH \rightarrow [=Si–O–Si=] \quad (6.8)$$

or

$$[\equiv Si–O–Si\equiv] + R–OH \rightarrow \equiv Si^{\bullet} + \equiv SiO^{\bullet} + R^{\bullet} + {}^{\bullet}OH \rightarrow \equiv Si–R + \equiv Si–O–OH \quad (6.9)$$

$$[\equiv Si–O–]\, Si^{\bullet\bullet} + R^{\bullet} \rightarrow [\equiv Si–O–]\, Si^{\bullet}–R, \quad (6.10)$$

where R = CH_3, C_2H_5, C_3H_7, etc.

In more detail, structural changes, as the alcohols become more complex, may have the following character:

(1) Ethyl alcohol (ethanol) as a result of thermal and mechanical degradation is represented by the following radical groups:

$C_2H_5OH \rightarrow CH_3 - CH_2 - OH \rightarrow H_3C^{\bullet} + H_2C^{\bullet\bullet} + {}^{\bullet}OH \rightarrow H_2C^{\bullet\bullet} + H_2C^{\bullet\bullet} + H_2O$,

in the interaction of which with the active centers of the surface of quartz in the complex, the following combinations of compounds are possible:

$$\equiv Si–OH + H_3C^{\bullet} + H_2C^{\bullet\bullet} + {}^{\bullet}OH \rightarrow \equiv Si–O–CH_3 + H_2C^{\bullet\bullet} + H_2O \quad (6.11)$$

$$=Si^{\bullet\bullet} + H_3C^{\bullet} + H_2C^{\bullet\bullet} + {}^{\bullet}OH \rightarrow =Si=CH_2 + H_2C^{\bullet\bullet} + H_2O \quad (6.12)$$

$$=Si^{\bullet\bullet} + H_2C^{\bullet\bullet} \rightarrow =Si=CH_2 \quad (6.13)$$

$$\equiv SiO^{\bullet} + H_3C^{\bullet} + H_2C^{\bullet\bullet} + {}^{\bullet}OH \rightarrow \equiv Si–OH + H_2C^{\bullet\bullet} + H_2C^{\bullet\bullet} + {}^{\bullet}OH \rightarrow \equiv Si–OH + 2H_2 + 2C + H_2C^{\bullet\bullet} + {}^{\bullet}OH \quad (6.14)$$

$$=Si^{\bullet\bullet} + H_2C^{\bullet\bullet} \rightarrow =Si=CH_2 \quad (6.15)$$

$$\equiv SiO^{\bullet} + {}^{\bullet}OH \rightarrow \equiv Si–O–OH \quad (6.16)$$

$$\equiv Si^{\bullet} + H_3C^{\bullet} + H_2C^{\bullet\bullet} + {}^{\bullet}OH \rightarrow \equiv Si–OH + H_3C^{\bullet} + H_2C^{\bullet\bullet} \quad (6.17)$$

$$\equiv Si^{\bullet} + H_3C^{\bullet} + H_2C^{\bullet\bullet} + {}^{\bullet}OH \rightarrow \equiv Si{-}H_3C + H_2C^{\bullet\bullet} \quad (6.18)$$

$$=Si^{\bullet\bullet} + H_2C^{\bullet\bullet} \rightarrow =Si{=}CH_2 \quad (6.19)$$

$$\equiv Si^{\bullet} + H_3C^{\bullet} + H_2C^{\bullet\bullet} + {}^{\bullet}OH \rightarrow \equiv Si{-}CH_3 + H_2C{-}OH \rightarrow \equiv Si{-}CH_3 + H_2O + C + H \quad (6.20)$$

$$= Si^{\bullet\bullet} + H_3C^{\bullet} + H_2C^{\bullet\bullet} + {}^{\bullet}OH \rightarrow =Si{=}CH_2 + H_2C^{\bullet\bullet} + H_2O \rightarrow =Si{=}CH_2 + H_2 + C + H_2O \quad (6.21)$$

$$\equiv Si^{\bullet} + H_3C^{\bullet} + H_2C^{\bullet\bullet} + {}^{\bullet}OH \rightarrow \equiv Si{-}OH + H_3C^{\bullet} + H_2C^{\bullet\bullet} \rightarrow \equiv Si{-}OH + H_3C^{\bullet} + H_2 + C \quad (6.22)$$

$$\equiv Si^{\bullet} + H_3C^{\bullet} \rightarrow \equiv Si{-}CH_3 \quad \text{and} \quad \equiv SiO^{\bullet} + H_3C^{\bullet} \rightarrow \equiv Si{-}O{-}CH_3 \quad (6.23)$$

Since the results of IR spectroscopy did not show the presence of ${}^{\bullet}CH_3$ groups in the spectrum, this complex is most likely to decay with formation of the final products $=Si{=}CH_2$, H_2, C. The presence of free carbon on the surface of quartz is started by the results of electron diffraction and its amount is measured by the absorption weight method.

Thus, the final result of these interactions is the formation of silanol groups, unbound (adsorbed) water, carbon, and Si=C double bonds, which predetermines the possibility of formation of polymer films on the surface of particles.

The possibility of polymerization is also associated with the fact that during the grinding process the mixture is heated. And increasing the temperature to 350–500°C in the presence of a catalyst, which can be SiO_2 and C, ensures the development of the dehydration process of alcohols with formation of olefins:

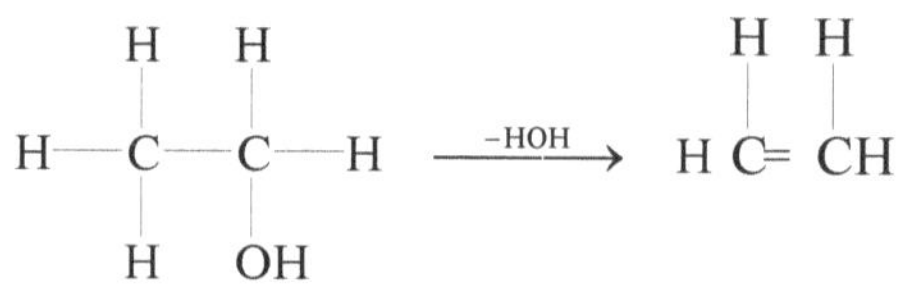

(6.24)

in which the presence of a double carbon bond determines the polymerization process—the sequential addition of other molecules to one olefin molecule due to the double bond breaking according to the scheme:

$$n\ H-\overset{\displaystyle H}{\overset{|}{C}}=\overset{\displaystyle H}{\overset{|}{C}}-H \rightarrow \underset{\text{polyethylene}}{(-CH_2-CH_2-)_n} \qquad (6.25)$$

This process runs via both ionic and radical mechanisms. The resulting radicals, due to their high activity, not only combine with each other, but also interact with whole molecules according to the chain reaction scheme.

(2) When using butyl alcohol (butanol) as a modifier, the interaction with the surface of quartz, on the one hand, follows all the above schemes and, on the other, is complicated due to intensification of the polymerization process and carbonization of particles as a result of partial destruction to carbon.

$$C_4H_9OH \rightarrow CH_3 - CH_2 - CH_2 - CH_2 - OH \rightarrow R - 3\ (\text{-}R''\text{-}) - OH, \qquad (6.26)$$

where $R = CH_3$, $R'' - CH_2$.

The group R″ forms double bonds with active centers on the surface of quartz and the possibility of the formation of large amounts of free carbon.

$$= Si^{\cdot\cdot} + R + 3(\text{-}R''\text{-}) + {}^{\cdot}OH \rightarrow =Si=R'' + 3H_2 + 3C + H_2O \qquad (6.27)$$

(3) When using polyhydric alcohols, the reactions follow the scheme:

with ethylene glycol $C_2H_4\ (OH)_2$

$$OH\text{–}CH_2\text{–}CH_2\text{–}\ OH \rightarrow OH - 2\ (\text{-}R''\text{-})\ \text{-}OH \qquad (6.28)$$

$$2\ (=Si^{\cdot\cdot}) + 2\ (\text{-}R''\text{-}) + 2({}^{\cdot}OH) \rightarrow 2(=Si=R'') + O + H_2O \qquad (6.29)$$

$$2\,(\equiv Si^{\bullet}) + 2\,(\text{-}R''\text{-}) + 2\,(^{\bullet}OH) \rightarrow 2\,(\equiv Si\text{–}R''\text{–}OH) \tag{6.30}$$

and with glycerin $C_3H_5\,(OH)_3$

$$\underset{\substack{|\\ OH}}{OH\text{–}CH_2 - CH\text{–}} - CH_2 - OH \rightarrow \underset{\substack{|\\ R''' - OH}}{OH - 2\,(\text{-}R''\text{-})} \text{–}OH, \tag{6.31}$$

where R″ - CH group with a carbon triple bond, the presence of which to a greater extent ensures the polymerization process on the surface of quartz. The amount of free carbon on the surface is insignificant. The formation of a polymer film and a decrease in the amount of free carbon on the surface of quartz, when processing it with polyhydric alcohols, have been confirmed experimentally.

For unsaturated alcohols, in particular, vinyl alcohol, all the considered reactions also take place, and the polymerization process during the mechanical processing of quartz is accelerated due to the presence of a double bond in the initial compound. The use of more complex organic compounds as additives in the mechanical processing of quartz leads to the creation of formations that are complex in shape and structure on the surface of a quartz particle as a result of the initial deep destruction of both the inorganic part of the material and its organic component. Then, active organic radicals are grafted to the reaction centers of the fresh surface of a quartz particle. In the case of using acrylic acid, the process is generally as follows. Acrylic acid ($CH_2{=}CHCOOH$) is the simplest unsaturated carboxylic acid, which polymerizes well due to the presence of a double bond, and its rupture stimulates the formation of various polymeric compounds. As in the cases discussed above with alcohols, the surface reaction centers $\equiv SiO^{\bullet}$ and $\equiv Si^{\bullet}$, arising from the destruction of quartz particles, initiate the graft polymerization of acrylic acid monomers and the grafting of its radicals, arising from the destruction of the initial materials.

A possible variant of the radical polymerization of the surface of a quartz particle with acrylic acid during mechanical degradation in a planetary centrifugal mill is the following scheme:

$$\begin{array}{l} \equiv Si{-}O{-}Si\equiv + n\,CH_2{=}CHCOH \\ \quad \rightarrow \equiv Si{-}O{-}(\underset{\displaystyle COOH}{\underset{|}{CH_2{-}CH}}{-})n + \equiv Si{-}(\underset{\displaystyle COOH}{\underset{|}{CH_2{-}CH}}{-})n \end{array} \qquad (6.32)$$

and

$$\begin{array}{l} \equiv Si{-}O{-}Si\equiv + nCH_2{=}CHCOH \rightarrow \\ \quad \rightarrow \equiv Si{-}O{-}(\underset{\displaystyle CO}{\underset{|}{CH_2{-}CH}}{-})n + \equiv Si{-}OH \end{array} \qquad (6.33)$$

Thus, a film of carbon-containing polymer is formed, cross-linked with the surface of a quartz particle and encapsulating it. The structure of the polymer surface layer, according to the obtained results of IR spectroscopy, contains additional crystalline inclusions with Fe. In addition to the polymer cross-linked to the surface of the particle, a homopolymer of different structure and composition is also formed. The final result is a new complex modified structure being transformed depending on the amount of substance, type and time of exposure. In such diverse compounds, a carbonaceous matter can appear as an independent phase, and is in chemical bonding with the surface of quartz, forming organo-elemental compounds. With long-term processing, the decomposition of the organic modifier, as shown experimentally, goes to carbon, which is possible as a result of the decomposition of the $COOH \rightarrow C + H_2O$ group. In this case, the CH_2 group of the polymer is additionally closed on the reaction centers of the surface of quartz $Si^{\bullet}$ or $Si^{\bullet\bullet}$, as well as on the inclusions of iron, forming organometallic complexes.

Inclusions of Fe with various forms of presence: both in metal (cluster), oxidized form, and in the form of compounds with silicon and carbon [14] play a significant role in silicon-carbon composites. The interaction of iron with the quartz surface occurs primarily via the hydroxyl groups of silanol centers with formation of iron silicates [15]. The second way of interaction is due to the incorporation of iron atoms into the polymer chain, grafted organic radicals to the surface of quartz. Thus, on the surface of a particle containing particles of implanted iron, a heterogeneous structure of a nanoscale scale with a stable potential difference between the inclusions of iron and carbon is formed [16].

When using more complex polymer modifiers, in particular polystyrene ($[-CH_2-CH-C_6H_5]_n$), the process of destruction and phase formation of the powder material proceeds with the formation of a large number of active radical groups: $=CH_2$, $\equiv CH$, which provide modification of the surface of quartz according to the above scheme, which clearly follows from the obtained results of electron microscopic studies, IR and Mössbauer spectroscopy. The main part of the polymer (C_6H_5 group) destructs to carbon and hydrogen. The final result of grinding quartz particles, transforming the structure of the surface layers before the appearance of polysilicic acid, the destruction of organic additives and the grafting of the resulting radicals to the reaction centers of the surface of a particle with inclusions of iron and reduced carbon is schematically presented in Fig. 6.1, where the particle after mechanochemical modification consists of a quartz core and a heterogeneous nanostructured surface layer.

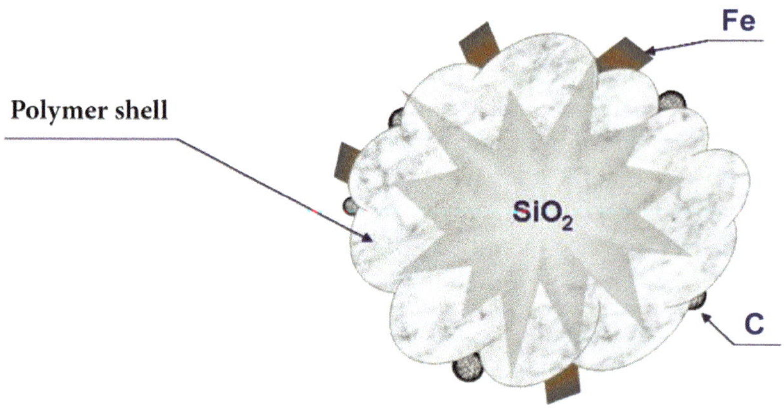

Figure 6.1 Scheme of the modified particle with all the constituent elements of the structure of the surface layers.

When polystyrene was used as a modifier on the surface of a quartz particle, free carbon was found in the greatest amount in various morphological forms of its manifestation. The observed experimental phenomenon is conditioned by the fact that during mechanical action there take place not only the processes of breaking bonds of various levels, but also radiation of various kinds and emission of electrons, accelerating the process of destruction and subsequent polymerization of organoelement compounds.

Various physical processes accompanying mechanical activation have been studied in sufficient detail by a number of researchers in specially designed works. In this case, much attention is paid in them to the phenomenon of exoelectronic emission, i.e., electron emission [17, 18]. Electron emission can be caused by an increase in the temperature of a pre-excited solid, its surface treatment in a gas discharge or irradiation with ultraviolet and X-rays, as well as mechanical effects on air: grinding, scratching, crushing, deformation. In addition, there is a "chemical emission" in the course of action of corrosive gaseous media on metals and alloys, resulting in the formation of surface films (metal corrosion). This process is investigated with a tip counter. The emission process, lasting during the action of the external source is followed by after-emission, which decreases according to a power law in proportion to t^{-a}, and the indicator depends on temperature. So, if quartz is processed at room temperature, the value of $a = 0.8$, and at 52°C $a = 7$.

There are several hypotheses explaining electron emission [18, 19], but in all cases of external influence this phenomenon is related transition of electrons to lower energy levels, i.e., the decrease in the electronic work function due to formation of an unstable surface phase. For non-metallic systems, the polar structure of the crystal is a necessary but not sufficient condition for the onset of emission.

The effect of exoelectron emission is most pronounced on piezomaterials, in particular, on quartz [18]. The energy of fast electrons when splitting quartz reaches hundreds of kiloelectron-volts. As the decrease of after-emission with time obeys the power law, the initial velocity of electrons escaping from the surface can be much higher. It was experimentally started that the surfaces of quartz, along which destruction occurred, have mosaic-spaced positively and negatively charged regions. Fused quartz, as well as silicate and organic glass do not exhibit electron emission. Crystal quartz also gives a noticeable glow when splitting it in the dark, i.e., a luminescence effect is revealed.

When considering the mechanisms of chemical reactions occurring on a freshly formed surface, it is necessary to take into account that the reaction proceeds on the radiating surface. With simultaneous treatment of solid inorganic materials and organic compounds in a mechanical reactor, chemical interaction processes proceed under the influence of local irradiation of the flow of exoelectrons. Polymerization reactions in a gaseous medium of a monomer in contact with a freshly formed radiating surface proceed as follows: radiation generates free radicals in the bulk of the monomer, which then initiate chain growth, i.e., bulk polymerization of the monomer or graft polymerization on the radiating surface. According to the calculations given in [20], the radiation effect achieved by electrons generated in a single act of adhesion failure is comparable to the action of electrons produced in an accelerator and having a power of about 1 MeV and a dose rate of 150 rad/s when the source acts during the day. The effect of the radiative action of electrons is explained by the concentration of energy in the thin surface layer, while the energy from external sources penetrates into the volume. The advantage of the action of electrons emitted by freshly formed surfaces is that they do not cause destruction of the substance they irradiate in the bulk. This implies the possibility of practical application of the emission of fast electrons in the control of mechanochemical processes and the formation of new structures and compounds, preferably on the surface of particles.

At a certain threshold destruction rate, in addition to electron emission, electromagnetic oscillations and radio waves are observed. Radio emission is pulsed. The cause of radio emission is the gas-discharge process in the resulting gap between the tearing-off surfaces, where a kind of accelerator arises due to the existence of high-intensity fields. The electron, accelerated by the field, is gaining sufficient speed for impact ionization of oncoming molecules and the formation of avalanches of positive and negative ions, i.e., gas discharge microplasma. The motion of particles in a plasma causes oscillations of various frequencies from acoustic to radio frequencies with a wavelength of the order of decimeters. Surface electrification is observed due to the remaining and gradually dissipating charge.

An important property of fast electrons emitted by a freshly formed surface is their ability to generate free radicals in the gas phase and on the surface of the irradiated bodies and thereby produce effects similar to those of radiation. This makes it possible to obtain graft polymerization on the surface of a radiating polymer and homopolymerization in the volume of monomer in contact with it.

The fact of the influence of electron emission on the processes and the change in the structure of the material during mechanochemical processing is considered in many works, but the role of the piezoelectric effect is mentioned in some cases, but not specifically considered. However, the relationship of these two phenomena is noted by many researchers.

Questions

1. What main structural changes take place on the surface of particles during the grinding of quartz?
2. Show the role of silylenic, silanone, peroxide, and other centers in the formation of surface structures of ground quartz.
3. Describe the role of defects and electrification of the surface of a deformable particle in the formation of silicic acid and polymerization of the surface of quartz.
4. How are the processes of grafting various alcohol groups on the surface of deformable quartz with the formation of polymer films?
5. Describe the features of structural changes when using complex organic compounds as additives for mechanical processing of quartz and obtaining composite systems.
6. What are the conditions for formation of a carbon-containing polymer film cross-linked with the surface of a quartz particle and encapsulating it during the mechanical processing of a mixture of quartz-organic modifier?
7. What is the role of physical phenomena: radiation of various kinds and electron emission in accelerating the processes of destruction and polymerization of organoelement compounds?
8. Describe the phenomenon of exoelectronic emission during grinding of quartz as a result of the electric field in the cracks of the destroyed body.

9. Describe the scheme of the polymerization process with simultaneous processing in the mechanical reactor of solid inorganic materials and organic compounds (radiation, formation of free radicals, growth of the polymer chain on the radiating surface of the particles).

6.2 The Piezoelectric Effect of Quartz Is Part of the Process of Modifying the Surface of Particles with Organic Compounds

The piezoelectric effect is the ability of some crystalline materials to generate an electric charge proportional to the mechanical voltage. The same materials have an inverse piezoelectric effect, i.e., are deformed in proportion to the applied voltage. For manifestation of this effect, in any case, it is necessary that at least one direction in the medium would be polar without a center of symmetry of the unit cell. Specifically, quartz in both the α-modification and β-modification, having polar directions in the crystal lattice, exhibits a piezoelectric effect [21]. α-Quartz crystals refer to the trigonal-trapeziohedral class of 3:2 trigonal system, and β-quartz crystals refer to the hexagonal-trapeziohedral 6:2 class of the hexagonal system. The elements of minimal symmetry of α-quartz are the axis of the third order (optical) and three axes of the second order perpendicular to it (electric), forming angles of 120° between them. Perpendicular to these axes are three axes, called mechanical.

The structure of quartz is based on chains of SiO_4 tetrahedra located along a third-order symmetry axis (along the optical axis). In a crystal structure, each Si ion with a positive charge of +4e is tetrahedrally surrounded by four O ions, each of which has a negative charge of −2e, and each O ion binds two Si ions. With full charge compensation, the cell is electrically neutral. When an external force is applied, the ions are displaced and a positive charge appears on one surface of the crystal, and a negative charge on the other, i.e., polarization occurs. For α-quartz, polarization is possible in three directions, for β-quartz—only in one direction. According to the materials considered in [20], this is a consequence of the fact that in α-quartz crystals by the piezoelectric method five types of uniform deformation can be obtained: compression or tension in the direction of the X and Y axes and shift in the YZ, ZX and XY planes. Moreover, the first three of them are carried out if the electric field is directed along the X axis, and the other two if the field acts in the direction of the Y axis. On the contrary, a mechanical displacement of the crystal elements occurs under the action of an external electric field.

So, the mechanical displacement of charged particles in the volume of the crystal lattice and the electric polarization are interrelated, as a result of which the crystal is a charged electric capacitor (or electric battery). The associated charges of spontaneous polarization create an external electric field (Fig. 6.2). The discharge of such a capacitor is bound, first, due to the internal conductivity taking place in any dielectric, and second, the bound charges of spontaneous polarization can be compensated by almost always available free charges in the atmosphere. In this case, the spontaneously polarized state of the dielectric is not lost, but it is compensated by the oppositely oriented field of free charges.

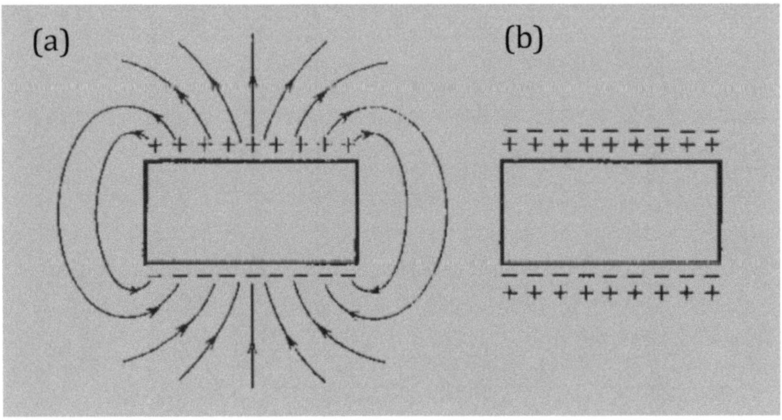

Figure 6.2 Spontaneously polarized dielectric in the absence of neutralizing charges (a) and when neutralized by external free charges (b).

To describe the piezoelectric effect, the intensity of the arising polarization is taken as a measure, the value of which is proportional to the charge (Q). The main characteristic of the piezomaterial is d—piezoelectric module, which is determined as

$$d = D/T = S/E, \quad (6.34)$$

where D is electrical induction (numerically equal to the total charge Q per unit area), T mechanical stress, S expansion or contraction, and E the electric field strength generated by the mechanical voltage.

Another characteristic is the piezoelectric coefficient $-g$ (mV/n):

$$E = -g\,T \quad (6.35)$$

or –e:

$$T = -e\,E \quad (6.36)$$

These characteristics are related through the dielectric constant by the ratio

$$g = d/\varepsilon \quad (6.37)$$

The most used characteristic is the coefficient of electromechanical coupling $-k$, which characterizes the fraction of mechanical energy converted into electrical energy and vice versa. For quartz, the value of k = 0.1 and is calculated according to the system of equations:

$$I_x = d_{11}t_{xx} - d_{11}t_{yy} + d_{14}t_{yz} \quad (6.38)$$

$$I_y = -d_{14}t_{zx} - 2d_{11}t_{xy}$$

$$I_z = 0,$$

where t denotes mechanical stresses and d piezoelectric modules.

The values of piezoelectric modules [21] for α-quartz in the SI system are equal to:

$$d_{11} = 2.31 \times 10^{-12}\ \text{C/N},\ d_{14} = 0.85 \times 10^{-12}\ \text{C/N}.$$

The intensity of the arising polarization is recorded in a matrix form:

	t_{xx}	t_{yy}	t_{xz}	t_{yz}	t_{zx}	t_{xy}
I_x	d_{11}	d_{11}	0	d_{14}	0	0
I_y	0	0	0	0	$-d_{14}$	$-2-d_{11}$
I_z	0	0	0	0	0	0

Using the calculated data on the values of the acting force at the point of contact of the grinding balls and the quartz powder being processed [18], it is possible to estimate the possible values of the polarization intensity and, consequently, the surface charge density and electric field strength around the quartz particle when grinding it in a mechanical reactor. With current efforts at the point of impact along the electrical axis *X*, according to [20], equal to ~ 4.5×10^9 N/m^2 (normal pressure) and 7.2×10^9 N/m^2 (tangential pressure), a charge appears on the surfaces of the crystal, perpendicular to the *X* axis, determining the polarization intensity:

$$I_x = d_{11}t_{xx} = 2.31 \times 10^{-12} \times 4.5 \times 10^9 = 103.5 \times 10^{-4} \text{ C/m}^2 \quad (6.39)$$

At a simultaneous action in the *YZ* plane of shifting (tangential) stresses of 7.2×10^9 N/m^2, the polarization intensity increases:

$$I_x = d_{11}t_{xx} + d_{14}t_{yz} = 2.31 \times 10^{-12} \times 4.5 \times 10^9 + 0.85 \times 10^{-12} \times 7.2 \times 10^9$$
$$= 164.7 \times 10^{-4} \text{ C/m}^2 \quad (6.40)$$

With the acting force of the same magnitude along the *Y* axis (i.e., if at the moment of impact a particle of quartz is turned to the ball direction Y), a charge equivalent to the first case but opposite in sign occurs on the same crystal surfaces. Qualitatively, the level of polarization intensity is maintained. The maximum polarization takes place under the action of shear stresses in the XY plane and its value corresponds to

$$I_y = -2d_{11}t_{xy} = -2 \times 2.31 \times 10^{-12} \times 7.2 \times 10^9 = -332.64 \times 10^{-4} \text{ C/m}^2 \quad (6.41)$$

Using the obtained polarization values and, consequently, the induced charge density (σ), the potential difference in the direction of the applied voltage was calculated. Its values are determined as

$$U = q/C = \sigma h/\varepsilon\, \varepsilon_o, \quad (6.42)$$

where h is the thickness of the crystal, ε is the dielectric constant, and ε_0 is the electric constant equal to 8.85×10^{-12} f/m.

For the three cases of polarization considered above, the voltage values for a crystal with a thickness of 100 μm are equal to

$$U_x = 103.5 \times 10^{-4} \times 10^{-4}/4.6 \times 8.85 \times 10^{-12} = 2.54 \times 10\ 5 = 254\ \text{kV} \tag{6.43}$$

$$U_{x(yz)} = 164.7 \times 10^{-3} \times 10^{-4}/4.6 \times 8.85 \times 10^{-12} = 4.046 \times 10^5 = 404.6\ \text{kV} \tag{6.44}$$

$$U_y = 332.64 \times 10^{-3} \times 10^{-4}/4.6 \times 8.85 \times 10^{-12} = 8.85 \times 10^5 = 885\ \text{kV} \tag{6.45}$$

Accordingly, the maximum electric field strength ($E = U/d$) inside and around charged particles 100 μm thick will be 2540, 4046 and 8850 MV/m, respectively, for each of the polarization directions.

According to the first Curie law, the polarization effect under compression or tension along the X axis does not depend on the size of the crystal, but is determined only by the acting forces. Therefore, as the crystals are crushed, the intensity of its polarization perpendicular to the X direction will not decrease. According to the second Curie law, the magnitude of the piezoelectric charge, due to the acting forces along the Y axis, depends on the size of the crystal (the relationship is directly proportional) and its size will decrease as it is crushed. Thus, as grinding progresses, there takes place self-organizing ordering of the direction of the electric field in the bulk of the mixture of quartz crystals being ground and deformed.

However, the voltage of the induced field caused by polarization decreases as the size of the crystals decreases in the process of grinding, and as the size of the crystals decreases to 10 and 1 μm, the maximum values of the effective force will be 25.4, 40.46, 88.5 kV and 2.54, 4.046, 8.85 kV, respectively. The values of intensity and energy of the induced electric field created by polarization of particles of different sizes, i.e., at different stages of grinding quartz in the mill also change. The calculated values of these characteristics are given in Table 6.1. The energy of the electric field of quartz particles is calculated by the equation for a plane capacitor $W = \varepsilon\varepsilon_0(V/d)^2S\,d/2$.

Accordingly, the volume density of the electric field energy is determined as

$$u = \varepsilon\,\varepsilon_0 E^2/2 \quad (6.46)$$

The volume density of the electric field energy is equivalent to the force acting on the unit of the charged surface of the deformed quartz crystal. As it is crushed, this characteristic for one particle of quartz decreases, but highly dispersed particles are combined into aggregates that already have a total electric field.

Table 6.1 Characteristics of the electric field induced by the piezoelectric effect of a quartz particle of different sizes

Particle size, μm	Polarization $I \times 10^3$, C/m^2	Potential difference U, kV	Tension E, MW/m	Electric field energy density u, W
100	103.5	254	2540	131.3
	164.7	404.6	4046	333.2
	-332.64	885	8850	1594.25
50	103.5	127	1270	32.82
	164.7	202.3	2023	83.3
	-332.64	442.5	4425	398.6
20	103.5	50.8	508	5.25
	164.7	80.9	809	13.32
	-332.64	177.0	1770	63.77
10	103.5	25.4	254	1.31
	164.7	40.46	404.6	3.32
	-332.64	88.5	885	15.9
1	103.5	2.54	25.4	0.031
	164.7	4.046	40.46	0.033
	-332.7	8.85	88.5	0.159

According to the calculated data obtained on quartz and sodium chloride [22], which also has a strong electrical effect under mechanical action, relative to the size of the zone under the impact of the ball, the contact area is $S = 3.8 \times 10^{-2}\,R^2\text{cm}^2 = 0.95$ mm (with *R*, the radius of the ball, being 5 mm). The thickness of the particle layer for the same exposure conditions is 1.1 mm. Then the volume of the powder in the contact zone is 1.045 mm^3, and the amount of quartz particles with a size of 100 microns in the contact zone will reach 250 pieces. Of these particles, only a part is deployed in the direction of the acting forces by the polarization axes *X* and *Y*. Even if this part of the oriented particles is only 10% of the total, i.e., 25 pieces, the electric field caused by the piezoelectric polarization of each particle in the interaction contact volume (1.045 mm^3) reaches tens of MV/m (25–370 MV/m). Thus, the remaining non-polarized by mechanical action particles are in a strong electric field and receive a charge due to the effect of electrostriction—deformation in an electric field, which does not depend on the direction of the active field, and its magnitude is proportional to the square of the applied voltage [23]. As a result, a charge is formed on the surface of such quartz particles, but of a smaller value (voltage of several tens of volts) than with the piezoelectric effect.

Quartz is also a pyroelectric, i.e., a charge caused by thermal expansion of the crystal appears on the surface of quartz particles, which especially sharply increases at temperatures preceding the polymorphic transition ($\alpha \leftrightarrow \beta$) from 400 to 573°C. The calculated temperatures at the contact between the ball and the ground silica powder make up 1187°C [22]. The temperature values at the contact are calculated under the condition that all the energy transferred by the ball to the powder being ground goes into heat. However, part of the energy goes to the piezoelectric effect and other structural changes in the deformable particle, so the actual temperature at the contact will be much lower. The pyro effect will actually be determined by the background temperature, which is 100–300°C, depending on the duration of grinding. The magnitude of such a charge induced by thermal deformation is small, and the difference of the induced potential is several volts (or tens of volts).

According to the calculated data, the time of interaction between the ball and the volume of particles in the contact zone is 32 μs. The time of removal of the piezoelectric effect after the cessation of force is 2–3 times longer [24]. In addition, with an increase in the frequency of the pulse effect of the mechanical load on the crystalline material, a polarization effect does not take place between two successive blows. Deformation of the structural elements of the crystal is transferred from elastic to plastic, and thus the effect of charge polarization is "fixed". As a result, the intensity of the induced electric fields increases and becomes more stable. At this stage, the electric polarization, i.e., an ordered arrangement of charged particles that form the structure of a solid under the action of an external mechanical action, leads to the formation of dipoles. Depending on the displacement, the degree of distortion (imperfection) in the structure and its stability over time change; this is how the dipole centers (electron–hole) of the exchange-coupled interaction are formed. The result of these structural changes is the ferromagnetism of particles induced by an external mechanical action [25, 26].

Currently, all types of external influences (mechanical, electrical, magnetic, temperature) are considered together and in interrelationship. Each of the external fields can cause the rest of the fields in the crystal, the degree of influence of which is determined by the level of the corresponding material properties. For piezomaterials, the relationship between mechanical, electrical, thermal, and magnetic properties is most pronounced and is particularly important for practical use.

Questions

1. What are the features of the structure of the crystal lattice of quartz provide manifestation of the piezoelectric effect?
2. What is the relationship of the mechanical displacement of charged particles in the volume of the crystal lattice and electric polarization: the formation of a charged electrical capacitor?
3. What are the main characteristics of the polarization process and indicators of the properties of the piezomaterial?
4. Estimate the possible values of the polarization intensity, surface charge density and electric field strength around the quartz particle when grinding it in a mechanical reactor?
5. Describe the formation of the pyroelectric properties of quartz.
6. What is the relationship of defects in the structure of particles, electric polarization and formation of dipole centers (electron–hole) with an external mechanical effect on quartz particles?

6.3 Simulation of the Process

The change in the structure of the crushed quartz particle at all hierarchical levels (from the particle volume to distortions within the crystal lattice) is determined by the type and degree of kinetic energy dissipation) transferred to the particle at the time of the impact of the ball and fixed in it as the elastic energy of the deformed crystal. All processes and phenomena that occur as a result of an impact and manifest themselves in both organic and inorganic media are summarized in the literature data in Table 6.2. All of these processes take place under pulsed loading conditions in mechanical reactors. Practically at every stage, the appearance of electric charges occurs, either explicitly or as a side effect, which is most pronounced for piezomaterials, which include quartz. Already at the stage of elastic compression (stress), the lattice is polarized and charges appear on opposite faces of the crystal. In this case, the component of the thermal effect for piezomaterials should decrease under equivalent conditions of mechanical action, since most of the energy impulse is spent on the displacement of charges and formation of an electric field around the particle.

Table 6.2 Processes and states that determine dissipation of the impact energy in a mechanical reactor

Types of processes and states
Triboplasma
Elastic stress
Vibrational state in a nonequilibrium system
Vibrational state in equilibrium (hot spots) - thermal radiation
Breaking the surface and forming a fresh surface
Transport effect (reactions between solids)
Plastic deformation, migration of lattice defects
Electronic emission
Surface electric charges
Generation of acoustic waves and radio waves
Luminescence
X-radiation

This process is non-equilibrium, developing in an open system with a constant redistribution of energy, the simplest scheme of which is presented in Fig. 6.2 and, therefore, it can be adequately described only by non-equilibrium thermodynamics of irreversible processes [27]. The redistribution of energy and the change of physical phenomena accompanying destruction, changing the Gibbs potential, also changes the kinetics of chemical processes during grinding. The solution of problems of nonequilibrium thermodynamics in mechanochemical processes is a field of research that deserves special attention, and for the formulation and solution of such problems, first of all, it is necessary to develop an optimally simplified model of the process, including many active factors. This section of the work is devoted to creation of just such a model of the process, the final stage of which is to obtain a quartz particle with a modified surface.

Based on the considered concepts of the piezoelectric effect in a quartz crystal and the calculated values of the electric field that sets up around deformable crystals in a mechanical reactor, a model for the formation of quartz-based nanostructured systems can be constructed as an electromagnetic response of a system to pulsed mechanical loading. Modification of the surface of a splitting particle is the result of the process of accelerated movement of radical ions (charged particles) of destroyed organic additives and their interaction with the charged surface of crystalline quartz. The final result is the formation of a composite particle with nanoscale formations on its surface. A number of side effects, well enough described in the scientific literature, and which have their influence on the modification process should be taken into account:

(1) Formation of local deformed and stressed areas with potential difference with respect to undeformed areas [27–29].
(2) Formation of charged defects on the surface and in the crystal volume (broken bonds, vacancies, dislocations, electron–hole pairs) and their displacement in the volume and on the surface of particles in the electromagnetic field and the field of elastic stresses [6, 27, 30–32].

(3) Formation of hot electrons (with energy greater than the average energy of electrons in the equilibrium state, that is, the formation of a two-temperature system) and the emission of electrons both external and internal [33, 34].

(4) The formation of electron–hole centers with an exchange coupling in the surface layers, i.e., manifestation of induced ferromagnetism in deformed crystals [35].

(5) Rearrangement of electron shells of atoms and formation of an additional zone of surface states caused by mechanical action on the crystal, which leads to a change in the thermodynamic characteristics of deformed particles and an increase in their chemical activity [27, 31, 36, 37].

(6) Ionization of the components of the gas atmosphere near the charged surface of deformed crystals, displacement, and sorption of ions of the corresponding sign in the areas of oppositely charged surface of quartz particles.

(7) The drift of charged particles in the electromagnetic field of the deformed surface of a quartz particle—nonlinear diffusion waves [38].

(8) Diffusion of charged vacancies and divacancies in the gradient field of elastic compression, accumulation of defects in the surface layer and an abrupt increase in density in the particle volume, "circulation of voids" and formation of new structures [39].

(9) Ultrasonic dielectric relaxation in piezoelectrics and the electric field accompanying the ultrasonic wave; acousto-electric waves, irradiation of surface layers, formation of new defective electron–hole centers [40].

Simulation of the process with the division into main and related factors is necessary for formation of ideas about the controllability over the process of mechanochemical synthesis (MCS) of nanocomposite systems. The impulse nature of the impact (at contact time $n \times 10$ μs.) determines the features of the high-speed process: ultrafast mass transfer and non-equilibrium conditions for the physicochemical effects and reactions. The number of pulses is determined by the ratio M_p/M_b and the exposure time. Process modeling is a sweep of the sequence of the process components in time and space, the main stages of which are presented below.

1. The area of mechanical loading (compression) is determined by the ratio of the size of the ball (10^{-2} m) and the initial size of ground quartz crystals (10^{-4} m) randomly oriented (statistical distribution with respect to the rotation of the axes providing polarization of the crystals) in the deformation zone relative to the direction of the current loads (Fig. 6.3a). A particle can be considered individually (as a compressed piezocrystal) or in an ensemble of particles of different packing densities concentrated and destroyed in the impact zone (Fig. 6.4b). In the experiments, they had a ball with a diameter of 10^{-2} m and an ensemble of particles of size $n \times 10^{-4}$ m.

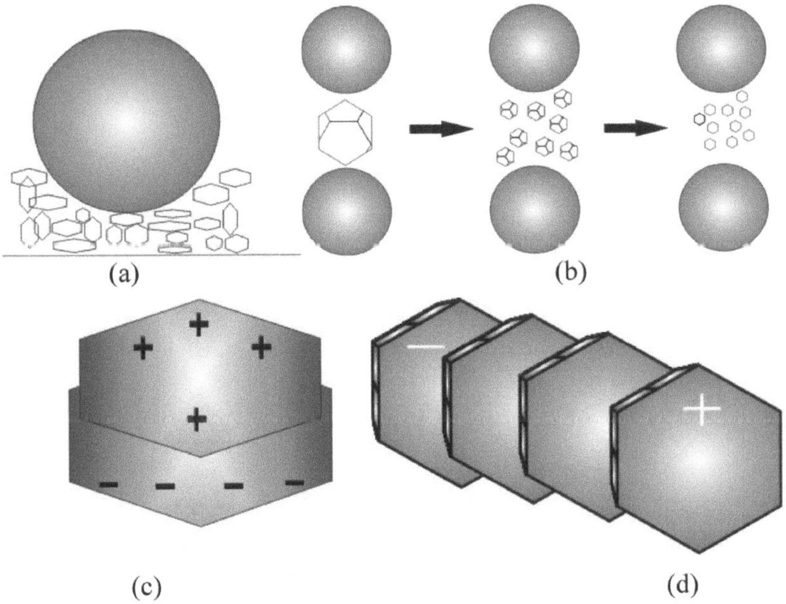

Figure 6.3 Model of the process of a quartz particle modification during mechanochemical treatment. The first stage is the piezoelectric effect during the mechanical activation of quartz particles.

2. Figures 6.3c,d show elastic compression and polarization of particles, charges on opposite surfaces of crystals, electric

field in the space between oppositely charged surfaces, repulsion and reversal of charged particles at like-charged surfaces, attraction of opposite surfaces with charge compensation in the ensemble volume and preservation of the external fields of ordered particles, agglomeration and formation of a dendrite-like structure of an ensemble of particles (i.e., an element of self-organization in the amount of ground particles).

3. Under the current load, which exceeds the elastic limit (σ > σ_{el}), there starts multiplication and movement of dislocations, annihilation of oppositely charged defects and fusion of dislocations with formation of pores and cracks with subsequent destruction of the crystal. Emission of electrons with their acceleration in the field of a crack and in the field of oppositely charged surfaces of polarized crystals (Fig. 6.4a).

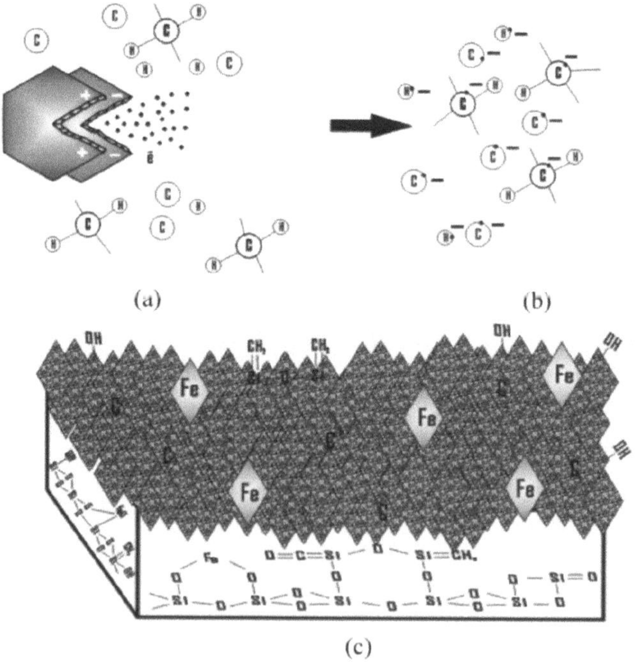

Figure 6.4 Model of the process of a quartz particle modification during mechanochemical treatment. The second stage—exoemission (a), ionization (b) and modification of the surface of quartz (c).

4. Irradiation of the surface of nearby crystals with a flux of fast electrons with the formation of radical ions SiO^- in the surface layers (Fig. 6.4b) and formation of radical ions in a gaseous medium. Accelerated movement of radical ions to the charged surface of crystals with the opposite sign.
5. Interaction of atmospheric radical ions with charged centers (grafting) in the field of a polarized charged surface of a deformed SiO_2 crystal (Fig. 6.4c). Structural heterogeneity and the presence of iron inclusions, the constant change (drift) of the surface charge ensures the presence of micro (nano-) electromagnetic fields, within which the movement of charged particles of ion radicals and directional surface modification by carbon-containing compounds occurs (Fig. 6.4d).

As a result of the periodic impulse effect of the grinding balls on the mixture of powders, the above listed actions (operations) are repeated several times. The destruction leads to an increase in the dispersion of particles and the heterogeneity of the system on average by volume of the mass of powder. Between two successive blows, the powder particles are in rotational motion with acceleration of $20g$ in the field of centripetal force. The interaction between particles is determined by the induced electromagnetic forces and manifests itself in a change in their orientation with each other with separation in space, or, conversely, as formation of conglomerates. With an increase in the dispersion of the system with each pulse of influence, the role of surface electromagnetic effects is enhanced. The interaction between the charged particles of the processed powder naturally must obey the laws of self-organization of planetary systems related to centrifugal and centripetal forces.

The presented scheme of the process (in the sequence of its development) allows us to quite clearly explain a number of experimental effects. First of all, it is the occurrence of stressed siloxane bonds, which are recorded by IR spectra after processing in ground quartz. They actively break up into groups (centers) with unsaturated valence and interact with gas-phase radicals in the process of MCT. Stressed bonds are a reflection of the polarization of charges in a crystal cell of quartz and formation of electric dipoles. Consequently, the gradient of fixed elastic-plastic deformations leads to a fairly stable formation of an electric field around the deformed particle and regions of ferromagnetic interaction.

Reactions on structural fragments (clusters) with strained siloxane bonds and with uncompensated valence occur with participation of negatively charged ion-radicals of the gas-phase medium, obtained as a result of the effect of the flux of emitted electrons on the radicals of degraded organic compounds-modifying material of the surface of quartz particles. Reactions run mainly on the positively charged surfaces of a polarized quartz crystal. Thus, in most cases, the structure and morphology of the surface of quartz particles change directly, this clearly following from the results of electron microscopic studies.

Electrification of a polarized crystal, emission of electrons and the presence of local oppositely charged centers in the defective areas of the particle surface, which are subject to drift under the influence of the general electric field of the crystal, affect the polymerization processes and the structural ordering of the polymer film on the surface of quartz, in which iron particles are embedded, which ensures manifestation of magnetic properties of modified particles.

Ultrafine particles of iron rubbed in the grinding process are exposed to a flow of electrons, which increases their activity and contributes to formation of chelate compounds with radical silanone groups (i.e., Si–O–Fe).

The main ion radicals obtained as a result of destruction of organic compounds in the process of mechanochemical treatment and irradiation by a stream of electrons emitted from the surface of split quartz crystals are: H_3C^-, H_2C^-, C^-, H^-. The more complex the initial compound and the longer the MCT, the greater the destruction and increase in the number of C^-, H^- end-products of destruction, which actively interact with the surface. In this case, H^- participates in formation of silanol centers, i.e., in formation of a monolayer of silicic acid and its polymerization, and accumulation of C^- provides carbonization of the surface.

Thus, the process of modifying a quartz particle is due to three interrelated effects:

(1) local surface breakage of bonds and localized charges related to defects on the surface and in the surface layer of the particle
(2) volumetric changes for the ensemble of crystals piezoelectric effect, i.e., polarization and the formation of an electric field with a voltage of up to 104 MV/m
(3) volumetric changes for the entire system—exoemission of electrons, ionization of the gaseous medium and surface

Charged particles (ion-radicals) rapidly moving in an electric field of charged crystals with a potential difference of several tens and hundreds of kV, depending on the particle size, interact with reaction centers on the surface, forming new structures, changing the particle morphology and properties.

Thus, in all cases of mechanical processing, the electromagnetic response of the system takes place, manifested primarily in the formation of electric charges and dipoles on the surface and in the volume of the particle. For piezomaterials to which quartz refers, the electric fields (with a strength of up to 10^4 MV/m) created by the mechanical compression of crystals in the impact zone significantly enhance all the electromagnetic effects of a deformable crystal, including exoelectronic emission, and accelerate the movement of electrons in the induced electric field of the system, determining the direction of their movement.

Irradiation of the induced electron flux of the surface of particles and the environment leads to ionization of radicals and atoms, the rooted motion of which in the electric field of piezoelements contributes to sorption and their grafting to the reaction centers on the freshly formed surface of splitting particles. The interaction of the field of centrifugal force in the PCM and the electromagnetic fields of deformable particles is the basis of self-organization of the structure on the scale of the ensemble of particles, as well as in the volume and on the surface of each of the constituent particles of the system.

Quartz modification is the result of interrelated repetitive processes: crystal deformation → polarization → electric charge on the surface → crystal splitting → electron emission → ionization of the surface and atmosphere → accelerated diffusion in the electric field of an ensemble of crystalline particles and along their surface → sorption and grafting of ions to charged centers surface → formation of an organometallic layer on the charged surface of a quartz particle. The pulsed nature of the force action ensures the constant renewal of the active charged surface of the particles and the pulsed excitation of the electromagnetic field of the generated electron flow. A quartz particle in a mechanical reactor works as a pulsed electromagnetic field generator.

The processes of chemical interaction of the surface of a deformable quartz particle and organic compounds from the atmosphere of a mechanical reactor are subjected to the periodicity of excitation and attenuation of electromagnetic fields in a deformable system of piezoceramic powders. The accelerating and directing action of the electric field on the process of grafting organic radicals on the surface of quartz and the formation of nanoscale objects on the surface of a particle is a component condition of the MCT of the systems under study.

Questions

1. What are the processes and conditions that determine dissipation of the impact energy in a mechanical reactor and manifest themselves in both organic and inorganic media?
2. Describe the model of the formation of nanostructured systems during the mechanochemical treatment of quartz—as the electromagnetic response of the system to pulsed mechanical loading.
3. Describe the process of modifying the surface of a splitting particle and formation of a composite particle with nanoscale formations on its surface—as a result of accelerated movement of radical ions of degraded organic additives and their interaction with the charged surface of crystalline quartz.
4. What are the main effects that have an impact on the modification process during mechanochemical treatment?
5. Describe the model and the main stages of the grinding process and modification of solid particles during processing in a mechanical reactor.
6. What is the sequence of stages of the periodic pulsed impact of grinding balls on a mixture of powders?
7. What is the role of electromagnetic forces arising from the destruction of particles and the formation of conglomerates?

8. What are the main interrelated effects that determine the process of modifying a quartz particle during mechanochemical treatment?

References

1. Radtsik V. A. Paramagnetic centers on the surface of quartz splitting. Interaction with CO and N_2O molecules. *Kinet. Catal.*, 1979. vol. XX, no. 2. pp. 448–455. (In Russian).
2. Radtsik V. A. Paramagnetic centers on the surface of quartz splitting. Interaction with H_2 and D_2 molecules. *Kinet. Catal.*, 1979. vol. XX, no. 2. pp. 456–460. (In Russian).
3. Heinicke G. *Tribochemisrty*. Mir, Moscow, 1987. 584 p. (In Russian).
4. Koroleva S. M., Milov A. D., Sukhoroslov A. A. Formation and paramagnetic relaxation of E′ - centers in mechanically activated quartz. *Chem. Phys.*, 1990. vol. 9, no. 2. pp. 258–263. (In Russian).
5. Radtsik V. A. Formation of free radicals in the interaction of groups with materials $(\equiv Si{-}O{-})_2Si{\equiv}O_2$ H_2, CH_4, C_2H_6. *Chem. Phys.*, 1995. vol. 14, no. 2. pp. 416–427. (In Russian).
6. Radtsik V. A. Reaction intermediates on the surface of solids. *Chem. Phys.*, 1995. vol. 14, no. 8. pp. 125–135. (In Russian).
7. Ailer R. *Chemistry of Silica*. Mir, Moscow, 1982. Part 1, 416 p.; Part 2, 712 p. (In Russian).
8. Radtsik V. A., Berestetskaya I. V., Kostritsa S. N. Silanone groups $(\equiv Si{-}O)_2Si{=}O$ on the surface of silica. Registration by IR spectroscopy. *Kinet. Catal.*, 1998. vol. 39, no. 6. pp. 940–945. (In Russian).
9. Radtsik V. A. Registration by IR spectroscopy in the overtone region of the $(\equiv Si{-}O)_2Si{=}O$ and $(=Si\text{-}0)_2Si<0>C=0$ groupings on the silica surface. *Kinet. Catal.*, 2001. vol. 42, no. 1. pp. 53–61 (In Russian).
10. Berlin A. A., Wolfson S. A., Enikolopyan N. S. *Kinetics of Polymerization Processes*. Chemistry, Moscow, 1978. 240 p. (In Russian).
11. Molchanov V. I., Selezneva O. G., Zhirnov E. N. *Activation of Minerals*. Nedra, Moscow, 1988. 208 p. (In Russian).
12. Chaykina M. V. *Mechanochemistry of Natural and Synthetic Apatites*. Publishing House of the Siberian Branch of the Russian Academy of Sciences, Geo Branch, Novosibirsk, 2002. 223 p. (In Russian).
13. Strelko V. V., Kanibolotsky V. A. Classification of reactions involving the surface of dispersed silicas and the study of the replacement of hydrogen associated with surface silicon atoms. *Colloid J.*, 1971. vol. 33, no. 5. pp. 750–754. (In Russian).

14. Grigoriev T. F., Vorsina I. A., Barinova A. P., Boldyrev V. V. Mechanochemical synthesis of dispersed layered composites based on kaolinite and higher carboxylic acids. *Rep. Acad. Sci.*, 1995. vol. 341. pp. 66–68. (In Russian).
15. Ketegenov, T. A., Urakaev, F. Kh., Tyumentseva, O. A., Mofa, N. N., Mansurov, Z. A. Synthesis of iron silicates on the surface of quartz particles in the process of their mechanical processing. *Rep Natl. Acad. Sci. Republic Kazakhstan*, 2003. no. 2. pp. 66–72. (In Russian).
16. Mofa N. N., Mansurov Z. A. Mechanochemical synthesis of surface nanostructures, the method of obtaining new composition materials for different purposes. *International Conference "Mechanochemical Synthesis and Sintering"*. Novosibirsk, Russia, June, 2004. pp. 67–68.
17. Exoelectronic emission. *Collection of Works*, ed. Kobozev N. I. Translation by Krylova I. V. Foreign literature, Moscow, 1962. 187 p. (In Russian).
18. Deryagin B. V., Krotova N. A., Smilga P. M. *Adhesion of Solids*. Science, Moscow, 1973. 280 p. (In Russian).
19. Nassenshteyn Kh. Electron emission from the surface of solids after mechanical treatment and irradiation. *Proc. Exoelectronic Emission*, ed. Kobozev N. I., Translated by I. V. Krylova. Foreign literature, Moscow, 1962. pp. 72–95. (In Russian).
20. Krotova N. A. *About Bonding and Sticking*. Publishing House of the ANSSR, Moscow, 1960. 237 p. (In Russian).
21. Smagin A. G., Yaroslavsky M. I. *Quartz Piezoelectricity and Quartz Resonators*. Energy, Moscow, 1970. 488 p. (In Russian).
22. Ketegenov T. A., Tyumentseva O. A., Urakaev F. Kh., Mansurov Z. A. Modeling the reaction of the interaction of the material of grinding bodies with the substance being processed in mechanochemical reactors. *Rep. Natl. Acad. Sci. RK.*, 2003. no. 1. pp. 67–72. (In Russian).
23. Shuvalov L. A., Urusovskaya A. A., et al. *Modern Crystallography. Physical Properties of Crystals*/Reference manual. Nauka, Moscow, 1981. vol. 4. pp. 165–186. (In Russian).
24. Shunin V. M., Nabatov S. S., Smetanina L. A., Yakushev V. V. Electrical response of flexible piezo-composite films to high-speed impact of a spherical body. *Chem. Phys.*, 1995. vol. 14, no. 2. pp. 159–165. (In Russian).
25. Zyryanov V. V. Para- and ferromagnetic resonance of intrinsic defects in non-magnetic dielectrics after MT. *Abstracts of the 5th All-Union Meeting "Modern NMR and EPR Methods in Solid State Chemistry"*. Chernogolovka, 1990. pp. 108–109. (In Russian).

26. Venevtsev Yu. N., Gagulin V. V., Lyubimov V. N. *Ferromagnetics*. Science, Moscow, 1982. 224 p. (In Russian).

27. Prigogine I., Stengers I. *Order from Chaos*. Progress, Moscow, 1986. 432 p. (In Russian).

28. Kochurova N. N., Rusanov A. I., Myrzakhmetova N. O. Jones–Ray effect and surface electrification. *Rep. USSR Acad. Sci.*, 1991. vol. 31, no. 6. pp. 1425–1427. (In Russian).

29. Khoroshun L. P., Maslov B. P., Leshchenko P. V. *Prediction of the Effective Properties of Piezoactic Composite Materials*. Naukova Dumka, Kiev, 1989. 208 p. (In Russian).

30. Davison S., Levin J. *Surface (Tamm) States*. Mir, Moscow, 1973. 232 p. (In Russian).

31. Avvakumov E. G. *Mechanical Methods of Activation of Chemical Processes*. Science, Novosibirsk, Sib. Branch, 1986. 300 p. (In Russian).

32. Ahmed-zadeh H. D., Baptismansky V. V., Zakrevsky V. A., Tomashevsky E. E. Paramagnetic centers formed during the destruction of silicon dioxide. *Phys. Solid.*, 1972. vol. 14, no. 2. pp. 422–426. (In Russian).

33. Maradudin A. *Defects and Vibrational Spectra in Crystals*. Mir, Moscow, 1968. 275 p. (In Russian).

34. Zaitseva L. A. Influence of a layer of hot electrons near the surface on the thermodynamic energy of semi-bounded solids. *Chem. Phys.*, 2002. vol. 21, no. 2. pp. 15–25. (In Russian).

35. Zyryanov V. V. Model of the reaction zone during mechanical loading of powders in a planetary mill. *Inorg. Mater.*, 1998. vol. 34, no. 12. pp. 1525–1534. (In Russian).

36. Khodakov G. S. *Grinding Physics*. Science, Moscow, 1972. 307 p. (In Russian).

37. Steinike, U., Mechanically induced reactivity of quartz and its connection to the real structure. *Izv. SB AS USSR. Ser. Chem. Sci.*, 1985. vol. 3. pp. 40–47. (In Russian).

38. Alekseev Yu. V., Kolotyrkin Ya. M., Popov Yu. A. On the diffusion of divacancies in a crystal under strong compression and shear. *J. Phys. Chem.*, 1989. vol. LXIII, no. 11. pp. 3059–3063. (In Russian).

39. Zhugin Yu. N. Behavior of α-quartz at high dynamic and static pressures: New results and representations. *Chem. Phys.*, 1995. vol. 14, no. 1. pp. 69–74. (In Russian).

40. Balakirev M. K., Gilinsky I. A. *Waves in Piezocrystals*. Science SB, Novosibirsk, 1982. 239 p. (In Russian).

Chapter 7

Mechanochemical Synthesis of Disperse Composition Systems of Different Purpose

7.1 Composition Systems Quartz Core–Polymer Shell

Chapter 5 reviewed the results of studies on mechanochemical processing, activation, and modification of quartz, which showed that when obtaining a composite powder material consisting of particles of quartz encapsulated in organometallic shells, the composition and structure of the modified surface are largely determined by the quality of the base material, i.e., the degree of crystallinity and the presence of impurities in quartzite, as well as the choice of modifier and modes of the MCT. The modifiers were mainly different alcohols. Then the range of modifying additives has been expanded. A 10% aqueous solution of ammonia (NH_4OH), polyvinyl alcohol $(C_2H_3OH)_n$, succinic acid ($C_4H_4O_6$), urea $CO(NH_2)_2$, and hydrous silicic acid—$H_2SiO_3 \bullet n(H_2O)$ were used. The studies were carried out on quartz of two deposits of Tekturmy and Aktas, which consist of 98.8% and 98.9% of silicon dioxide with a slight difference in the content of impurities [1–3].

Mechanochemical Synthesis of Composite Materials
Zulkhair A. Mansurov, Nina N. Mofa, Tlek A. Ketegenov, and Bakhtiyar S. Sadykov

ISBN 978-981-4800-88-4 (Hardcover), 978-1-003-12081-0 (eBook)
www.jennystanford.com

They differ mainly in the density of macroscopic structure. But some difference in the content of impurities and in the microstructure of the material showed a number of distinctive features in dispersion and modification of quartz particles.

For testing the modes of mechanochemical synthesis and modification of the powder in order to obtain polymer-inorganic particles for concrete practical use, quartz sand from the Kuskuduk deposit with quartz content of 81.3% and 18.7% of microcline $K(Si_3Al)O_8$ was taken. Also, this sand contains various elements that saturate the surface and are dissolved in the volume of particles. According to the results of spectral analysis, it contains from 0.1% to 1.0% of impurities of iron, magnesium, calcium, sodium, the presence of which imparts a corresponding color to quartz particles (Fig. 7.1a). The structure of the surface layers of the colored particles is layered, which is confirmed by the results of electron microscopic and microdiffraction analyzes (Fig. 7.1b,c). The microdiffraction pattern of "kikuchi lines," light and dark stripes, is characteristic of the layered structure of the surface of the particles. This material can be said to have a naturally modified surface. Impurities can act as active centers in the process of mechanochemical modification of the surface.

(a) (b) (c)

Figure 7.1 Optical (a), electron microscopy (b) and microdiffraction (c) of quartz particles.

The mechanochemical treatment (MCT) of quartz material was carried out in a centrifugal planetary mill (CPM) with an acceleration of 20*g* with a working chamber volume of 450 mm^3, a working chamber rotation speed of 1200 rpm and a platform rotation speed of 700 rpm, energy consumption of 2.2 kW/h.

When grinding, the grinding time, the ratio of the mass of the material and the mass of grinding bodies (M_p/M_b) were varied. Carbon and nitrogen-containing modifying additives were 10% aqueous ammonia (NH_4OH), polyvinyl alcohol $(C_2H_3OH)_n$, succinic acid ($C_4H_4O_6$), and urea $CO(NH_2)_2$, which during mechanical impact get destructed and polymerize on the surfaces of quartz particles being crushed.

After 10 min of mechanical treatment of quartz, 96% of the powder was particles up to 12 microns. A longer processing time of up to 30 min provided an almost complete transfer of the material to this level of dispersion (Table 7.1). After grinding in the presence of modifiers, the dispersion of the powder significantly decreased, the proportion of larger particles appeared, the number and size of which depend on the time-strength processing conditions and the type of modifier. According to the results obtained in [1, 4, 5], the most effective, from the point of view of formation of modified particles with the nanolayer structure of the surface layers, is the treatment with the ratio of the mass of powder to be ground to the mass of balls (M_p/M_b) equal to 1/2, optimal processing time depends on the modifier. So, with the use of polyvinyl alcohol, the processing time should not exceed 20 min, with the participation of an aqueous solution of ammonia (ammonia) 10 min treatment is preferable, and modifying with urea gives the best results for longer processing times up to 30 min. In this case, the most polydisperse powder is obtained with a large proportion of larger particles, which are the composite formation of a quartz + polymer layer. In the presence of succinic acid, the highest dispersion powder was obtained. After 10 min of the MCT, over 96% of the powder consists of particles up to 12 microns in size. With the increase in the duration of the MCT dispersion increases.

An indirect confirmation of the conclusions made from the results of the measurement of dispersion is the data for determining the bulk density of the powder. Figure 7.2 presents the dependences of the bulk density of quartz powder on the treatment time (MCT). The increase in the time of treatment leads to a decrease in bulk density. However, the level of bulk density values increases slightly with the use of modifiers.

Table 7.1 Dispersibility of particles of quartz powder modified with MCT depending on the time of treatment and type of modifier

Material	The percentage of particles, %					Crystallite size
		Dispersibility, μm				
	Time, min	0–12	12–23	23–35	>35	L, Å
Quartz	10	96	4			2900
	20	97	3			2350
	30	99	1			2140
Quartz+5% $(C_2H_3OH)_n$,	10	61	28		11	2470
	20	77	17		6	1960
	30	86		14		2090
Quartz+5% $(NH_2)_2CO$,	10	77.8		20.8	1.4	3400
	20	77.8		14.1	13.2	3210
	30	70.4		22.3	7.3	2940
Quartz+5% $(C_4H_4O_6)$	10	96.6	2.1	1.3		3150
	20	98.3		1.7		2790
	30	98.6		1.4		2530

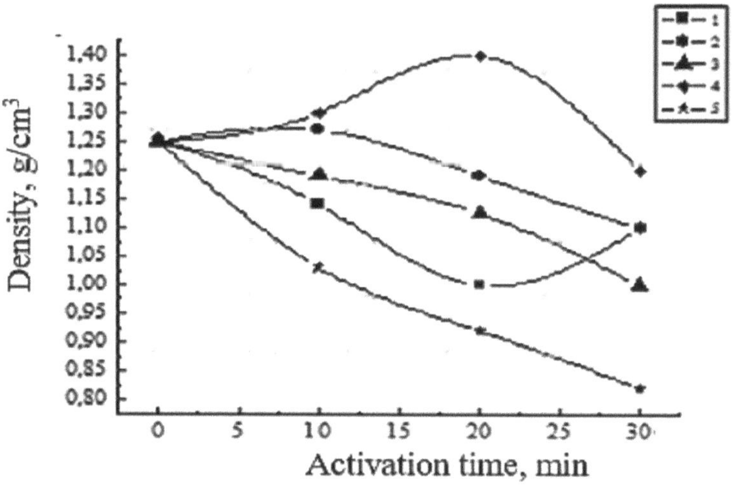

Figure 7.2 Dependence of changes in the bulk density of quartz on the time of the MCT without modifier (1) and in the presence of 5% urea (2), polyvinyl alcohol (3), aqueous ammonia solution (4) and succinic acid (5).

The highest values are obtained when treating quartz with an aqueous solution of ammonia, then when using urea, and finally, polyvinyl alcohol. The results obtained are due to a change in the polydispersity of the powder, which determines the packing density of the particles when the measuring vessel is filled with powder as well as the formation of a modified layer on the surface of the particles during grinding. The extreme nature of the dependence of the bulk density of quartz on the time of treatment with an ammonia solution indicates that at 20 min the most significant changes occur in the structure and morphology of the powder particles. Low values of the bulk density of the powder, modified with succinic acid, are associated both with its high dispersion and simultaneously with the presence of a significant contribution from the modified polymerized part of the powder.

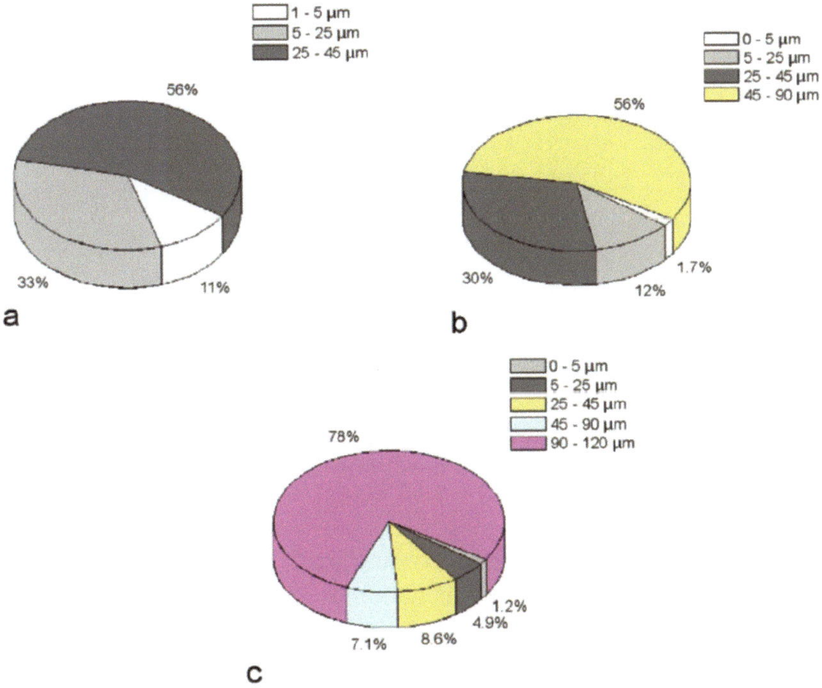

Figure 7.3 Particle size distribution for a quartz powder modified with liquid ammonia solution during MCT with M_p/M_b ratio equal to 1/2 (a), 1/3 (b) and 1/4 (c).

Both an increase in the processing time and a change in the ratio of the mass of the powder to be ground to the mass of the balls (M_p/M_b) to 1/3 and 1/4, namely, with an increase in the number of balls, leads to a decrease in the bulk density and an increase in dispersion. As an example, in Fig. 7.3, for comparison, the data on the dispersion of quartz powder after 20 min of the MCT with ammonia with a different M_p/M_b ratio are given. For all other cases of MCT, patterns are preserved, only quantitative indicators change.

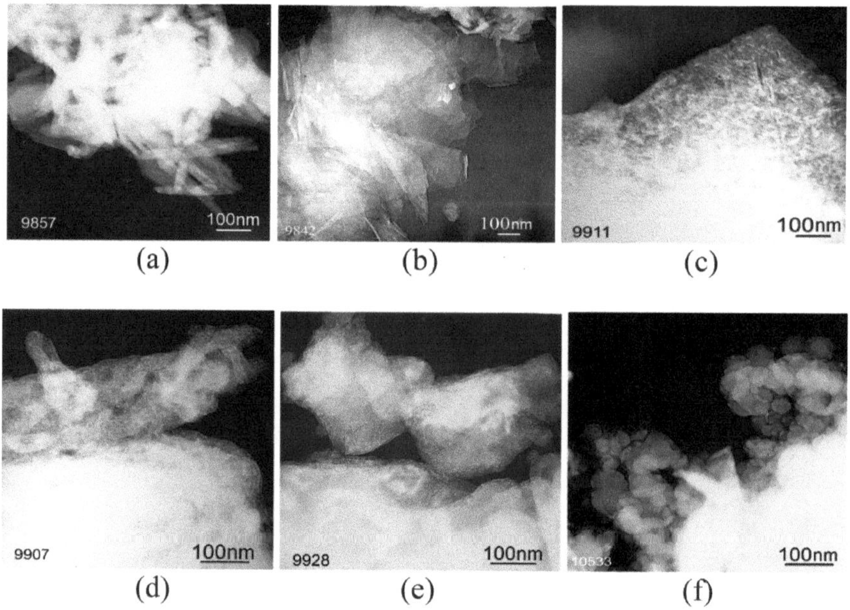

Figure 7.4 Electron-microscopic images of quartz after manual abrasion (a) and after MCT for 20 min without (b) and in the presence of modifiers: (c) –$(C_2H_3OH)_n$; (d) –$(NH_2)_2CO$; (e) –NH_4OH; (f) –$(C_4H_4O_6)$.

The change in the structure and state of the surface layer of a particle during mechanical action is clearly demonstrated by electron microscopic images. Even with manual abrasion, the surface layer of particles is destroyed (Fig. 7.4a). With intensive grinding of quartz particles in the mill, the destruction of the surface layer is deeper and its structure becomes loose

(Fig. 7.4b). Processing of quartz particles in the presence of polyvinyl alcohol leads to formation of a sufficiently dense polymer film on the particle surface (Fig. 7.4c). When modifying with urea, a loose polymer layer is formed on the surface of the particle (Fig. 7.4d).

Polymerization of the surface of quartz particles also occurs in case of an MCT with an aqueous solution of ammonia (Fig. 7.4e), and the presence of particles of varying degrees of dispersion is fixed. Another feature of the particles modified with nitrogen-containing compounds was noted—their high sensitivity to the effects of the electron beam. The morphological peculiarity of quartz particles obtained by MCT in the presence of succinic acid is the formation of nanostructured formations in the surface layer (Fig. 7.4f), which are also characterized by high sensitivity to the electron beam. A modified particle is a composite formation: a crystalline base—quartz with a different level of imperfection, and a structurally modified surface layer of a particle with polymer films of various configurations [5, 6].

The difference in activity during modification of quartz particles in the course of MCT process is determined by both the state and structure of the surface layer and the degree of imperfection in the volume of the particle. One of the indicators of the imperfection of particles is the size of the constituent crystallites. The smaller the crystallite size, i.e., the greater the length of their interface, the higher the imperfection of the sample. During mechanical processing, the longer the force effect and the more grinding balls (i.e., the more shocks occur on the particle of the ground powder), the smaller the crystallite size due to accumulation of dislocations and vacancies in the material. Along with the accumulation of dislocations, their annihilation occurs, so linear dependencies are not always the case (Table 7.1). So, when grinding quartz for 10 min and M_p/M_b = 1/2, the crystallite size was 2900 Å, and after treatment for 30 min and M_p/M_b = 1/4, it decreased to 1900 Å. The MCT in the presence of modifiers has a different effect on the crystallite size. Treatment with polyvinyl alcohol under similar conditions reduces the size of the crystallites, to 2470 Å and to 1840 Å, respectively.

In the presence of an aqueous solution of ammonia, urea and succinic acid, changes in the structure of particles during MCT do not occur unambiguously, with a small exposure time and a ratio of $M_p/M_b = 1/2$, the crystallite size increases markedly, respectively, for the above modifiers up to 3020, 3400 and 3150 Å and, conversely, decreases with more rigid (with an increase in the number of grinding balls) and prolonged processing to 1500, 1760 and 1870 Å. The reason for the growth of crystallites may be the sink of dislocations generated by mechanical (force) impact on the sample surface in areas where the particle interacts chemically with organic modifier compounds.

In the process of modifying quartz particles during MCT, a significant role is played by the piezoelectric effect of quartz, which determines the directionality and efficiency of the process of interaction between the surface of the destruction of quartz and the environment [7, 8]. During mechanical action on a quartz particle, the electromagnetic response of the system takes place, manifested primarily in the formation of electric charges and dipoles on the surface and in the volume of the particle. For piezomaterials, the electric fields created by the mechanical compression of crystals in the impact zone significantly enhance all the electromagnetic effects of a deformable crystal, including exoelectron emission [9, 10]. Irradiation induced by the electron flux of the surface of particles and the environment leads to ionization of radicals and atoms, the rooted motion of which in the electric field of piezoelements contributes to sorption and their grafting to the reaction centers on the freshly formed surface of the splitting particles, and ultimately contributes to formation of a polymerized surface layer. All processes of chemical interaction of the surface of a deformed quartz particle and organic compounds from the atmosphere of a mechanical reactor occur under conditions of pulsed mechanical action and therefore are subjected to the periodicity of excitation and attenuation of electromagnetic fields in a deformable system of piezoceramic powders, which determines the specificity of the formation of layered structures on the surface of particles to be ground and the nonlinear dependence of indicators of structural changes with time.

Questions

1. What distinctive features of dispersion and modification of quartz particles are observed depending on the content of impurities and micro-composition of mineral raw materials, on the modes of mechanochemical processing and modifying additives?
2. Describe the features of composite formations of quartz+polymer layer when modifying particles with succinic acid and aqueous ammonia.
3. What morphological features of quartz particles modified by nitrogen-containing compounds are a consequence of their high sensitivity to the effects of the electron beam?
4. Describe the features of the change in the defect structure and production of nanostructured formations in the surface layer of quartz particles in the presence of succinic acid and polyvinyl alcohol in the MCT.
5. What electrophysical phenomena during MCT of quartz, which cause ionization of radicals and atoms, contribute to sorption and their grafting to reaction centers on the freshly formed surface of splitting particles, and to the formation of a polymerized surface layer?

7.2 Composite Systems Inorganic Core–Polymer Shell, Obtained by Mechanochemical Treatment of Calcite and Wollastonite

In addition to quartz, other minerals—calcium carbonate (i.e., calcite or marble) with a $CaCO_3$ content of 95.7% and quartz—4.3% and wollastonite powder, which is calcium metasilicate Ca $[SiO_3]$ and contains impurities of quartz up to 14% and calcium carbonate up to 11% were taken. Nitrogen- and carbon-containing compounds aqueous ammonia, polyvinyl alcohol, and urea were used as modifiers for encapsulation in a polymer shell.

Dispersion of calcite powder changes depending on the parameters of the MCT (time and M_p/M_b ratio), as well as on the type of modifiers. The results of measuring the dispersion showed that marble (i.e., calcite) was already well dispersed for 10 min at a ratio of M_p/M_b = 1/2 (Table 7.2). The increase in the number of grinding balls, and, consequently, the number of blows on the powder leads to agglomeration of small active particles, as a result of which the fraction of the fine fraction decreases. Similarly, the change in dispersion occurs with an increase in processing time to 20 and 30 min. With a longer MCT, the aggregates are destroyed and the fraction of the highly dispersed powder fraction increases [11, 12].

Table 7.2 Dispersity of calcite particles after different conditions of MCT of samples and the type of modifiers

		Percentage ratio,%		
		Dispersity, μm		
Material, M_p/M_b	**Time, Min**	**≤11.6**	**22.3–34.8**	**34.8–58**
$CaCO_3$, 1/2	10	92.7	7.3	
$CaCO_3$, 1/3	10	84.3	15.7	
$CaCO_3$, 1/4	10	62.2	37.8	
$CaCO_3$, 1/2	20	81.3	17.3	1.4
$CaCO_3$, 1/2	30	88.6	9.3	2.1
$CaCO_3$, 1/2	40	98.6	1.4	
$CaCO_3$+5%NH_4OH, 1/2	10	95.5	4.5	
$CaCO_3$+5%NH_4OH, 1/2	20	97.0	3.0	
$CaCO_3$+5%NH_4OH, 1/2	30	99.0	1.0	
$CaCO_3$+5%PVA, 1/2	10	87.3	119	0.8
$CaCO_3$+5%PVA, 1/2	20	90.1	9.9	
$CaCO_3$+5%PVA, 1/2	30	95.2	3.6	1.2
$CaCO_3$+5%$(NH_2)_2CO$, 1/2	10	96.4	3.6	
$CaCO_3$+5%$(NH_2)_2CO$, 1/2	20	96.6	3.4	
$CaCO_3$+5%$(NH_2)_2CO$, 1/2	30	98.0	2.0	

Note: PVA, polyvinyl alcohol.

The use of modifiers during MCT of calcite increases the degree of dispersion of the powder, i.e., first of all, aqueous solution of ammonia, polyvinyl alcohol and urea with respect to marble behave like surfactants (probably due to the presence of water and C/CO which release in the process of mechanical destruction of organic compounds).

If dispersion of the powder increases as a result of the MCT (Table 7.2), the bulk density decreases (Fig. 7.5). After processing for 20 min, the powder is distinguished by some increase in bulk density, with a general tendency to decrease with both an increase in the number of grinding balls and the processing time. For calcite powder, modified by various compounds (in the amount of 5%), the density values are reduced even more. The density of the powder, crushed in the presence of polyvinyl alcohol and urea is noticeably lower than when just activated or treated with an ammonia solution. The lower density of the powder treated with polyvinyl alcohol and urea is indirect evidence that organic particles of lower density than marble are formed on the surface of the particles, as a result of which the total density of the modified powder decreases.

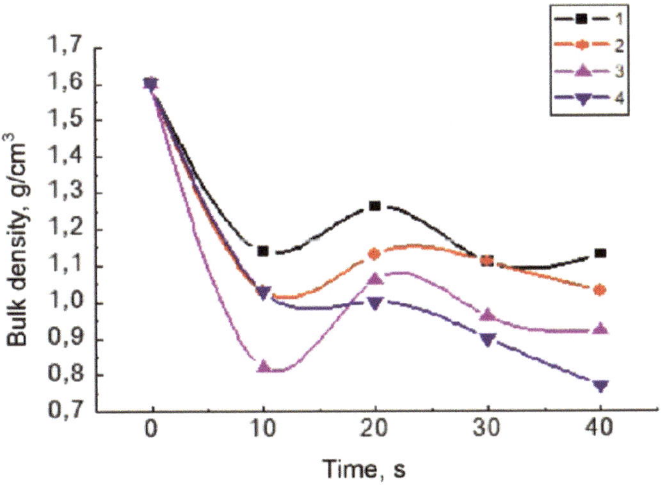

Figure 7.5 Dependence of changes in the bulk density of marble ($CaCO_3$) on the time of MCT without a modifier (1) and in the presence of a 5% aqueous solution of ammonia -NH_4OH (2), polyvinyl alcohol -$(C_2H_3OH)_n$ (3) and urea -$(NH_2)_2CO$ (4).

The features of the morphology of the surface of calcite particles modified in MCT are illustrated by electron microscopic images (Fig. 7.6).

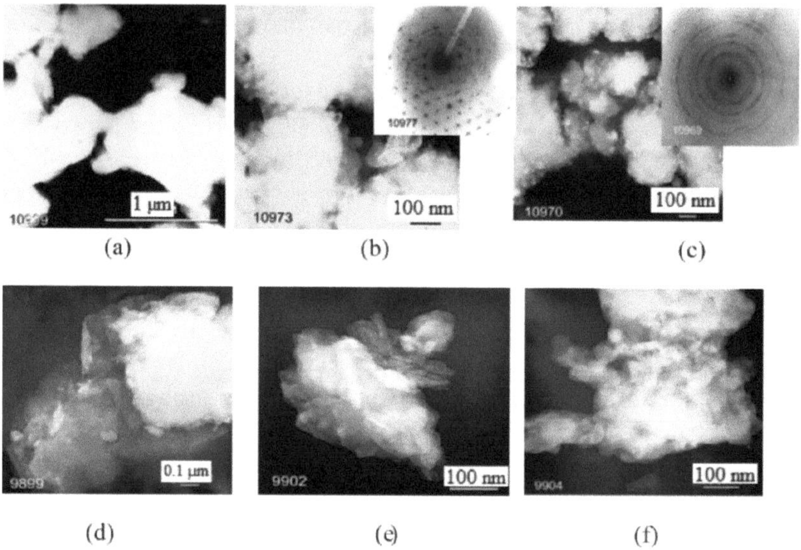

Figure 7.6 Microstructure and microdiffraction of dispersed calcite particles in the presence of succinic acid (a), liquid ammonia (b), mixtures of succinic acid with liquid ammonia (c), with polyvinyl alcohol (d) and urea (e, f) for 20 (a–e) and 30 min (f).

Calcite particles modified with succinic acid, compared with quartz, are distinguished by a more dense surface layer and high polycrystalline particles (Fig. 7.6a). When modifying with an aqueous solution of ammonia (liquid ammonia), the surface layer of the particles becomes, on the contrary, looser with a high level of structurization (Fig. 7.6b) and, according to the microdiffraction pattern, besides polycrystalline particles, single-crystal formations of cubic modification are observed.

Electron microscopic images of calcite after the MCT in a mixture of succinic acid and ammonia demonstrate the complexity of the structural interactions between these modifiers and their products (water, ammonium group and succinic anhydride) with mineral particles. With sufficiently long processing of the mineral, a large number of small single-crystal formations are observed (Fig.7.6c).

In addition, the scale of the structure of the modified surface layers of the particle increases.

When grinding calcite with polyvinyl alcohol and urea, there takes place the formation of polymer films of various thickness, enveloping the particles and formation of conglomerates of such particles. The longer the treatment, the more complex compositions are formed with the high dispersion of their component parts (Fig. 7.6f).

Thus, the results of electron microscopic studies have shown that during MCT of both quartz and calcite, there occurs not only particle grinding, but also transformation of the internal structure and the surface layer of the particles. Modifying nitrogen- and carbon-containing additives are destructed in the process of MCT and actively interact with the surface of the ground particles. As a result of this interaction, film or loose formations on the surface of the particles are produced and the powder material is obtained from particles of a multilayer composite structure.

Under similar conditions, experiments were carried out on the MCT of wollastonite samples in the presence of nitrogen and carbon-containing organic compounds (10% aqueous solution of ammonia (NH_4OH), polyvinyl alcohol $(C_2H_3OH)_n$ and urea $(NH_2)_2CO$). Wollastonite is calcium metasilicate $Ca[SiO_3]$ with chemical composition 48.3% CaO and 51.7% SiO_2 and refers to the family of pyroxenoids of the subclass of chain silicates. Wollastonite crystals have a needle-like shape and the particles retain this structure during subsequent grinding, so it is an effective reinforcing material and provides enhanced strength to composite systems [13–16].

When using wollastonite as a filler in composites for high strength of "cross-linking" the inorganic filler and the base (matrix), it is required that on the surface of the particles there should be formed structural elements of a nanoscale level, ensuring their strong contact with the matrix material. To this end, film structures are formed on the surface of wollastonite particles, providing not only the necessary contact strength, in particular, with organic systems, but also qualitatively new properties of the system: its mechanical, thermal, electrical, and other properties of the final synthesis product. Thus, modification of powder particles directly in the grinding process is one of the directions of mechanochemical processing of inorganic materials [17]. The task is to choose a modifier that would provide a directional change in the structure and properties of the material.

The results of changes in the dispersity of wollastonite powder particles after MCT are presented in Table 7.3. The main part of the powder consists of particles up to 12 μm in size. As a result, with the increase in processing time, the fraction of the finely dispersed fraction increases (up to 92%). At the same time, the particles agglomerate and an elongated form with a size of up to 116 μm is observed in the powder.

Table 7.3 Dispersion of particles of wollastonite samples after different conditions of MCT

		Percentage ratio, %			
		Dispersity, μm			
Material	**Time, min**	**0–11.6**	**11.6–23.2**	**23.3–58.0**	**58–116**
W*	Initial	79.6	13.0	7.4	
	10	85.3		5.5	9.2
	20	86.4		4	9.6
	30	91.8	4.9	3.3	
W+5%NH_4OH	10	78.7		21.3	
	20	92.7		6.6	0.7
	30	97.9			2.1
W+5% $(C_2H_3OH)_n$	10	94.3		0.8	4.9
	20	74.0		22	4.0
	30	93.9		5.5	0.6
W+5% $CO(NH_2)_2$	10	76.6	18.6	4.8	
	20	74.3		22.3	3.4
	30			99.8	0.2
W+5% $(C_4H_4O_6)$	10	52.27	24.15	21.61	1.97
	20	43.62	28.43	25.74	2.21
	30	50.0	27.0	20.71	1.59
W+5% $H_2SiO_3n(H_2O)$	10	47.96	28.62	23.42	
	20	53.69	25.2	21.11	
	30	58.83	30.05	10.67	

*W, wollastonite

An aqueous solution of ammonia is an effective surfactant that promotes the grinding of particles. However, at 10 min of processing simultaneously with grinding, agglomeration of particles occurs rather intensively with the formation of a fraction (23–58) μm, which constitutes up to 21.3% of the powder volume.

When using polyvinyl alcohol, succinic acid and urea as modifiers in the MCT process, along with grinding solid particles of inorganic powder and destruction of organic modifiers, there takes place polymerization and production of composite formations from fine particles of wollastonite powder and organic polymers. These formations are represented in the powder by a larger fraction, the largest amount of which, when using polyvinyl alcohol, is observed after 20 min of treatment. In case of MCT of wollastonite with urea, the longer the treatment, the more particle (up to 99.8%) fractions with a size of (23–58) μm are produced. When using succinic and silicic acid, the effect of enlarging wollastonite particles is most pronounced. Significant changes in the dispersion of the powder depending on the time of MCT are not observed.

The observed difference in the dispersion of the powder after the MCT is indicated in the results of measuring the bulk density. If the bulk density of wollastonite powder in the initial state is 1.2 g/cm^3, as a result of the MCT it decreases to 1.08 g/cm^3. At the same time, after MCT of wollastonite with silicic acid, urea, polyvinyl and ammonia, it slightly increases, varying within 1.12–1.14 g/cm^3, depending on the type of modifier. The highest value is noted for urea-modified powder. These results are due to the polydispersity of the powder, contributing to the high packing density of particles in a measured volume. The use of succinic acid as a modifier, on the contrary, leads to the most significant decrease in the bulk density of wollastonite. So after 30 min of processing, it reaches 0.7 g/cm^3. These changes may be due to a significant contribution of the organic component in the volume of the modified wollastonite powder.

During MCT, there takes place not only the destruction of particles, but also accumulation and redistribution of defects in the particle volume. The change in the imperfection of particles can be estimated from the size of the crystallites. With an increase in processing time, defects get accumulated and the size of crystallites decreases in wollastonite particles. X-ray analysis of amples of wollastonite powder showed that if in the initial state the crystallite size is 1087 Å, after the MCT it decreases to 775 Å. However, in the presence of carbon and nitrogen-containing modifiers during the MCT process, a significant increase in crystallite size is observed, especially in the first 10 min. With longer treatment, it is possible to reduce them.

Table 7.4 Change in the size of crystallites of wollastonite powder particles as a result of MCT with different modifiers

	Crystallite size, L, Å		
	Time of MCT (t_{mct}), min		
Modifier	**10**	**20**	**30**
	905	840	775
NH_4OH	2350	2420	2390
$(C_2H_3OH)_n$	1980	1550	1160
$(NH_2)_2CO$	1500	1710	1890
$(C_4H_4O_6)$	1110	973	895
$H_2SiO_3n(H_2O)$	910	754	740

Ammonia-modified particles have the largest crystallite size (L) (Table 7.4). When treated with urea, an increase in the time of MCT leads to an increase in the size of crystallites, but to a lesser extent than with ammonia, and when using polyvinyl alcohol, an inverse relationship is observed, and this relationship is very pronounced. Quantitative analysis showed that with an increase in processing time, the urea content in the sample decreases, this indicating its destruction during the MCT process and interaction with the surface of solid inorganic particles, which results in surface modification. The growth of crystallites is related to the drain of defects to the surface, where chemical processes of interaction of surface layers with modifier additives begin to occur, causing diffusion of dislocations and vacancies into the reaction zone. With the increase in the MCT time, the accumulation of defects prevails over their realization in surface transformation, as the growth of new polymer layers occurs on the already modified surface layer. Of all the modifiers considered, the exception is silicic acid. In its presence, there is a decrease in the size of the crystallites during the MCT of wollastonite, this indicating the different nature of the interaction between the modifier and the dispersible powder.

Thus, during MCT, wollastonite proved to be quite active in changing the dispersion and transforming the internal (bulk) structure of the particles. The results of the research indicate that the selection of modifiers and processing modes can be used to obtain powder systems of different dispersion and imperfection with a modified structure and composition of the surface layer of particles.

Dispersion with a simultaneous production of polymer formations on the surface of the wollastonite particle, which lead to its enlargement, is clearly represented by the results of electron microscopic studies. As is known, wollastonite is characterized by an elongated needle-like particle shape. The samples of wollastonite powder used in the experiment also consist of needle-shaped particles (Fig. 7.7) of various thickness and length. In addition, they contain impurities of silicon dioxide, small particles of which have an irregular shape.

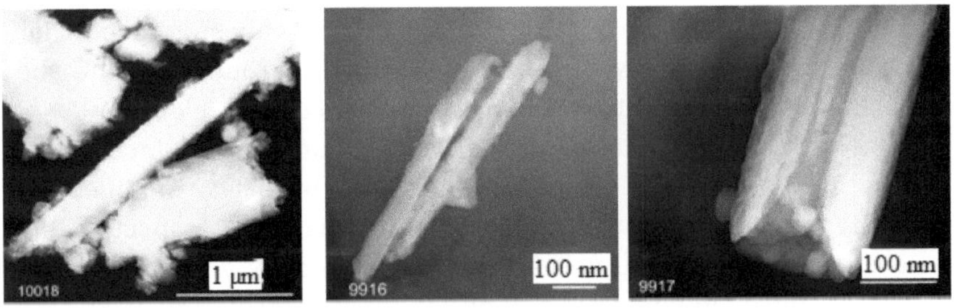

Figure 7.7 Electron microscopic images of wollastonite samples in the initial state.

The MCT of wollastonite without modifiers results in grinding and splitting of particles, preserving their needle shape. Silicon dioxide impurities are dispersed to an amorphous state and envelop wollastonite particles. As a result, conglomerates of different structure are formed (Fig. 7.8). In this case, the surface layer is loosened and activated, therefore, there takes place transformation and displacement of surface structural formations under the action of the electron beam.

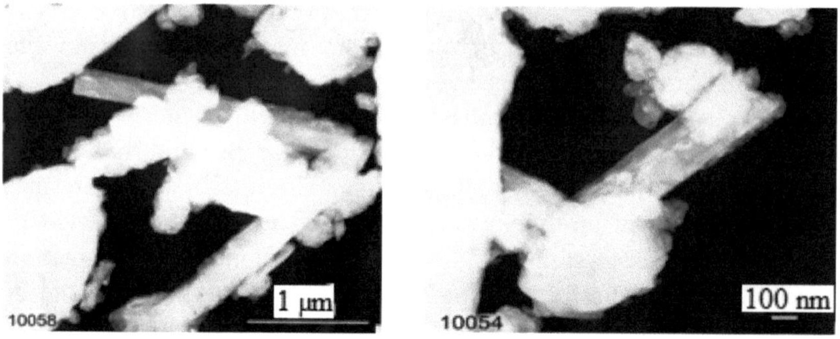

Figure 7.8 Electron microscopic images of wollastonite samples after MCT.

During the mechanochemical treatment of wollastonite powder in the presence of ammonia, destruction of the particles proceeds more intensively, and loose porous formations of a rounded shape are formed at the edges (Fig. 7.9), in the formation of which, apparently, calcium carbonate and quartz impurities took part. Under the influence of the electron beam, the particles of the sample often transform and take the form of spheroids this indicating a high activity of the modified particles. The noted fact of an increase in the activity of wollastonite powder particles after MCT with a nitrogen-containing modifier and the change in their structure under the influence of an electron beam confirms the previously obtained results on the modification of quartz [2, 6], when the high activity of particles, modified by MCT with an aqueous solution of ammonia, with respect to the electron beam. Thus, was also noticed there is a specific effect of the nitrogen-containing modifier on the structure of silicate compounds.

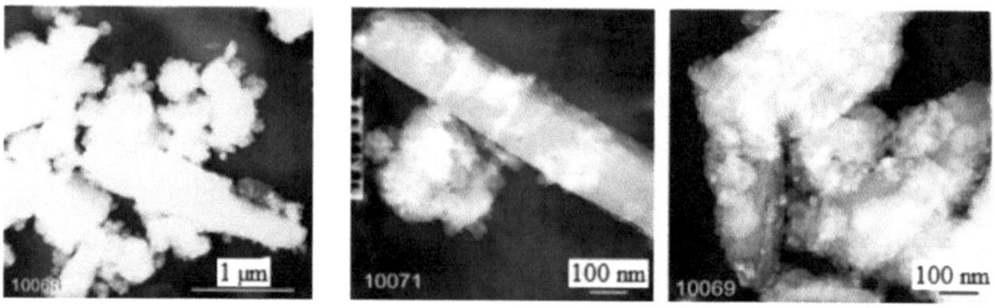

Figure 7.9 Electron microscopic images of wollastonite samples after MCT in the presence of aqueous ammonia.

In the case of wollastonite powder mechanochemically treated in the presence of polyvinyl alcohol, a polymer film is formed on the surface of the elongated particles, i.e., inorganic oxide particles are encapsulated in an organometallic film (Fig. 7.10a,b). Some formations are transformed under the influence of an electron beam with formation of a new structured particle surface (Fig. 7.10c).

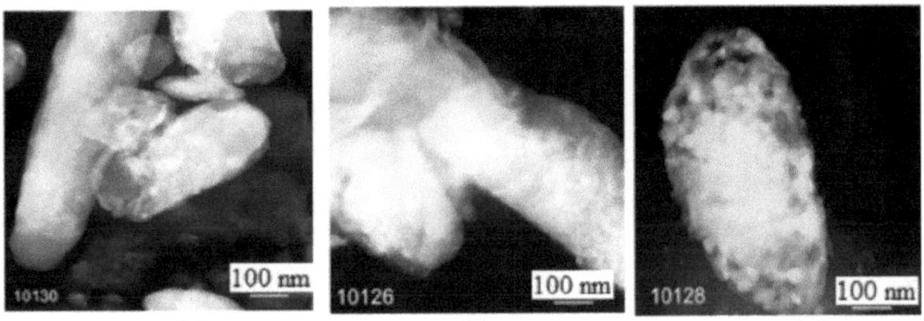

(a) (b) (c)

Figure 7.10 Electron microscopic images and microdiffraction of wollastonite samples after MCT in the presence of polyvinyl alcohol (a, b) and with prolonged exposure to the electron beam (c).

When processing wollastonite in the presence of succinic acid and urea, the process of modifying particles also manifests itself quite clearly in the transformation of the structure of the surface layer of particles (Fig. 7.11). With an increase in processing time, the contribution of the polymer formation to the morphology and structure of the particle increases. As a result, the proportion of the enlarged modified fraction of powder particles increases, which clearly follows from electron microscopic images (Fig. 7.11).

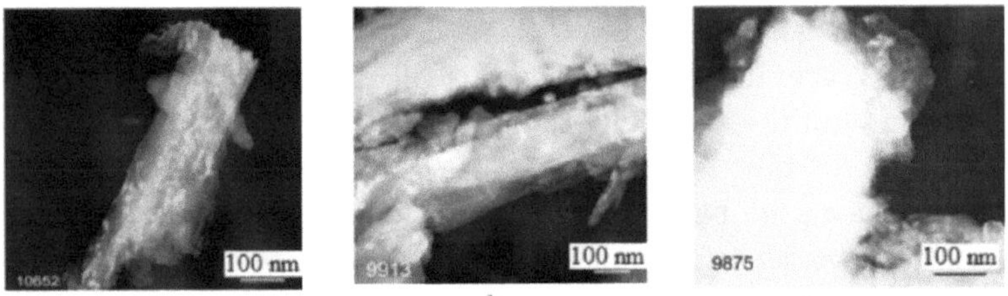

a b c

Figure 7.11 Electron microscopic images of wollastonite samples after MCT in the presence of succinic acid (a) and urea (b, c) at 10 min (a, b) and 30 min (c).

A distinctive feature of the modification of wollastonite with silicic acid is the looser structure of the surface layer of particles after the MCT. Over time, the particles are gradually enveloped in amorphous silicon dioxide, depending on the time of MCT and the modifying film grows layer-by-layer (Fig. 7.12).

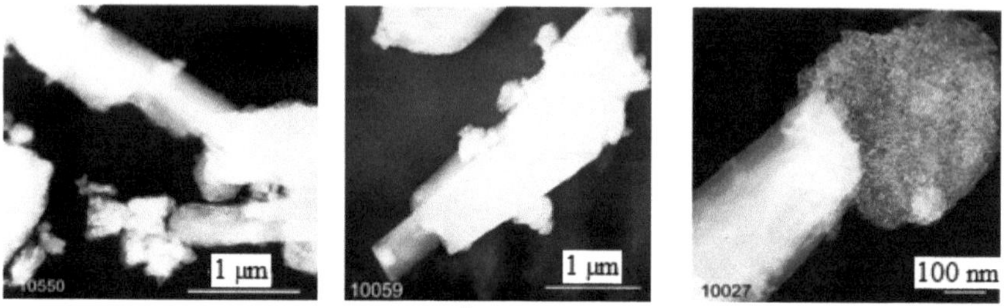

a b c

Figure 7.12 Electron microscopic images of wollastonite samples after MCT in the presence of silicic acid for 10 min (a), 20 min (b), and 30 min (c).

Wollastonite, encapsulated in silicone films, will be an effective reinforcing filler in the preparation of ceramic materials, in particular, insulating porcelain, the main component of which is silica [18]. Such a directional change in the surface structure of wollastonite particles will improve the compatibility of the filler and the matrix, which will improve the mechanical and dielectric properties of the composition.

Thus, the results of electron microscopic and X-ray diffraction studies have shown that MCT of wollastonite powder consisting of $CaSiO_3$ and containing mostly quartz impurities resulted in not only grinding of particles, but also accumulation and annihilation of defects, transformation of the internal structure and surface layer of particles. Modifying nitrogen- and carbon-containing additives are destructed in the process of MCT and actively interact with the surface of the ground particles. As a result of this interaction, the formation of film or loose formations are produced on the particles of minerals and a powder material of a composite structure is obtained. In addition to the general laws in transformation of the structure of particles of the material under study, there are individual features of the interaction of particles with various modifiers. Ammonia has a greater effect on the structure in the volume of the particles and increases the activity of the particles to external influences, while polyvinyl alcohol, succinic acid and urea actively participate in reactions on the surface of the particles and ensure their encapsulation into organoelement nanostructured shells of various density and fragmentation. Mechanochemical treatment of wollastonite in the presence of silicic acid provides the formation of a silicone capsule on the surface of particles.

The presented results of the studies indicate that by choosing appropriate modifiers and modes of mechanical processing it is possible to obtain powder systems of different dispersion and imperfection with a modified structure and composition of the surface layer of particles. The resulting composite powder systems consist of particles that are an inorganic mineral core and a polymeric encapsulating shell. Together, these structural adjustments change the energy state and reactivity of the material.

The obtained results on the modification of quartz, calcite (marble) and wollastonite in the process of MCT vividly illustrate the possibilities of this method in obtaining mineral fillers for composite systems based on polymers. The choice of the mineral-filler, a specific modifier for it and the processing modes of this system will be determined by the polymer base of the composite and what properties it is necessary to provide the synthesized material with. The fundamental approach in solving these problems will be discussed in the following sections on modifying mineral systems in the process of MCT and directed to changing their properties according to the functional area.

Questions

1. Describe the distinctive features of dispersion and modification of calcite particles depending on the content of impurities and the micro-composition of mineral raw materials, on the modes of mechanochemical processing and modifying additives.
2. What is the specificity of dispersion of calcite when using an aqueous solution of ammonia, polyvinyl alcohol and urea?
3. What morphological and structural features of the composite formations calcite + polymer layer are observed when the particles are modified during MCT with an aqueous solution of ammonia?

4. What are the conditions for production of film or loose formations on the surface of calcite particles during MCT?
5. What is the composition and structural features of the wollastonite mineral that provide reinforcing properties and increased strength to composite systems?
6. Describe the features of the process of grinding wollastonite, polymerization and production of composite formations of highly dispersed particles of wollastonite powder and organic polymers.
7. What features of changes in the defective structure of wollastonite during MCT with modifiers contribute to the growth of new polymer layers on the particle surface?
8. What is the specific effect of the nitrogen-containing modifier on the structure and activity of silicate compounds after MCT?
9. Describe the conditions for encapsulation of inorganic oxide particles in a dense organometallic film and formation of a loose structure of the surface layer of particles after MCT.
10. What are the conditions for the production of powder composite systems based on mineral raw materials of various dispersity and imperfection with a modified structure and composition of the surface layer of particles at MCT?

References

1. Mofa N. N. Mechanochemical processing is a progressive technological process of creating new composite materials. *(Review)/Collection of Chemistry and Chemical Technology. Modern Problems.* Kazak university, Almaty, 2004. pp. 163–198. (In Russian).
2. Mofa N. N., Shabanova T. A., Ketegenov T. A., Chervyakova O. V., Mansurov Z. A. Electron-microscopic investigations of quartz particles modified by mechanochemical processing. *Eurasion Chem. Tech. J.*, 2003. no. 5. pp. 297–303.
3. Mofa N. N., Ketegenov T. A., Mansurov Z. A., Soh D. W. Nanocomposite magnetic powder materials obtained using mechanochemical synthesis. *Trans. Electron. Mater.*, 2004. vol. 5, no. 1. pp. 24–33.
4. Mansurov Z. A., Mofa N. N., Shabanova T. A. Hybride, nano-structurized materials of special purpose on the basis of silicon dioxide. *Adv. Eng. Ceramics Composites*, 2011. vol. 484. pp. 230–240.
5. Mansurov Z. A., Shabanova T. A., Mofa N. N. *Synthesis and Technology of Nanostructured Materials: A Course of Lectures.* Kazak University, Almaty, 2012. 335 p. (In Russian).

6. Mansurov Z. A., Mofa N. N., Shabanova T. A. Mechanochemical treatment, structural features, properties and reactivity of SHS systems based on natural materials. Part 1: Mechanochemical synthesis of highly dispersed quartz-based nanostructured systems. *Eng. Phys. J.*, 2013. vol. 86, no. 4. pp. 793–800. (In Russian).
7. Smagin A. G., Yaroslavsky M. I. *Quartz Piezoelectricity and Quartz Resonators*. Energy, Moscow, 1970. 488 p. (In Russian).
8. Zhugin Yu. N. Behavior of α-quartz at high dynamic and static pressures: new results and representations. *Chem. Phys.*, 1995. vol. 14, no. 1. pp. 69–74. (In Russian).
9. Urusovskaya A. A. Electrical effects associated with plastic deformation of ionic crystals. *Uspekhi Fiz.*, 1968. vol. 96, no. 1. pp. 39–60. (In Russian).
10. Zaitseva L. A. Influence of a layer of hot electrons near the surface on the thermodynamic energy of semi-bound solids. *Chem. Phys.*, 2002. vol. 21, no. 2. pp. 15–25. (In Russian).
11. Molchanov V. I., Selezneva O. G., Zhirnov E. N. *Activation of Minerals During Grinding*. Nedra, Moscow, 1988. 208 p. (In Russian).
12. Boldyrev V. V., Avvakumov E. G. *Fundamental Principles of Mechanical Activation, Mechanosynthesis and Mechanochemical Technologies*. Publishing House of the Siberian Branch of the Russian Academy of Sciences, Novosibirsk, 2009. 343 p. (In Russian).
13. Tyulnin V. A., Tkach V. R., Eyrikh V. I., Starodubtsev N. P. *Wollastonite (A Unique Multi-Purpose Mineral Raw Material)*. Publishing House "Ore and metals", Moscow, 2003. 142 p. (In Russian).
14. Lei Y., Yan Q., Liu S., Yuan J. Papermaking of mineral fiber composites. *J. Sci. Ind. Res.*, 2010. vol. 69. pp. 215–220.
15. Sara M. Robinson, Martin F. Sheridan, Anthony G. Ferradino. *The Advantages of Wollastonite, a Non-Traditional Filler, in Fluorohydrocarbon (FKM) Elastomers*. Presented at a meeting of the Rubber Division, American Chemical Society Savannah, Georgia April 29 – May 1, 2002.
16. Mansurov Z. A., Mofa N. N., Chernoglazova T. V. *Wollastonite Is a Universal Reinforcing Filler of Composite Materials*. Kazak University, Almaty, 2018. 308 p. (In Russian).
17. Berlin A. A., Wolfson S. A., Oshmyan V. G., Enikolopov N. S. *Principles of Creating Composite Polymeric Materials*. Chemistry, Moscow, 1990. 238 p. (In Russian).
18. Kozlov G., Zaikov G. *Fractal and Local Order in Polymeric Materials*. Nova Science publishers, New York, 2001. 420 p.

Chapter 8

Mechanochemistry under the Conditions of Ultrasonic Treatment of Powder Systems

8.1 Ultrasonic Treatment of the Material Is a Way to Change the Structure and State and Obtaining Nanostructured Systems

Ultrasonic treatment is a variety of mechanochemical processing of materials [1–4]. For ultrasonic treatment (UST), devices with electro-acoustic emitters (magnetostrictive or piezoelectric) are used. Under the action of ultrasound (acoustic oscillations with a frequency higher than 20 kHz), different physical and chemical effects occur in the substance, which are used in various technological processes. Ultrasonic energy significantly affects chemical reactions (oxidation, polymerization, depolymerization, electrochemical, and other processes), accelerates many mass transfer processes (dissolution, extraction, impregnation of porous bodies, etc.), the course of which is limited by the diffusion rate. Along with the mechanochemical treatment of various substances in dynamic action mills for production of highly dispersed systems up to nanostructured materials, ultrasonic processing has been widely used in both scientific and practical terms [4–7].

Mechanochemical Synthesis of Composite Materials
Zulkhair A. Mansurov, Nina N. Mofa, Tlek A. Ketegenov, and Bakhtiyar S. Sadykov

ISBN 978-981-4800-88-4 (Hardcover), 978-1-003-12081-0 (eBook)
www.jennystanford.com

In the course of UST, most of the processes in liquids are accompanied by the phenomenon of ultrasonic cavitation and the occurrence of acoustic flows. Acoustic cavitation is an effective means of concentrating the energy of a low-density sound wave into a high energy density, related to pulsation and collapse of cavitation bubbles. The overall picture of formation of a cavitation bubble is presented in the following form (Fig. 8.1). In the rarefaction phase of an acoustic wave in liquid, a gap is formed as a cavity, which is filled with the saturated vapor of the liquid (Fig. 8.2). In the compression phase, under the action of increased pressure and surface tension forces, the cavity collapses, and steam is condensed at the interface. The gas dissolved in the liquid diffuses through the cavity walls, which then undergoes a strong adiabatic compression [4]. At the moment of bubble collapse, the pressure and temperature of the gas reach significant values—according to some data, up to 100 MPa and 5000–25000 K. After the cavity collapses in the surrounding liquid, a spherical shock wave propagates attenuating rapidly. As the explosion takes place in less than a nanosecond [3], it develops very high cooling rates exceeding 10^{11} K/s. Bubble explosions also lead to the appearance of streams in a liquid the velocity of which reaches 150 m/s.

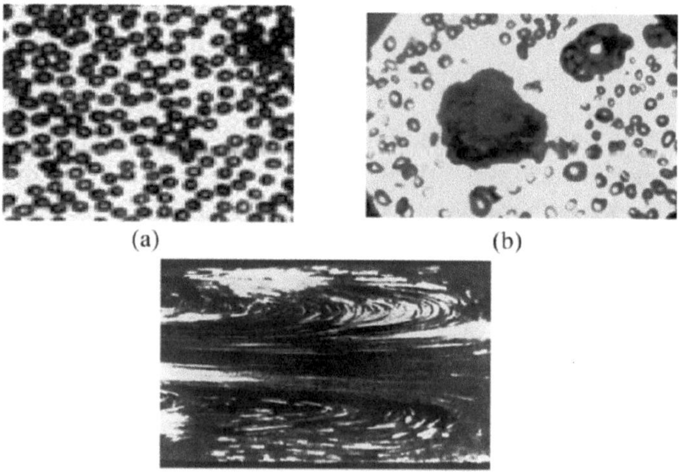

(a) (b)

(c)

Figure 8.1 Cavitation bubbles (a), their collapse (b) and flows (c) that appear in the working fluid during the collapse of cavitation cavities in the process of ultrasonic treatment of the fluid [1].

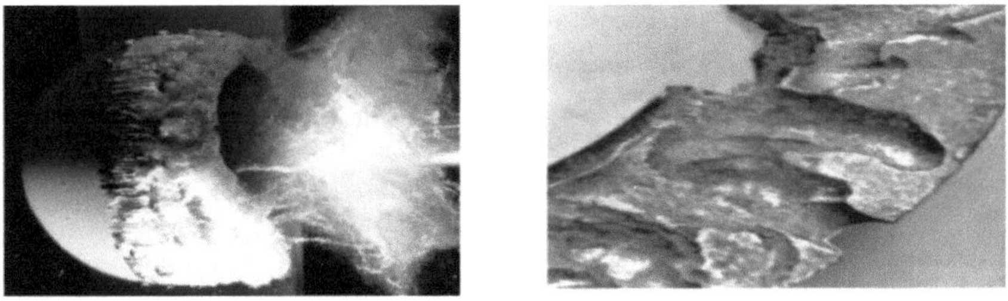

(a) (b)

Figure 8.2 The effect of the collapse of cavitation bubbles (a) and mechanical erosion of the surface of materials (b) during ultrasonic processing of the system [4].

The pressures created by the collapse of cavitation bubbles in liquids cause intense mechanical erosion of the surface of materials. These effects are widely used to remove various unwanted films and contaminants from the surface of materials, to accelerate chemical reactions, homogenization, and other purposes [7]. The high level of stresses created by ultrasonic vibrations in solids can lead to the development of fatigue phenomena and destruction.

Ultrasonic dispersion is a fine grinding of solid particles in a liquid medium (transition to a dispersed state), which allows obtaining highly dispersed, homogeneous, chemically pure powders, which cannot be produced by other methods—mechanical, chemical, thermal [8, 9]. The particles of the powder obtained by ultrasonic dispersion have a "smoothed" surface. Ultrasonic dispersion allows to obtain highly dispersed (average particle size—microns and micron fractions), homogeneous and chemically pure mixtures (suspensions) of solid particles in liquids. During ultrasonic dispersion of suspensions, the dispersion of the product is increased by several orders of magnitude compared with conventional mechanical grinding.

Dispersion is conditioned by irradiation of a liquid with solid particles in its volume by a sound field of a definite frequency and intensity. Substances are ground under the action of shock waves arising from the collapse of cavities, flows. The process of grinding starts at a certain ultrasound intensity exceeding the threshold value depending on cavitation strength of the liquid, the state of the solid phase surface, and the nature and magnitude of the interaction forces between the individual particles and solid phase [8]. With the increase in ultrasound intensity, the rate of dispersion increases; it also increases with brittleness and with a decrease in hardness and sticking of the particles together.

Ultrasonic dispersion is most effective when processing amorphous substances and aggregating substances such as soil and rocks, when splitting textured materials such as cellulose, glass wool, asbestos, and acting on plant and animal cells. Many natural minerals—kaolin, gypsum, mica, sulfur, graphite, etc., disperse quite easily, pure metals in a more difficult way. Thus, the authors of [10] created a laboratory setup and a method for synthesizing highly dispersed oxide powders, based on the combination of hydrothermal treatment of the starting materials with a simultaneous action of ultrasound of high power. This made it possible to obtain highly dispersed oxide powders. Other methods for producing nanoparticles with ultrasonic treatment have been developed [11, 12]. All these methods are based sonochemical reactions. Sonochemistry is also used to precipitate various nanomaterials (metals, metal oxides, semiconductors) on the surfaces of ceramic and polymeric materials. A uniform homogeneous coating layer is formed on the surface. Nanoparticles are attached to the surface via formation of chemical interactions with the substrate.

For production of nanoscale oxides and metal hydroxides, a solution of a metal salt (usually chloride) in an appropriate solvent is subjected to the action of powerful ultrasound in the presence of a base, such as, for example, an alkali metal hydroxide. Under these conditions, highly active radicals that make up the core of the nanoparticles are created inside the rapidly exploding cavitation bubbles. In such a sonochemical reaction, a solution of one mole of metal salt gives up to several hundred grams of a nano-product, having a size on the nanoscale from 5 to 60 nm, in a surprisingly short time, about 3–6 min [13]. Examples of compounds, nanoparticles of which can be obtained in this way are oxides: FeO, Fe_2O_3, Fe_3O_4, NiO, Ni_2O_3, CuO, Cu_2O, Ag_2O, CoO, CO_2O_3, and crystalline hydrates: $Fe(OH)_3$, $Co(OH)_3$, NiO(OH). Under ultrasonic action, chemical reactions are accelerated and conversion of metal salts or oxides to metal powder in relatively large quantities (1 mol) is completed within 5–10 min. The resulting powder consists of ultra-fine metallic or non-metallic particles in the nanoscale range (5–100 nm) [14, 15].

As shown by the results of the carried out research, depending on the parameters of ultrasonic vibrations (frequency, power, intensity) and the emitter design, substances can be destroyed or activated, promote synthesis, polymerization and dissociation, control the kinetics of the chemical or physicochemical process, including processes or individual stages of mass transfer [16–18]. Acoustic radiation can render a significant effect on homogeneous redox and electrochemical reactions both with ion recharging and with the release of a new phase. In industry, as a rule, powerful ultrasonic fields are used in cavitation mode. At the same time, relatively few studies related to the use of ultrasound of low intensity and power were carried out for the treatment of chemical and electrochemical systems. Therefore, the use of low-power ultrasonic fields in this area may be important for evaluating the mechanism of sound action on the chemical and physico-chemical characteristics of chemical processes and predicting the development of a certain type of reaction [19, 20].

Questions

1. What phenomena and processes accompany ultrasonic processing in liquids?
2. What is acoustic cavitation with an UST in liquids?
3. How does mechanical erosion of the surface of materials happens under the influence of pressure created during the collapse of cavitation bubbles in liquids?
4. What are the features of ultrasonic dispersion in comparison with conventional mechanical grinding?
5. Describe the dependencies of the process of destruction and activation of the substance on the parameters of ultrasonic vibrations (frequency, power, intensity) and design of the emitter.
6. Give examples of ultrasonic influence to accelerate chemical reactions and the conversion of salts or metal oxides as well as production of ultra-fine metallic or non-metallic particles in the nanoscale range.

7. Describe the ultrasonic influence, as a way to control the kinetics of chemical or physico-chemical processes of synthesis, polymerization and dissociation of a substance.

8.2 Changes in the Structure, Properties, and Modification of Quartz and Calcite under the Influence of Ultrasound

The studies were conducted on quartz from the Kuskuduk deposit (KQ) with a quartz content of 81.3% and microcline $K(Si_3Al)O_8$-18.7%, and Tekturmas deposit (TQ), which consists practically of quartz (98.84%) with a small amount (about 1%) of aluminosilicate impurities. The processing was carried out in an ultrasonic bath (Fig. 8.3), the electro-acoustic emitter was piezoelectric resonantly tuned transducers with a frequency of 40 and 100 kHz and a power of 50 W. The UST was carried out with time variation (for 20, 40, and 60 min) and frequency of exposure. Treating was carried out both at separate and simultaneous action of low and high frequencies of the ultrasonic range. The latter is advantageous in that cavitation bubbles arising at low frequencies receive great acceleration under the influence of hydrodynamic flows created by high frequency ultrasonic vibrations.

Figure 8.3 Ultrasonic bath

Ultrasonic treatment was conducted at a frequency of 40 kHz in distilled water (pH = 6.1). After introduction of 100 g of quartz powder in 300 ml of water, its pH increased by an average of 1.0, and after ultrasonic treatment it increased by another 0.5–1.15, depending on the purity of the quartz material and the processing time. With the introduction of 10% of liquid glass in water, its alkalinity increased to 10.9, and the introduction of quartz powder and ultrasonic treatment led to a slight decrease in pH (by 0.5–0.8). After dissolving 30 mg of silicic acid in water (i.e., when a 10% aqueous solution of silicic acid is obtained), pH of the medium decreased within 1 and after UST there is a slight increase in pH depending on the processing time and material purity (Fig. 8.4).

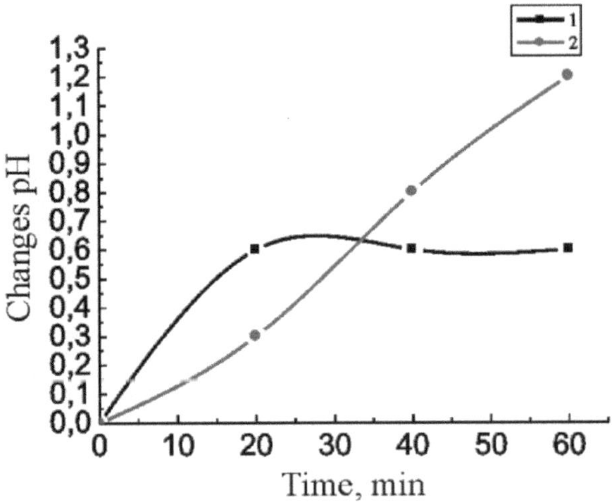

Figure 8.4 Dependence of changes in pH of a 10% aqueous solution of silicic acid on the time of ultrasonic treatment of quartz KQ (1) and TQ (2).

All changes in pH of an aqueous solution indicate the processes of destruction and dissolution of the surface layers of particles and the possibility of chemical reactions on their surface. Confirmation of the fact that chemical reactions on the surface of quartz particles occur during UST is a change in the crystallite size in comparison with the initial state of the particles (Table 8.1). The average crystallite size in the initial state for Kuskuduk quartz is 725 Å, and for Tekturmasky –915 Å. After UST within 60 min, the value of L decreased slightly, i.e., the imperfection of the particles increased, and after treatment in the presence of dissolved liquid glass and silicic acid, their values increased, which is a result of chemical reactions on the surface and the flow of defects into the reaction zone.

Structural changes both in the volume of particles and on their surface lead to a change in such a structurally sensitive characteristic of the material as the dielectric constant. In the initial state, the dielectric constant of the powders is ε = 2.27. After UST in water, an increase in the dielectric constant of quartz particles to ε = 7–8 is observed, which may be due to saturation of the surface layer of particles with hydroxyl groups.

Table 8.1 The size of crystallites and the broadening of X-ray lines of quartz samples (KQ) and (TQ) after 60 min of UST in water and 10% aqueous solutions of liquid glass and silicic acid

Samples, modifier	**Crystallite size, L, Å and lattice microdistortion**			
	Plane index (100), *d* = 4.25		**Plane index (101), *d* = 3.34**	
	L, Å	**ε-dist. 1 × 10–5**	**L, Å**	**ε-dist. 1 × 10–5**
KQ*	675	235	530	160
TQ**	846	250	987	220
KQ, $Na_2O \times nSiO_2 + mH_2O$	678	232	666	146
KQ, $H_2SiO_3 \cdot n(H_2O)$	786	250	763	180
TQ, $Na_2O \times nSiO_2 + mH_2O$	960	260	900	220
TQ,$H_2SiO_3 \cdot n(H_2O)$	1027	276	875	210

*KQ, quartzite from the Kuskuduk deposit; *TQ, Quartzite from the Tekturmas field.

After sonication in an alkaline and acidic media, changes in the dielectric constant of the quartz material are observed. After treatment in an acid medium (in the presence of silicic acid) for quartz TQ, there is practically no change in the values of dielectric constant, and treatment in an alkaline medium leads to a significant increase in ε by more than an order of magnitude (Fig. 8.5).

For KQ quartz, the dielectric constant changes are more significant. After the UST, in an acidic medium, ε increases to 14, and in an alkaline medium—to 50. The observed changes in ε values for quartz particles of this field are due to the fact that in the initial state they have a surface layer saturated with various impurities, which gives them a specific color, as shown in Chapter 7 (Fig. 7.1). In the process of UST, especially in an alkaline or acidic medium, this surface layer is destroyed and various compounds with hydroxyl groups are formed.

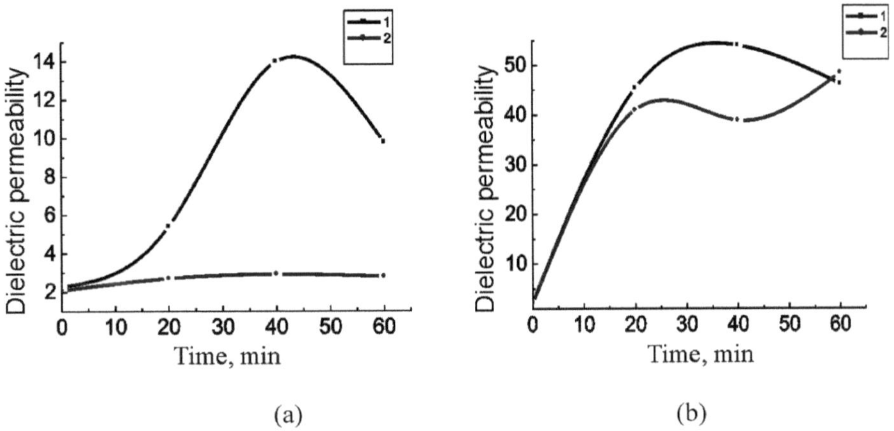

Figure 8.5 Dependence of dielectric permeability of quartzite powder of various deposits on the time of UST in 10% aqueous solution of silicic acid (a) and sodium liquid glass (b). Quartzite KQ (1) and TQ (2).

Thus, the presented results on measuring the pH of the treatment medium and the dielectric constant of the quartz powder material indicate a change in the state and structure of the particles after UST depending on the medium and the treatment time. At the same time, the degree of purity of the material being processed is important, since each compound in the composition of the processed ultrasonic powder is involved in this process. And the final result is a synergic effect from the presence of all the components in the mixture of compounds.

Electron-microscopic studies have shown that ultrasonic processing of quartz in an aqueous medium results, first of all, in erosion of the surface of particles, chipping and destruction of particles to smaller sizes (Fig. 8.6). The nature and degree of destruction depends on the processing time and composition of the aqueous medium. When silicic acid and sodium liquid glass are introduced into water, after UST, the particles have a nanostructured friable surface, especially when using liquid glass, i.e., when processing in an alkaline medium.

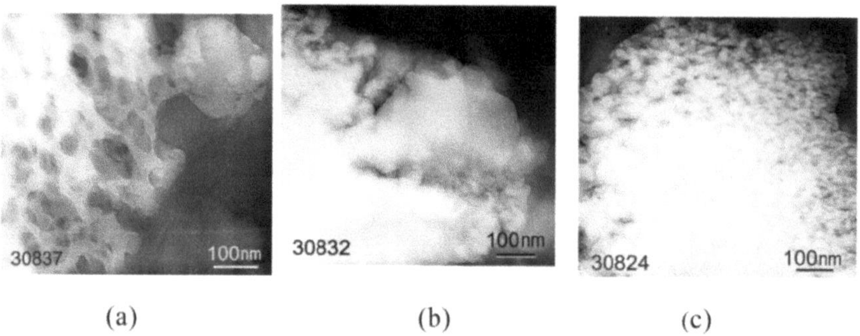

Figure 8.6 Electron microscopic images of quartz particles after UST in water (a) and in 10% aqueous solution of silicic acid (b) and sodium liquid glass (c).

To modify the surface layer of quartz particles and to obtain a polymer film that encapsulates a particle during UST, up to 10% butanol—C_4H_9OH (2), glycerine—$C_3H_5(OH)_3$, or urea—$(NH_2)_2CO$ were introduced into the aqueous medium. If, in the initial state, the particles of quartz have a clear cut, then after UST in the presence of these modifiers, their morphology changes significantly. Films are formed on the surface of the particles as a result of interaction of the surface of the particle with the medium in which the powder is treated (Fig. 8.7). If for particles treated with ultrasound in water, only erosion of the surface of the particles is observed, during processing in the presence of butanol, glycerol, and urea, polymer films are formed on the surface of the particles, which encapsulate the quartz particle. In the surface layer of the particle is likely the formation of compounds with the presence of urea and butanol ions in them are likely to be formed.

Compared to quartz, the particles of calcite used in the work under the conditions of UST mode are modified with butanol and glycerol, and are characterized by a dense surface layer and high polycrystallinity of particles (Fig. 8.8). When urea is modified, the surface layer of the particles becomes, on the contrary, looser with a high level of structure.

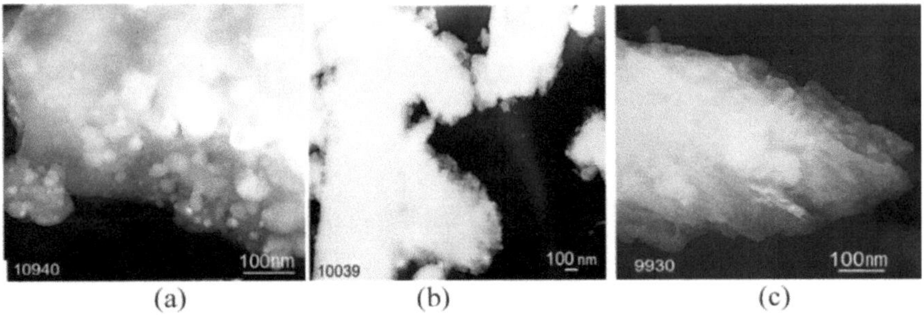

Figure 8.7 Microstructure of quartz particles after UST in 10% aqueous solutions of C_4H_9OH (a), $C_3H_5(OH)_3$ (b) and $(NH_2)_2CO$ (c).

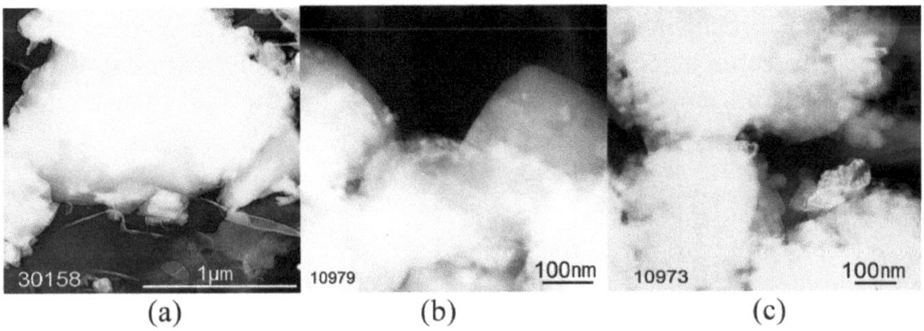

Figure 8.8 Microstructure of calcite particles after UST in 10% aqueous solutions of C_4H_9OH (a), $C_3H_5(OH)_3$ (b) $(NH_2)_2CO$ (c).

During joint ultrasonic treatment of a mixture of quartz and calcite (50:50), it was found that the system being processed is agglomerates, since the surface interaction of the quartz and calcite particles takes place (Fig. 8.9). The system is represented by aggregates of particles. Particle aggregates having a similarity of faceting have an internal ordering different from rounded particle aggregates (Fig. 8.9c).

As the surface of the particles changes during UST, an organic component is added, the dispersity, i.e., particle size, changes, and it all counts for the bulk density. This characteristic of the powder material can serve as a complex indicator of the change in particle size and the formation of surface structures in the UST process.

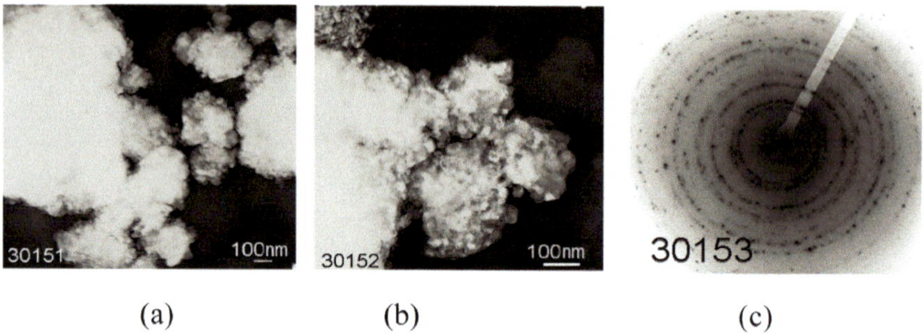

Figure 8.9 Microstructure and electron diffraction pattern of particles of a mixture of quartz and calcite after UST in the presence of C_4H_9OH (a) and $(NH_2)_2CO$ (b).

In case of measuring the bulk density of quartz with calcite in the ratio of 50:50 depending on the time of treatment with ultrasound, one can observe a tendency for the bulk density to decrease (Fig. 8.10).

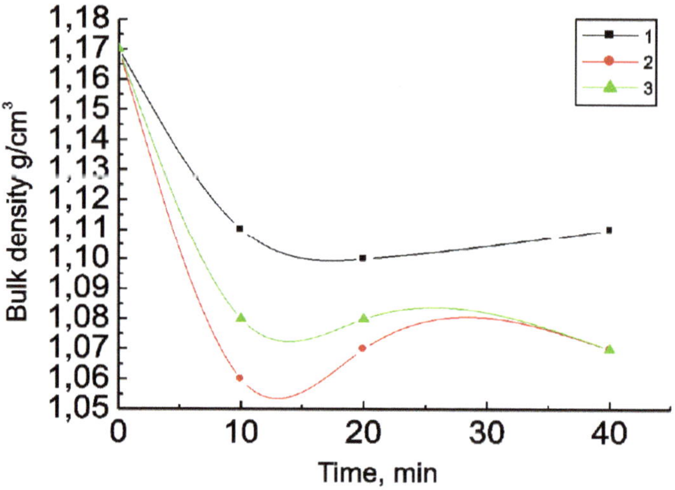

Figure 8.10 Dependence of the bulk density of mixtures of quartz and calcite in the ratio (50:50) on the time of treatment with ultrasound in water (1) in aqueous solution of $(NH_2)_2CO$ (2), and C_4H_9OH (3).

This is due to formation of organic layers on the surface of particles. In most cases, the dependence of this indicator for a mixture of quartz with calcite, on the time of UST is not linear. Thus, the values of the bulk density of mixtures of quartz and calcite in the ratio (50:50) depending on the processing time for UST in water and with addition of butanol or urea decrease from 1.17 to 1.06 g/cm^3. The decrease in bulk density is due to the fact that either the dispersion of the powder increases or a thicker organic film appears on the surface. Organic matter has a lower density, and bulk density is a total indicator of changes in the dispersion, state and structure of particles.

Thus, the presented investigation results on the structure and morphology of quartz and calcite particles after UST showed significant changes in both the structure, morphology of the particles, and the values of the bulk density of the treated powder, depending on the time of treatment and the composition of the aqueous medium.

Questions

1. According to the physicochemical characteristics of the medium and the material being processed, can chemical reactions on the surface of quartz particles (or other materials) during UST be assessed?
2. What features of the changes in the dielectric constant of a quartz material are observed after sonication in an alkaline and acidic medium?
3. Describe the dependence of changes in the state and structure of particles after an UST on the medium, processing time and the degree of purity of the processed material.
4. What should be the composition of the aqueous medium during UST for obtaining particles of polymer films encapsulating a quartz particle on the surface?
5. What are the differences in the surface layer of calcite particles compared to quartz when modifying the powder in the UST mode?
6. Describe the bulk density as a total indicator of changes in the dispersion, state and structure of particles encapsulated in an organic film during ultrasonic treatment.

8.3 Changes in the Structure, Properties, and Modification of Wollastonite under the Influence of Ultrasound

The mechanochemical treatment of wollastonite brand FW 200 in the mode of ultrasonic (US) exposure was carried out at frequencies of 40 and 100 kHz and, at the same time, ultrasound exposure at low and high frequencies in order to detect resonant effects. The processing time and the composition of the aqueous solution in which the wollastonite powder was processed varied. As the modifying additives of wollastonite during UST, the same compounds that were used in the processing of quartz and calcite—butanol, glycerin, and urea, as well as ammonia—were taken.

For the treated ultrasonic particles of wollastonite powder, bulk density measurements were performed. The values of these characteristics for the powder after the UST in water at different frequencies of influence and different treating times are presented in Fig. 8.11. The bulk density of the powder after the UST increases, and the highest values are recorded after processing at a frequency of 40 kHz. After 40 min of UST, a slight decrease in density is observed. The bulk density can increase due to formation of a highly dispersed fraction of the treated powder, contributing to its compaction in a measured volume. An abnormal change in the bulk density after 40 min of treatment may be conditioned by the peculiarity of the state of the surface layer of the particles.

The role of the surface layers is enhanced after ultrasonic treatment in water with 10% butanol. In this case, the influence of high-frequency oscillations increases (Fig. 8.11b). A UST time of 40 min remains a special point in the duration of treatment, also leading to an increase in the bulk density with participation of alcohol groups in modification of wollastonite particles. The effect of particle modification is enhanced during UST with participation of glycerol (Fig. 8.11c). First of all, this is indicated in a decrease in the bulk density of almost two times (from 0.7 to 0.4–0.3 g/cm^3), which may be due to formation of a polymeric encapsulating hydrophobic film on the surface of the particle.

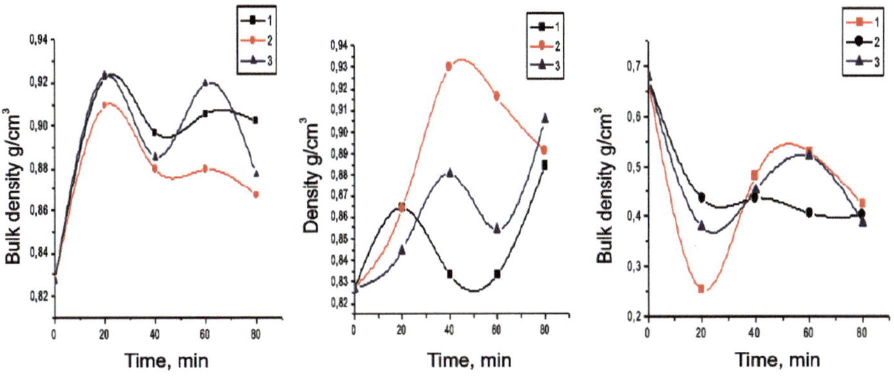

(a) (b) (c)

Figure 8.11 Dependence of the bulk density of wollastonite powder particles on the time of UST at frequencies of 40 (1), 100 (2) and simultaneously at 40 and 100 kHz (3). UST was carried out in distilled water (a), with the participation of butanol (b) and glycerin (c).

After the UST in the presence of ammonia, the bulk density of the powder increases to 0.88 g/cm^3 (Fig. 8.11a). After the UST with urea, the bulk density of the powder varies from 0.7 to 0.88 g/cm^3 (Fig. 8.11b). The role of the frequency of the UST during the 40–80 min of exposure is weak, i.e., there takes place stabilization of characteristics and, consequently, the structure of the material under such processing conditions. In all cases of UST treatment of wollastonite the structure of the surface layer of particles changes.

The change in the surface layer of particles up to their encapsulation clearly follows from the results of electron microscopic analysis of wollastonite after ultrasonic treatment in various aqueous solutions. After treatment in water, the particles have an average size of up to 7–10 μm in length and 1 to 5 μm in width. Splitting of particles along the length under the influence of ultrasound is observed (Fig. 8.12b). The particles after treatment for 40 min are highly sensitive to the electron beam with formation of Moiré transitions, this indicating stratification of the particles (Fig. 8.12c).

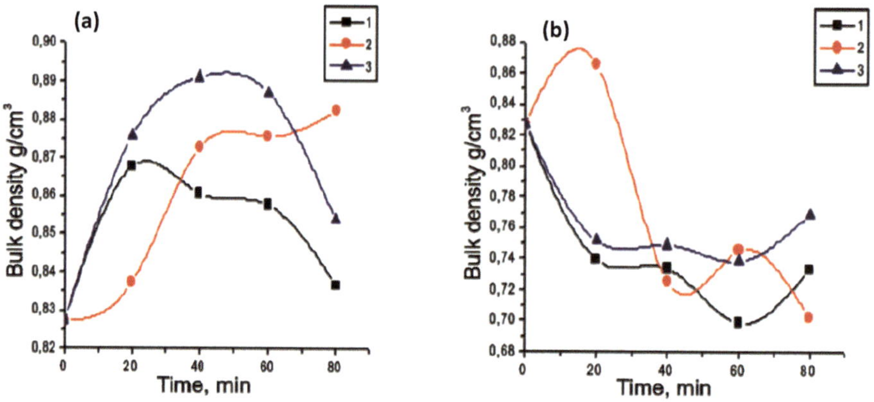

Figure 8.12 Dependence of the bulk density of wollastonite powder particles on the time of the UST at frequencies of 40 (1), 100 (2) and simultaneously at 40 and 100 kHz (3). UST was carried out with the participation of ammonia (a) and urea (b).

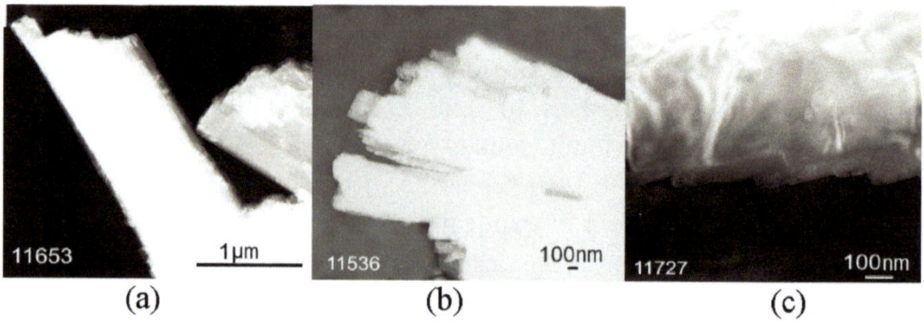

Figure 8.13 Electron microscopic images of wollastonite particles in the initial state (a) and after the UST in water for 20 min (b) and 40 min (c) at 40 kHz.

When introducing 10% butanol into water and carrying out UST of wollastonite powder in such a medium, the particles disperse more intensively, especially at 40 kHz for 40 min, the UST (Fig. 8.13).

After a longer treatment and at a higher frequency of ultrasonic exposure, the buildup of the polymer layer on the surface of the particle is more pronounced. In this case, according to the microdiffraction pattern, the bulk of the particles retain the crystal structure. After UZT in the presence of glycerin, the process of modifying particles and building up the polymer layer occurs to an even greater degree, especially with 40 min of treatment (Fig. 8.14).

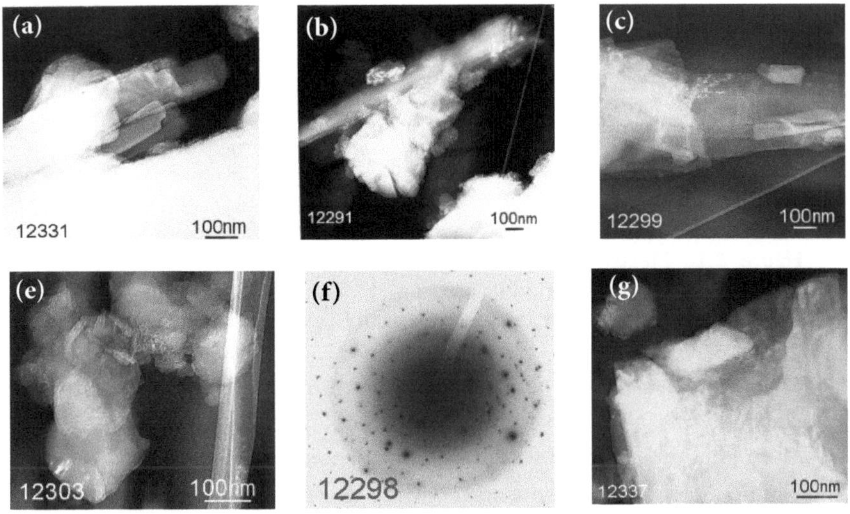

Figure 8.14 Electron microscopic images of wollastonite particles after UST in a 10% butanol solution in water for 20 min (a), 40 min (b), and 60 min (c) at 40 kHz and at 100 kHz for 40 min (d, e) and 60 min (f).

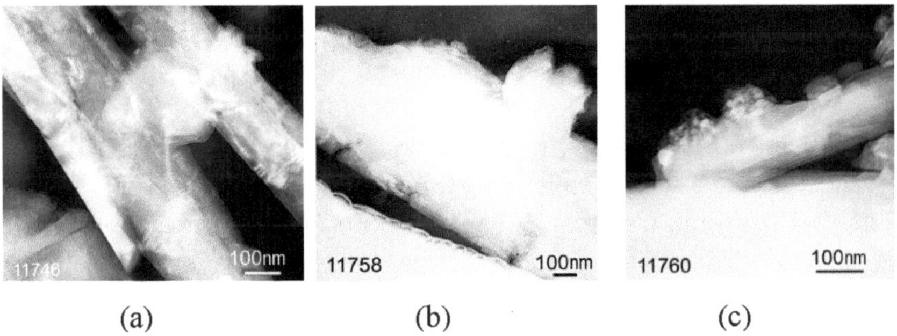

Figure 8.15 Electron microscopic images of wollastonite particles after UST in 10% glycerin in water for 40 min at 40 kHz (a), 100 kHz (b) and combined effects of 40 and 100 kHz (c).

Ultrasonic treatment in the presence of urea leads to greater structuring of the surface layer of particles (Fig. 8.16), aggregation of fine particles and transformation of the structure of their surface with prolonged action of the electron beam on the object of study.

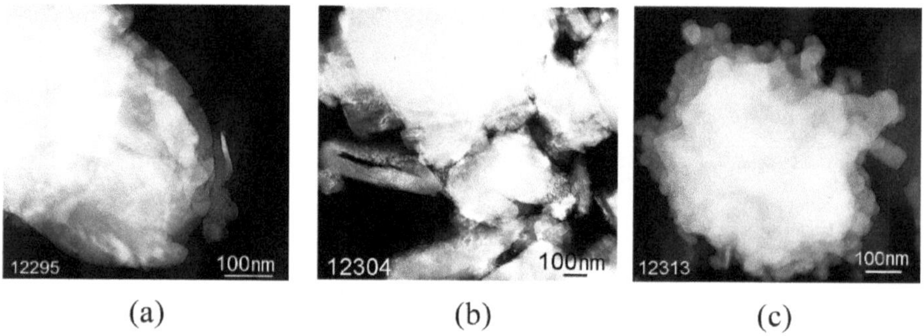

Figure 8.16 Electron microscopic images of wollastonite particles after UST in a 10% urea solution in water at 40 kHz for 40 min (a), 60 min (b), and with prolonged electron beam exposure (c).

After the UST, X-ray structural studies of wollastonite were carried out, and the change in defects and crystallite size was determined depending on the ultrasonic treatment modes (Table 8.2). It has been stated that the maximum change in the defective structure, and, consequently, the activity of wollastonite particles, falls on 40 min of a UST with simultaneous exposure to low and high frequencies and in the presence of butanol in water. In comparison with the samples after the MCT, during ultrasonic processing, the imperfection of particles is less pronounced, but anisotropy is more pronounced in the deformation and accumulation of defects along different planes of the crystal lattice. Microdistortion is more fixed in the plane 400. Wollastonite needle-like particles split along this plane.

Ultrasonic treatment of wollastonite in the presence of 10% glycerin leads to a significant increase in the size of the crystallites of the treated particles (Table 8.3) at the same time with a significant degree of lattice distortion compared with particles treated simply in water and in the presence of butanol. These data indicate that the modified particles of wollastonite retain a high degree of imperfection, i.e., high activity. Similar results were obtained when using urea as a modifier. Thus, when modifying, an encapsulation of the energy state of the substance occurs, which can be realized in subsequent chemical reactions.

The change in the state of the wollastonite powder after the UST was estimated by measuring its dielectric constant, since this characteristic is very sensitive to the change in defectiveness and the presence of new structural elements on the surface of the particles. It has been stated that UST in water for 40 min at frequencies of 40 and 100 kHz leads to a slight increase in the dielectric constant of wollastonite powder (Fig. 8.17).

Table 8.2 The size of crystallites and microdistortions of the wollastonite crystal lattice in samples depending on UST conditions: frequency, time and processing medium

Frequency, kHz; UST time, min; modifier	Crystallite size L, Å and microdistortions ε-dist. × 10–5 along various crystallographic planes							
	(400) d = 3.84		(002) d = 3.519		(320) d = 2.978		Average value	
	L, Å	ε-dist. 1×10–5	L, Å	ε-dist. 1×10–5	L, Å	ε-dist. 1×10–5	L, Å	ε-dist. 1×10–5
40 kHz, 40 min H_2O	990	290	1048	255	862	275	967	273
100 kHz 20 min H_2O	998	275	1083	246	905	261	995	261
100 kHz 40 min H_2O	957	316	1057	248	880	270	965	278
100 kHz 60 min H_2O	1084	318	1075	259	882	271	1014	283
100 kHz 80 min H_2O	1011	263	1045	251	886	270	981	261
(40+100) kHz 40 min, H_2O	978	303	1051	256	846	283	958	281
(40+100) kHz 20 min, C_4H_9OH	1026	306	1107	265	848	231	994	267
(40+100) kHz 40 min, C_4H_9OH	964	325	1048	267	863	279	958	290
(40+100) kHz 60 min, C_4H_9OH	996	305	1062	272	867	210	975	262

Table 8.3 The size of crystallites and microdistortion of the crystal lattice of wollastonite after UST in the presence of glycerin, depending on the frequency and treating time

Frequency, kHz; UST time, min	Crystallite size L, Å and microdistortions ε-dist. ×10–5 along various crystallographic planes							
	(400)		(002)		(320)		Average value	
	L, Å	ε-dist. 1×10–5	L, Å	ε-dist. 1×10–5	L, Å	ε-dist. 1×10–5	L, Å	ε-dist. 1×10–5
40 kHz, 20 min	1155	297	1332	266	790	313	1092	292
40 kHz, 40 min	1024	322	1081	292	653	313	919	309
100 kHz, 40 min	1277	274	1167	296	951	282	1131	284
100 kHz, 60 min	1259	319	1344	290	851	315	1151	308
40+100 kHz, 40 min	1330	403	1316	347	786	330	1144	360

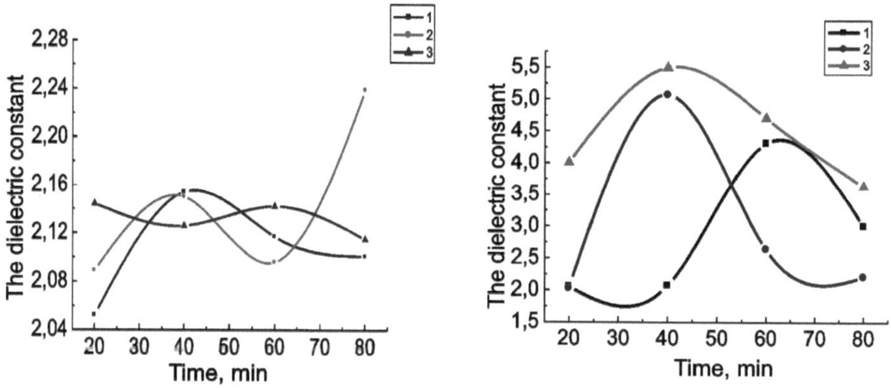

Figure 8.17 The dependence of the dielectric constant of wollastonite on the time of the UST in water (a) and in aqueous solution with 10% glycerol (b). UST frequency 40 (1), 100 (2) and combined effects of 40 and 100 kHz (3).

With an increase in the duration of treatment at different frequencies, the patterns in the change in the dielectric constant differ. And only after simultaneous exposure to different frequencies, the values of the dielectric constant stabilize. After UST in the presence of butanol, the level of values is almost the same as after treatment in water. The presence of glycerin in an aqueous medium leads to a 2–2.5-fold increase in the dielectric constant of wollastonite, especially after co-processing with low and high frequencies (Fig. 8.18).

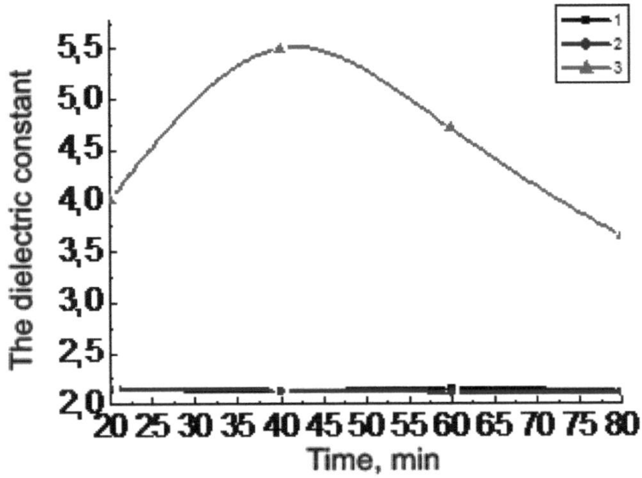

Figure 8.18 Dependence of the dielectric constant of wollastonite on the time of the UST in water (1) and in 10% aqueous solutions of butanol (2) and glycerin (3). UST with combined frequencies of 40 and 100 kHz.

The results indicate an increase in the activity of wollastonite after the UST. The treatment of wollastonite with modifying additives contributes to an increase in the values of dielectric constant. The most significant change in this characteristic occurs with introduction of alcohols into the aqueous medium in which the UST of wollastonite is carried out, due to formation of a polymer layer on the surface of the treated particles.

As in case of MCT, during ultrasonic treatment in the presence of ammonia and urea, the dielectric constant of wollastonite is significantly reduced compared with modified butanol and glycerin (Fig. 8.19). After treatment at a frequency of 40 kHz, the values ε of modified wollastonite vary in the range of 2.10–2.14. As the frequency rises to 100 kHz, the dielectric constant decreases to 2.06–2.03. At joint processing with frequencies of 40 and 100 kHz, intermediate values of ε are obtained, which increase with time of the UST.

The data obtained indicate that the polarization of wollastonite as a dielectric after modification with compounds containing amine groups (liquid ammonia and urea) is due to a different mechanism of polarization of the dipole structures of the substance compared to the modified glycerin.

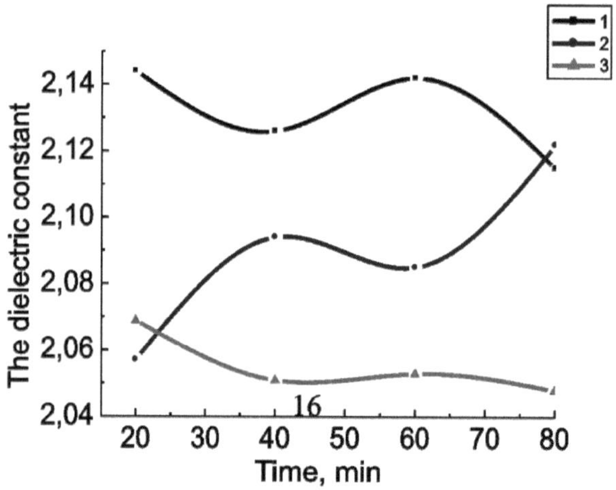

Figure 8.19 Dependence of the dielectric constant of wollastonite on the time of the UST in water (1), in an aqueous solution with 10% ammonia (2) and from 10% urea (3). UST with combined frequencies of 40 and 100 kHz.

The treatment of wollastonite with ultrasound in water with 10% glycerin and 10% urea led to the most significant structural changes in the surface of the particles, which manifested itself in an increase in the dielectric constant (Fig. 8.20). Moreover, the most significant changes are observed after treatment for 80 min at a frequency of 40 kHz.

From the experimental results discussed above, it follows that the choice of modifier, the presence of the optimal amount of it in the aqueous medium, as well as the choice of ultrasonic treatment modes, provide the production of a powder material with particles encapsulated in nanoscale organometallic shells of different composition. Depending on the level of indicators of the properties of modified during UST wollastonite particles when using it as a filler of composite systems, we should expect to obtain a composite, in particular on a polymer base, with the necessary set of properties.

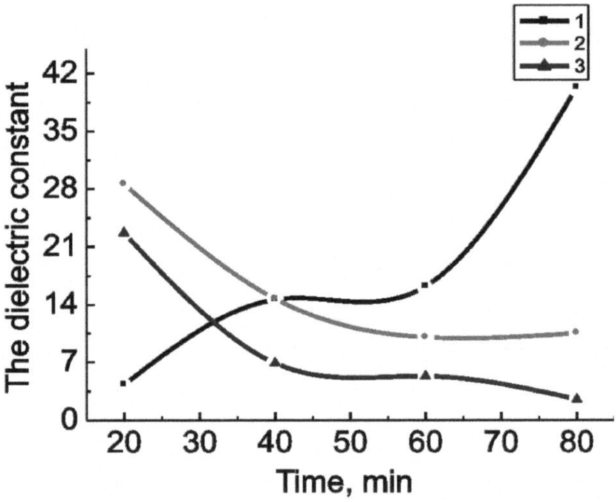

Figure 8.20 Dependence of the dielectric constant of wollastonite on the time of an UST in an aqueous solution with 10% glycerin and 10% urea. The frequency of the UST 40 (1), 100 (2) and 40 + 100 kHz together (3).

Questions

1. What features of changes in bulk density depending on the time of the UST of wollastonite are associated with the state of the surface layer of particles?
2. What is the peculiarity of the modification of wollastonite particles during UST with the participation of glycerin, ammonia, and urea?
3. Describe the lengthwise splitting of wollastonite particles as a result of UST under the action of ultrasound and their high sensitivity to an electron beam with the formation of moiré transitions.
4. What are the features of the structure of the surface layer of wollastonite particles in the result of the UST in the presence of glycerol, liquid ammonia and urea?
5. Describe the changes in the defect structure of wollastonite particles depending on the time and frequency of UST and the use of a modifying additive for encapsulating the energy state of a substance during UST.
6. What is the effect of modifying additives during UST of wollastonite on the dielectric constant of the powder as a result of the formation of a polymer layer on the surface of the treated particles?

7. Give a rationale for the choice of modifier, the optimal amount of it in the aqueous medium under ultrasonic treatment conditions, to obtain a powder material with particles encapsulated in nanoscale organometallic shells of different composition, and with the necessary set of properties.

8.4 Simulation and Quantum-Chemical Calculations of the Formation of Surface Compounds in the Mechanochemical Synthesis of Hybrid Powder Nanocomposite Systems

To predict the structure and properties of wollastonite particles as a result of modifying various organic compounds of particles in the process of mechanochemical synthesis of nanocomposite powder materials of complex composition, inorganic core and organic surface nanostructured layer, the most promising method is using quantum chemical calculations and modeling the process of formation of surface compounds when modifying wollastonite particles.

8.4.1 Methods for Calculating the Electronic Structure of Molecules and Solids, Simulation and Quantum Chemical Calculations of the Formation of Compounds on the Surface of Wollastonite Particles

Quantum mechanics and statistical physics in principle allow us to give an exact mathematical description of any experimentally observed molecular interactions. The basis of modern quantum chemistry is the Schrödinger equation for stationary states of a many-electron system. Usually, an adiabatic approximation is used, i.e., they assume that the nuclei are fixed and one can solve equations only for the motion of electrons. The task of calculating the electron levels of molecules is reduced to solving the single-electron Schrödinger equation. However, even in this case, the Schrödinger equation can be solved exactly for one-electron systems only. Therefore, various semi-empirical and non-empirical methods are used in quantum chemical calculations. The most common of these is the self-consistent field method (SCF) or the Hartree–Fock method. In this method, it is assumed that each electron moves in the field of atomic nuclei and the effective averaged field of other electrons.

Semi-empirical calculations are currently carried out in the CNDO, INDO and NDDO valence approximations [21]. In these approximations, the calculation is carried out only for valence electrons, and the electrons of the inner shells are included in the skeleton of the molecule; the minimum basis is used; a significant part of the Coulomb integrals is neglected. The inaccuracy of the calculation can be partially compensated by the successful selection of parameters.

In recent years, the most widely used of the semi-empirical methods are MNDO-like methods (NDDO approximation), which include MNDO, AM1 [21, 22], PM3, PM6 [23, 24]. All these methods are slightly different from each other and give approximately the same (quite satisfactory) results. A detailed overview of the application of MNDO-like methods to various problems is given, for example, in [25]. The peculiarity of the AM1 method is a somewhat better description of intermolecular interactions, whereas in the PM6 method, parameters are developed for a larger number of elements, including non-transition metals [26, 27]. Computational methods of quantum chemistry allow us to calculate many physicochemical characteristics of molecules: bond lengths and bond angles, dipole moments, ionization potentials, energy (heat) of formation, force constants, etc. [28].

Many problems so far are not amenable to solution by ab initio methods even after their ultimate simplification [25]. It should be especially noted that due to limited machine time, most non-empirical calculations are possible only in small and medium-sized bases. But even if the calculation is possible in a sufficiently large basis, the exact solution of the Schrödinger equation will not be found, but only its solution in the Hartree–Fock approximation. Thus, non-empirical methods are still approximate. Modern non-empirical methods are used in the calculations of Gauss AO (basic sets of Gaussian-type orbitals), while it is possible to select basic sets. Each equivalent Slater autonomous area is represented by several Gaussian electron density distribution functions, which saves machine time when calculating one- and two-center integrals. In non-empirical methods, it is possible to calculate the second-, third-, and fourth-order correlation corrections of the perturbation theory ("MP2," "MP3," or "MP4"). The values of the correlation corrections depend on the quality of the basis set in which the SCF procedure is performed, that is, on the method of describing the virtual orbitals in the original wave function.

Density functional theory (DFT), i.e., the method of calculating the electronic structure of many-particle systems in quantum physics and quantum chemistry is currently used to calculate the electronic structure of molecules and the condensed state of matter [29]. This method is one of the most widely used and universal methods in computational physics and computational quantum chemistry. The density functional theory method largely solves the problem of calculating systems involving a large number of particles by reducing the problem of a many-body system with the potential of electron–electron interaction to a single-particle problem.

The main problem associated with the density functional theory method is that exact analytical expressions for exchange and correlation functionals are known only for the particular case of a gas of free electrons. Nevertheless, the existing approximations allow us to calculate a number of physical quantities with sufficient accuracy. In physical applications, the local density approximation is most common, in which it is assumed that the functional calculated for a certain point in space depends only on the density at that point.

In calculations for solid-state physics, the local density approximation with the basis of plane waves is still widely used. The calculations of the electronic structure of molecules require more complex expressions for functionals. In contrast to the density functional method in the quantum theory of a solid, the density functional as applied to molecular systems is constructed in the basis of atomic orbitals approximated by Gaussian functions. Thus, a large number of approximate functionals for calculating the exchange-correlation interaction was developed for problems of chemistry. Some of them contradict the approximation of a spatially homogeneous electron gas, but, nevertheless, in the limiting transition to electron gas, they should be reduced to a local density approximation. In the calculations of quantum chemistry, one of the most common is the type of exchange functional B3LYP [30], based on a hybrid functional, in which the exchange energy is calculated using the exact result obtained by the Hartree–Fock method. To calculate the electronic states of molecules and radicals of modified wollastonite, the density functional method in the B3LYP version was used.

All the above-described quantum-chemical methods of molecular modeling are implemented in the software packages Gaussian 09, Mopac 2012, Gamess, which were used in the performance of this work. The calculations were performed on a multiprocessor computer and on a computing cluster of the KazNTU supercomputer. Quantum-chemical calculations of transition states in the elementary stages of reactions have become the usual method for determining activation energies and reaction rate constants. In combination with the solutions of kinetic equations, quantum chemical calculations make it possible to refine the mechanisms of chemical reactions, to determine intermediate and final products, to estimate experimentally undefined constants. The choice of one or another kinetic scheme of chemical reactions, which are most adequate to the set of experimental data, largely depends on the accuracy of the results of quantum chemical methods.

Wollastonite is a natural calcium silicate with the molecular formula $CaSiO_3$ (48.27% CaO, 51.73% SiO_2). It refers to the family of pyroxenoids of the subclass of chain silicates and in its crystal structure infinite silicon-oxygen chains with a period of repeatability of three tetrahedrons are connected by columns of calcium atoms, which are in two non-equivalent structural positions with distorted octahedral coordination. The repeatability of this structure determines its chain [31]. Due to this structure, the crystals of wollastonite acquire needle-shaped particles.

The structural features of wollastonite are schematically shown in Fig. 8.21. The structure is presented by chains of silicon-oxygen tetrahedra [SiO_2], which stretch along the axis and have a repeatability period equal to three tetrahedra [32]. In this case, two tetrahedra are directed in one direction, and the third—in the opposite direction. The main structural feature of wollastonite is the coordination of the periods of two large CaO_6 octahedra and three SiO_4 tetrahedra: The diortho group $[Si_2O_7]$ is linked with every second octahedron, and a bridge tetrahedron is thrown over the odd octahedron. In addition, each Ca atom is associated with two oxygen atoms belonging to tetrahedra located directly above and below the Ca atom [33]. Cationic ribbons formed by octahedra are a stable element of the wollastonite structure.

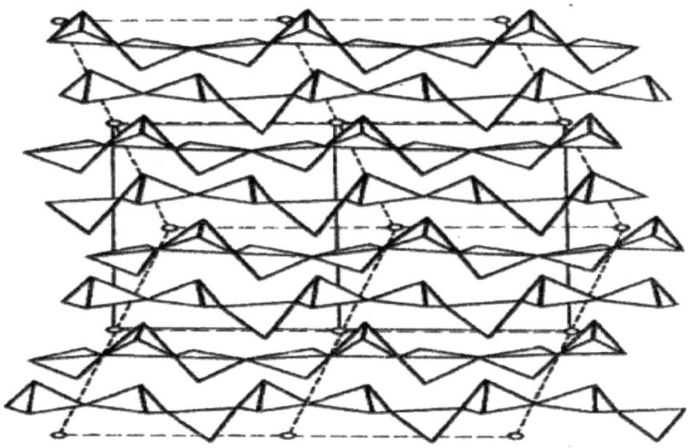

Figure 8.21 Scheme for the construction of wollastonite from silicon-oxygen chains. Wollastonite triclinic grids are shown by a dotted line [32].

Simulation of wollastonite for calculations was carried out from known crystallographic structural data. Figure 8.22 shows a fragment of the crystal structure of wollastonite and the main structural unit of the crystal lattice formed by silicon-oxygen tetrahedra.

For calculations, we chose minimal quasi-one-dimensional clusters consisting of three and six articulated silicon-oxygen tetrahedra with calcium atoms and free surface hydroxyl groups (Fig. 8.23).

A geometrically homogeneous silica matrix containing locally arranged fixed centers can be considered as a convenient and relatively accurate model for the study of molecular and chemical interactions. The presence of calcium atoms in the selected models corresponds to the generally accepted structure of wollastonite

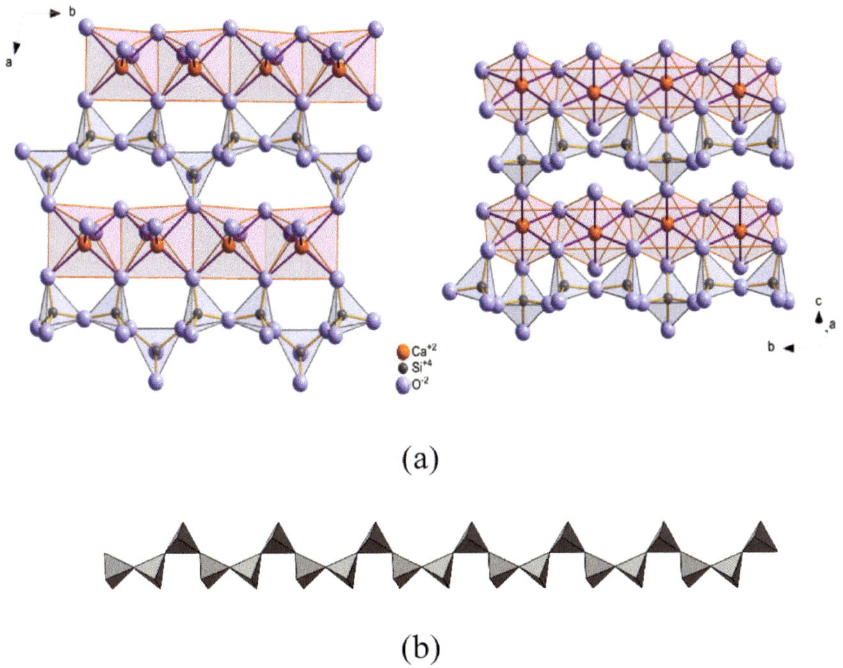

Figure 8.22 Structure of wollastonite in the directions of axes a and c (a) and its silica-oxygen chain (b).

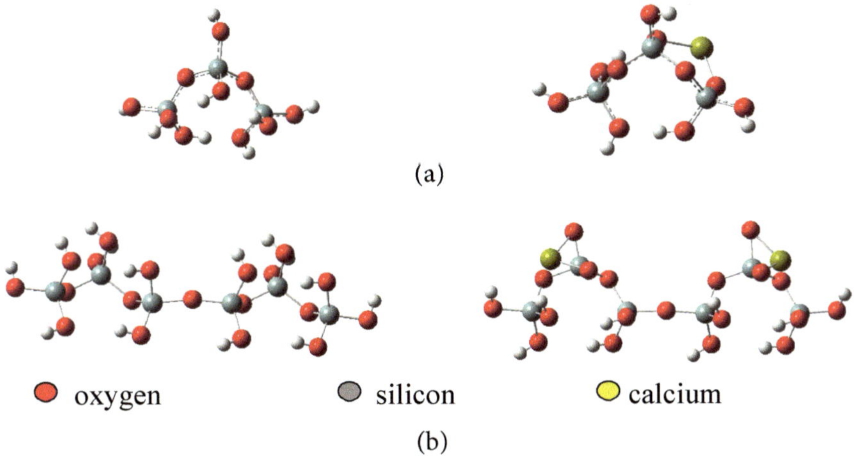

Figure 8.23 Silica-oxygen tetrahedra corresponding to one (a) and two (b) repetition periods in the wollastonite chain.

and its chemical composition. In quantum chemical calculations, the interatomic distances and bond angles corresponding to the crystallographic data were used as the initial approximation. The results of calculations of the energy of formation of surface complexes by the PM6 and B3LYP methods for three and six articulated silicon-oxygen tetrahedra turned out to be almost identical, this allowing us to use a minimal three-membered silicon-oxygen cluster with one repeatability period in modeling the interaction of adsorbents with the surface.

In studies of the reactivity of molecules interacting with fixed centers, it is very important to have reliable information about the nature and geometrical arrangement of the active groups on the surface of a solid. Quite often, even the same type of chemical reactions involving certain surface centers are carried out in completely different conditions. This may be due to the peculiarities of the electronic and geometric structure of the modifying reagents, the presence of electron and proton donor molecules in the reaction volume, the degree of hydration and hydroxylation of the surface of the initial silica, diffusion and other factors.

In the structure of wollastonite, silicon, oxygen and hydrogen atoms of siloxane bonds can act as reactive surface centers. The active centers are also surface hydroxyl groups, calcium ions and, in some cases, siloxane bridges. The concepts of the nature of the active centers, the structure of the hydrate cover and the surface properties of silicon-containing compounds were formed on the basis of numerous spectral and adsorption data, systematic studies of the activity of various groups in chemical reactions. The most significant advances in the study of the surface properties of such compounds are associated with a wide use of spectral methods for studying surface phenomena, especially vibrational spectroscopy [34, 35].

8.4.2 Adsorption Complexes of Butanol and Urea on the Surface of Silica and Wollastonite

As shown in Fig 8.21, wollastonite $CaSiO_3$ structure consists of 3D network of $[SiO_4]$ tetrahedra and $[CaO_6]$ octahedra. The directional change in the surface properties of dispersed silicas, which are widely used for solving many practical problems and in scientific research, is a very urgent and important problem. As a result of chemical surface modification, it was possible to create new practically important highly specific adsorbents, selective catalysts, active fillers of polymers, effective thickeners of dispersion media [36]. The possibilities of mechanochemical synthesis are even more extensive. In the process of grinding powders of inorganic compounds in the presence of modifying organic additives: linear and branched aliphatic alcohols, bifunctional molecules (amino alcohols, hydroxy acids, hydroxyperoxides, polyhydric alcohols), aromatic alcohols—various organic substances are formed on the surface of the particles.

Thus, the formation of hydrogen bonds between electronegative atoms or electrons of the molecules of the adsorbed substance and hydrogen atoms of silanol groups on the surface of silica plays the main role in the process of adsorption of molecules. The interaction of butanol molecules and amides with the surface of wollastonite can also be carried out via formation of hydrogen bonds with silanol groups or via coordination with the participation of calcium ions.

The studies show that adsorption of polar molecules of compounds with the help of H-bonds occurs most strongly on the surface silanol groups that are not connected by hydrogen bonds with neighboring OH groups. The strength of the hydrogen bond is determined by the offset of the IR absorption band related to the stretching vibrations of isolated OH groups. Stronger hydrogen bonds are usually formed by weaker bases, for example, with a nitrogen atom in a urea molecule, and weaker ones, with an oxygen atom of a butanol molecule. There are cases when there exist several centers in which the molecule is attached via hydrogen bonds to the surface. This situation is most characteristic of urea molecules, which has several potentially active groups in its composition. Such multiple hydrogen bonded surface complexes are much more stable. The most favorable molecular form during the adsorption of a substance on the surface of silica is precisely that which provides the formation of the greatest number of hydrogen bonds. Butanol molecules with the OH group can enter into a hydrogen bond, both as a donor and as a proton acceptor.

As a wollastonite powder containing a small amount of quartz was used in the work, processes on the surface of wollastonite and quartz particles were simultaneously considered when solving the problem of modeling the surface modification of particles used in the experiment powder.

To simulate the adsorption of butanol and urea on the surface of silica and wollastonite, quantum chemical calculations of the interaction of molecules with adsorption centers were carried out. The calculations were performed with full optimization of the geometry of model clusters using the B3LYP density functional method with the exchange potential in the 6–31/G basis set. The starting geometry of the simulated systems was obtained by the semi-empirical method PM6 [37, 38]. The calculated optimal values of bond lengths, valent, and dihedral angles are in good agreement with the experimental values of silica and wollastonite. The interaction energy was calculated by the formula

$$\Delta E = E_{AB} - (E_A + E_B).$$

Figures 8.24 and 8.25 show the cluster models of the adsorption of butanol on the surface of silica and wollastonite by the type of hydrogen bond with a single silanol group. As a result of the calculations, it was found that butanol binds more strongly to the surface when the hydrogen atom of the silanol group is the donor atom of the hydrogen bond.

The interaction energy reaches –14.93 kcal/mol. If the donor atom of the hydrogen bond is the hydrogen atom of the hydroxyl group of butanol, then the interaction energy is –8.66 kcal/mol.

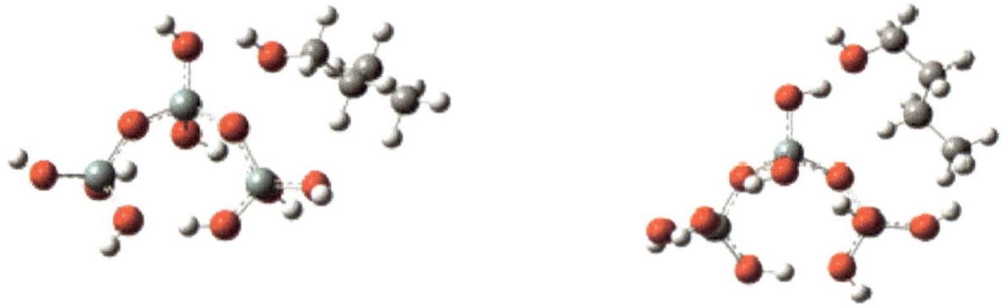

Figure 8.24 Models of adsorption of butanol on the surface of silica by the type of hydrogen bond.

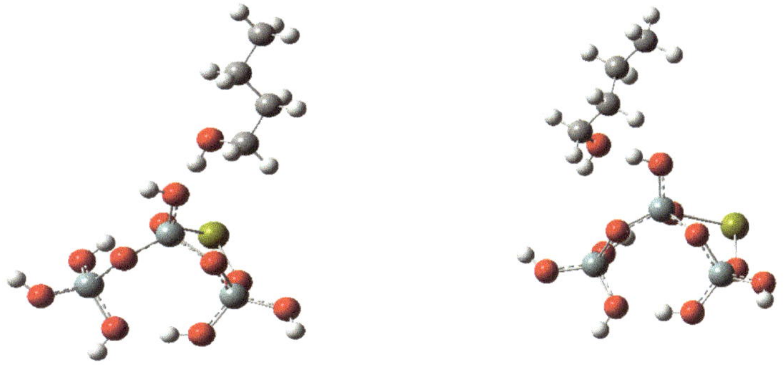

Figure 8.25 Models of adsorption of butanol on the surface of wollastonite by the type of hydrogen bond.

The presence of calcium ions in the vicinity of the silanol group prevents the formation of hydrogen bonds. As a criterion of the donor ability of the proton of the silanol group, its effective electrostatic positive charge can serve. According to the calculations, the charge on the hydrogen atom of a single silanol group is 0.426 AU in silica, and 0.405 AU in wollastonite, which does not result in energy gain during adsorption—the interaction energies take positive values –0.502 and 3.2 kcal/mole, respectively.

The direct interaction of butanol molecules with calcium ions leads to energy gain of up to –42.42 kcal/mol, i.e., butanol coordination by surface calcium ions is much stronger than via hydrogen bonding (Fig. 8.26).

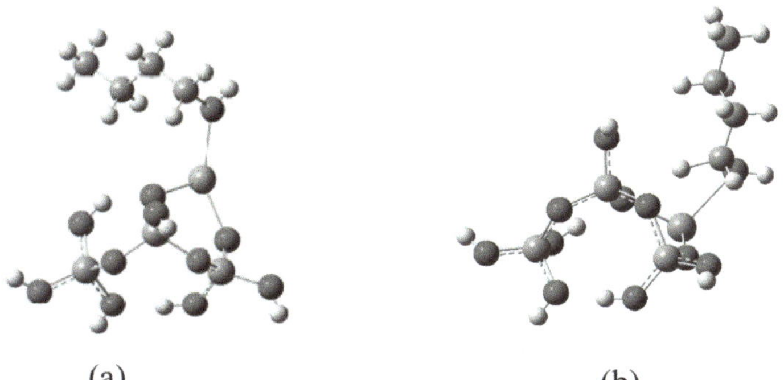

(a) (b)

Figure 8.26 Model of the interaction of butanol molecules with calcium ions on the surface of wollastonite: single point (a), two-point adsorption (b).

Even more durable binding of butanol occurs during the two-point coordination of the molecule—an oxygen atom to a calcium ion and a hydrogen atom to a bridging oxygen of siloxane bonds. In this case, the binding energy reaches −47.4 kcal/mol. This circumstance is important for explaining the effect of hydrophobization of the surface of wollastonite powders after mechanochemical activation in the presence of butanol. Strongly bound alcohol molecules form the hydrophobic shell of wollastonite particles.

Butanol and urea molecules differ in their donor-acceptor ability to form hydrogen bonds. Urea molecules predominantly act as a donor of hydrogen bond protons. Figures 8.27 and 8.28 show models of the interaction of urea molecules with the surface of silica and wollastonite with the formation of hydrogen bonds. For the surface of silica, the interaction energy of urea with silanol groups is −9.60 and 14.18 kcal/mol, which practically coincides with the energies of interaction of butanol with the surface of silica on the type of hydrogen bond.

The peculiarity of the urea molecule is that its structure contains a carbonyl group along with two amino groups. This circumstance determines the nature of the interaction of urea molecules with the surface of wollastonite—the electronegative oxygen atom of the carbonyl group is coordinated by positive calcium ions. In accordance with this, the energy of interaction of urea with the surface of wollastonite far exceeds the corresponding energy for butanol.

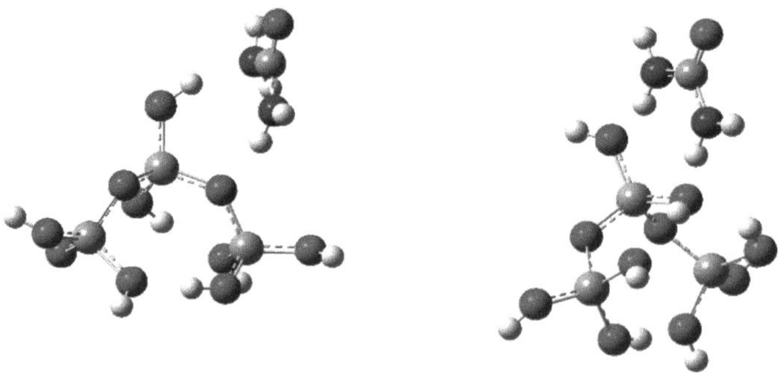

Figure 8.27 Models of urea adsorption on the surface of silica by the type of hydrogen bond.

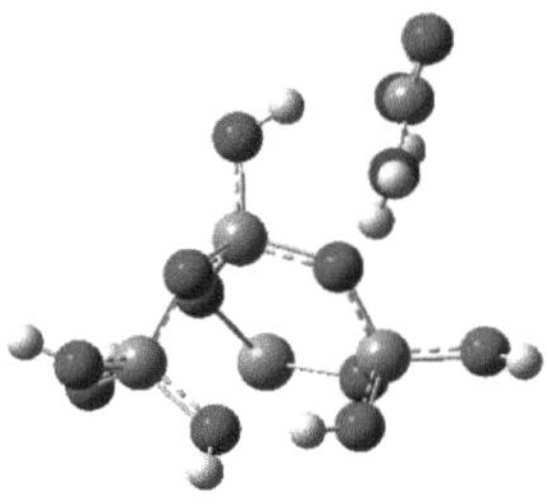
(a)

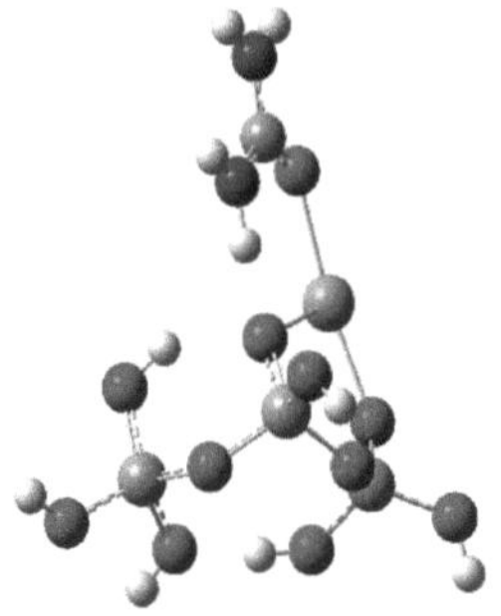
(b)

Figure 8.28 Models of urea adsorption on the surface of wollastonite by the type of hydrogen bond (a) and with the participation of calcium ion (b).

Thus, the adsorption energy for wollastonite is –20.08 and –61.99 kcal/mol. Such a high interaction energy is primarily due to formation of ionic C–O–Ca bond. The proton of the amino group forms a hydrogen bond with the bridging oxygen of the surface, which makes an additional contribution to the total interaction energy. In all the cases considered, the urea molecule tends to form two-center chemisorption bonds with the surface of silica and wollastonite.

Most reactions involving the active centers of the silica surface are heterolytic processes. The chemisorption of alcohols and amides is the simplest one-step method of introducing the corresponding surface ether and urethane compounds of a given structure. From an analysis of the known basic laws of reactions occurring in the surface layer of dispersed silicon dioxide, it follows that the rate of interaction is largely determined by the values of the effective charges on the atoms of the transition complex, including a fixed center and an attacking molecule. If the oxygen atom of the silanol group is attacked by a reagent, which contains a strong proton acceptor, the chemical reaction takes place under milder conditions.

To obtain complete information about the reaction mechanism, it is necessary to calculate the multidimensional potential energy surface (PES), i.e., calculate the dependence of the total energy on the coordinates of atomic nuclei. The most interesting and important in studying the reaction mechanism are stationary points on the PES—minima and saddle points, in which the derivatives of the total energy over all independent coordinates are zero. At the point of minimum of the total energy, the matrix of second derivatives has only positive eigenvalues, and at the saddle point it has one negative eigenvalue. The minima of the total energy correspond to stable structures and intermediates, and the saddle points correspond to transition states. The main difficulty in calculating the PES for organic reactions of polyatomic molecules is the need to calculate the total energy of the reactants for a very large number of points, which seems almost impossible. Therefore, when constructing PES, not all independent degrees of freedom usually vary, but only some of them, most often one or two geometrical parameters, which change most strongly during the reaction, and carry out optimization for all other degrees of freedom, i.e., find their optimal values (corresponding to the minimum total energy) for each PES point (the so-called relaxation scanning method). The transition from varying to optimizing for most of the degrees of freedom in constructing PES makes it possible to greatly reduce the cost of machine time and makes the calculation of PES practically possible for reactions involving fairly large organic compounds [38].

By the semi-empirical method of PM6, one-dimensional and two-dimensional PES sections were calculated for the reaction of electrophilic addition of butanol to the surface of silica. The reaction of the surface esterification of silica and wollastonite was simulated in accordance with the equation:

$$-Si-OH + C_4H_9-OH \rightarrow -Si-O-C_4H_9 + H_2O$$

Figure 8.29 presents the results of calculations of the interaction of butanol in the esterification process with the surface silanol group of silica and wollastonite with the formation of a free water molecule.

The process of electrophilic addition can be considered initially as a stage of electrophilic replacement of a proton in the silanol group of the surface with the formation of a bimolecular transition state having the structure of a trigonal bipyramid with a silicon atom in $3sp^3$ d hybridization. One of the vertices of the trigonal bipyramid is occupied by the oxygen atom of the silanol group of the surface, which uses one of its lone electron pairs to form the bond. The exclusion of the OH group from this complex returns the silicon atom to the sp3 state with the formation of a stable product bound to the surface by a siloxane bridge.

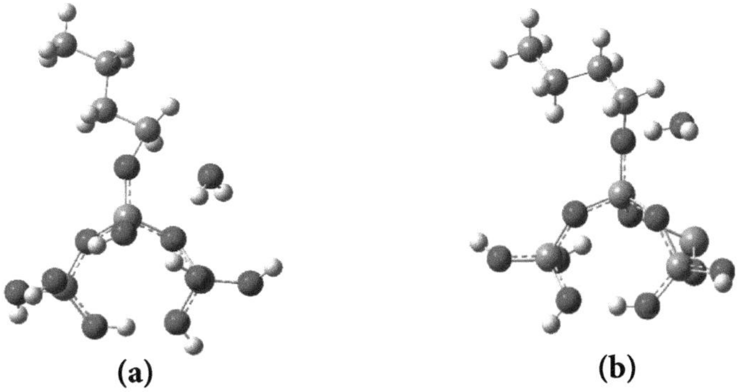

Figure 8.29 Models of the interaction of butanol in the process of esterification on the surface of silica (a) and wollastonite (b).

Figure 8.30 presents the results of calculations of PES for the coordinates of the esterification reaction. The reaction length is the bond length of the OH silanol group.

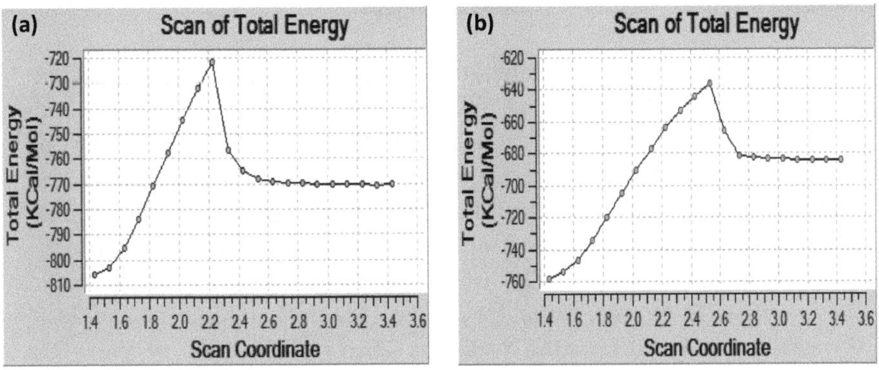

Figure 8.30 PES for the coordinates of the esterification reaction on the surface of silicon dioxide (a) and on the surface of wollastonite (b).

Calculations show that as the selected reaction coordinate increases, the proton of the silanol group approaches the hydroxyl group of butanol with the formation of intermediate $C_4H_9ON_2^+$. The subsequent detachment of the H_2O molecule leads to the direct interaction of the butyl group with the surface of silica with the formation of surface ether. The reaction mechanism for silica and wollastonite is the same, but the presence of calcium cation increases the activation energy from 84.3 kcal/mol to 121.9 kcal/mol. This result is shown in Fig. 8.31. It is consistent with a decrease in the donor activity of the proton of the silanol group in wollastonite, as was shown above.

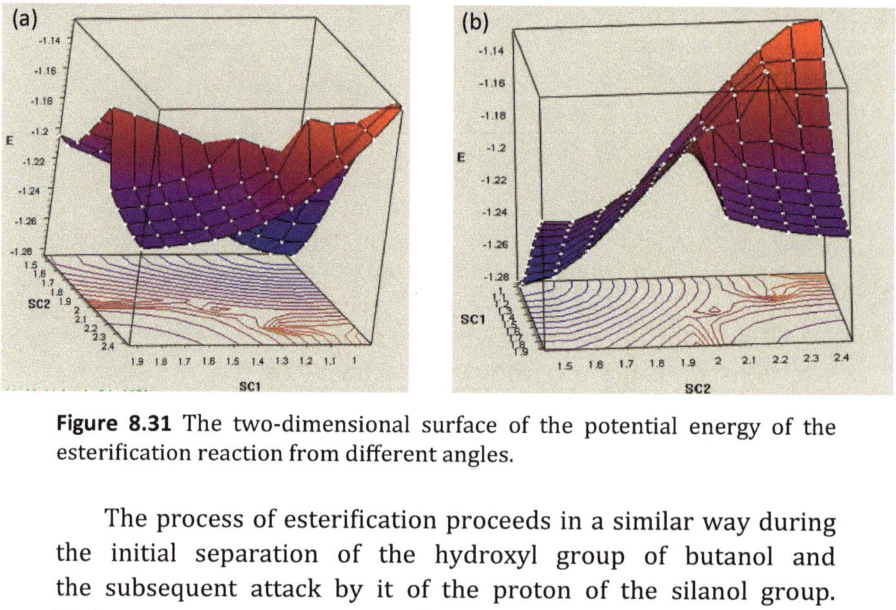

Figure 8.31 The two-dimensional surface of the potential energy of the esterification reaction from different angles.

The process of esterification proceeds in a similar way during the initial separation of the hydroxyl group of butanol and the subsequent attack by it of the proton of the silanol group. Both processes can proceed synchronously. Calculations of two-dimensional PES show the presence of several local maxima corresponding to transition states or intermediates during surface esterification. The minimum values on the energy scale correspond to the initial products (the left branch of the PES curve), and the reaction products—the right branch of the PES. The transition from the initial to the final products is accompanied by the overcoming of the potential barrier, the height of which corresponds to the activation energy. Different cross sections of multidimensional PES correspond to the results of the calculations presented in Fig. 8.31.

Another group of substitution or addition reactions proceeds with an attack of the nucleophilic reagent on the silicon atom of the silica surface. At the stage that determines the reaction rate in accordance with the experimental data, quasicyclic (mainly four-center) transition complexes are formed, followed by destruction of the existing bonds and formation of new compounds. In this case, the modeling of nucleophilic addition reactions was not performed. However, a preliminary assessment of the formation of surface ether by the mechanism of nucleophilic attachment shows that this mechanism is energetically more preferable in the presence of calcium cations. The energy gain in this case reaches about 13 kcal/mol.

The experimental data on the interaction of urea with silica, wollastonite and aluminosilicates are represented slightly. It is known that the reactions of urea with acids lead to the formation of urethates. Compounds with silicon oxide, including silica, are often considered as compounds exhibiting weak acidity. In this case, the reaction of interaction of urea with silica proceeds according to the equation

$$-Si-OH + (NH_2)_2CO \rightarrow Si-O-CONH_2 + NH_3$$

Figure 8.32 presents the results on calculations of the final products of the interaction of urea with silica with the formation of surface urethane and free ammonia and the PES scan curve using the PM6 method. The length of the C–NH_2 bond was chosen as the scan coordinate (reaction coordinate). The positions of the remaining atoms were automatically optimized.

The transition from the initial to the final products occurs with overcoming the potential barrier (activation energy) height of 51 kcal/mol. Under the conditions of mechanochemical activation, this barrier can be easily overcome.

The results of calculations of the reaction of the interaction of urea with wollastonite show that in the presence of calcium the formation energy of the reaction products is 0.18 kcal/mol, which is significantly lower than the energy of formation of the same products on the silica surface –5.96 kcal/mol.

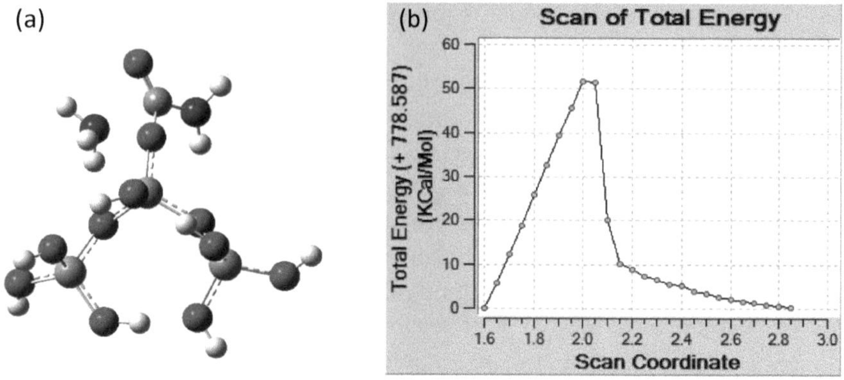

Figure 8.32 Model of the interaction of urea (a) and PES (b) for the coordinates of the reaction of ureterization on the surface of silica.

The mechanism of the reaction of ureterization is close to the mechanism of the reaction of esterification. Both reactions can proceed either with separation of the proton of the silanol group with formation of a (intermediate) charged intermediate, or with the separation of the OH group of butanol or NH_2 group of urea and their subsequent attack by the proton of the silanol group with the subsequent detachment of H_2O or NH_3 molecules and the formation of surface products.

Thus, the results of quantum chemical modeling and calculations of the surface compounds of butanol and urea indicate that these compounds are easily adsorbed on the surface of silica and wollastonite nanopowders both via the formation of hydrogen bonds and coordination with calcium cations. Mechanochemical activation favors the course of surface esterification and ureterization reactions with formation of an organic nanostructured layer.

Questions

1. Give the Schrödinger equation and other semi-empirical and non-empirical methods for the quantum chemical description of the stationary states of a multielectronic system.
2. What is the method of semi-empirical calculations in the valence approximations of CNDO, INDO and NDDO, AM1, PM3, PM6?

3. Describe the method of calculating the electronic structure (density functional) of particle systems in quantum physics and quantum chemistry to calculate the electronic structure of molecules and the condensed state of matter.
4. Simulation of wollastonite, as the main structural unit of the crystal lattice formed by silicon-oxygen tetrahedra.
5. What are the reactive surface centers in the structure of wollastonite that provide its high chemical activity?
6. What is the role of hydrogen bonds in the process of adsorption of molecules on the surface of silica and wollastonite?
7. How is the formation of strong hydrogen bonds through the nitrogen atom and weak ones through the oxygen atoms of alcohol molecules on the surface of silica and wollastonite carried out?
8. How do butanol molecules enter into a hydrogen bond with the OH group - as a donor, or as a proton acceptor on the silica surface?
9. Provide quantum-chemical calculations of the interaction of butanol and urea molecules with adsorption centers on the surface of silica and wollastonite.
10. How is the two-point coordination of butanol molecules with wollastonite calcium ions during the formation of the system bridged oxygen and siloxane bond, providing the effect of hydrophobization of the surface of wollastonite powders after mechanochemical activation in the presence of butanol?
11. What models exists of interaction of urea molecules with the surface of silica and wollastonite with the formation of hydrogen bonds?
12. Give the calculation of the energy of the esterification reaction on the surface of silicon dioxide and wollastonite and increased activation energy in the presence of calcium atoms.
13. Give the calculation of the products of interaction of urea with silica and wollastonite with the formation of surface urethane and free ammonia using the PM6 method.
14. What are the general provisions in the mechanism of the reaction of ureterization and esterification, which determine the effectiveness of the adsorption of butanol and urea on the surface of silica and wollastonite powders during mechanical action?

References

1. Novitsky B. G. *The Use of Acoustic Vibrations in Chemical-Technological Processes*. Chemistry, Moscow, 1983. 191 p. (In Russian).

2. Margulis M. A. *Basics of Sound Chemistry*. Chemistry, Moscow, 1984. 260 p. (In Russian).
3. Ginevsky A. S. *Aeroacoustic Interaction*. Mashinostroenie, Moscow, 1978. 178 p. (In Russian).
4. Kardashev G. A. *Physical Methods of Intensification of Chemical Technology Processes*. Chemistry, Moscow, 1990. 206 p. (In Russian).
5. Akulichev B. A. *Powerful ultrasonic fields*. ed. Rosenberg, L. D. Science, Moscow, 1968, Part 4. pp. 129–166. (In Russian).
6. Gershgal D. A., Fridman V. M. *Ultrasonic Technological Equipment*. Energy, Moscow, 1974. 178 p. (In Russian).
7. Goodwin G. *Ultrasound Equipment. Chemistry and Ultrasound/Trans. from English*. Science, Moscow, 1993. 161 p. (In Russian).
8. Mamaev Yu. A., Khrunina N. P. Experimental studies of the effectiveness of ultrasonic treatment of sand-clay gold-bearing rock after mechanical activation. *Mining Informational Anal. Bull.*, 2007. OB 9. pp. 357–367. (In Russian).
9. Urazovsky S. S., Polotsky I. G. On the dispersion by ultrasound. *Colloid J.*, 1940. no. 9. pp. 350–356. (In Russian).
10. Meskin P. E. Hydrothermal synthesis of highly dispersed powders based on oxides of titanium, zirconium and hafnium using ultrasonic and microwave effects: dissertation of Cand. chemical Sciences: 02.00.01./ Moscow Institute of Technology. Moscow. 162 p. (In Russian).
11. Flynn G. Physics of acoustic cavitation in liquids. *Physical Acoustics.*, ed., Meson W. Mir, Moscow, 1967. vol. 1, P. B. 138 p. (In Russian).
12. Pol V. G., Reisfeld R., Gedanken A. Sonochemical synthesis and optical properties of europium oxide nanolayer coated on titania. *Chem. Mater.*, 2002. vol. 14. pp. 392–398.
13. Kolenko Yu. V., Burukhin A. A., Churagulov B. R., Oleynikov N. N., Mukhanov V. A. Synthesis of nanocrystalline powders of various crystal modifications of ZrO_2 and TiO_2 by the hydrothermal method. *J. J. Neorgan. Chemistry*, 2002. vol. 47, no. 11. pp. 1755–1762. (In Russian).
14. Shabanova N. A., Popov V. V., Sarkisov P. D. *Chemistry and Technology of Nanodispersed Oxides*. "Akademkniga", Moscow, 2006. 309 p. (In Russian).
15. *Physical Bases of Ultrasonic Technology: A Collection of Works*. Publishing House of the Academy of Sciences of the USSR, Moscow, 1969. 688 p. (In Russian).

16. Ginberg A. M. *Ultrasound in Chemical and Electrochemical Engineering Processes*. Mashgiz, Moscow, 1962. 136 p. (In Russian).

17. Abramov, O. V., Kharbenko, I. G., Shvegla, S. *Ultrasonic Treatment of Material*. Mashinostroenie, Moscow, 1984. 243 p. (In Russian).

18. Novitsky B. G. *The Use of Acoustic Vibrations in Chemical-Technological Processes*. Chemistry, Moscow, 1983. 191 p. (In Russian).

19. Margulis M. A. *Sound and Chemical Reactions and Sonoluminescence*. Chemistry, Moscow, 1986. 288p. (In Russian).

20. Baranchikov A. E., Ivanov B. K., Tretyakov Yu. D. Sonochemical synthesis of inorganic materials. *Adv. Chem.*, 2007. vol. 76, no. 2. pp. 147–168. (In Russian).

21. Wada Y., Tsukada M., Fujihira M., Matsushige K., Ogawa K. et al. Prospects and problems of single molecule information devices. *Jpn. J. Appl. Phys.*, 2000. vol. 39. Part 1. No. 7A. pp. 3835–3849.

22. Clark T. *Computer Chemistry*: Trans. from English Mir, Moscow, 1990. 383 p. (In Russian).

23. Dewar M. J. S., Thiel W. Ground states of molecules. 38. The MNDO method. Applications and parameters. *J. Am. Chem. Soc.*, 1977. vol. 99, no. 15. pp. 4899–4907.

24. Dewar M. J. S., Thiel W. Ground states of molecules. 39. MNDO results for molecules containing hydrogen, carbon, nitrogen and oxygen. *J. Am. Chem. Soc.*,1977. vol. 99, no. 15. pp. 4907–4917.

25. Vul A. Y. Materials electronic technology, 1999. no. 3. pp. 4. (In Russian).

26. Stewart J. J. P. Optimization of parameters for semiempirical methods I. Method. *J. Comput. Chem.*, 1989. vol. 10, no. 2. pp. 209–220.

27. Stewart J. J. P. Optimization of parameters for semiempirical methods II. Applicatios. *J. Comput. Chem.*, 1989. vol. 10, no. 2. pp. 221–264.

28. Stewart J. J. P. A semiempirical molecular orbital program. *J. Comput.-Aided Mol. Des.*, 1990. vol. 4, no. 1. pp. 1–105.

29. Stewart J. J. P. Optimization of parameters for semiempirical methods III. Extensions of PM3 to Be, Mg, Zn, Ga, Ge, As, Se, Cd, In, Sn, Sb, Te, Hg, Tl, Pb, and Bi. *J. Comput. Chem.*, 1991. vol. 12, no. 3. pp. 320–341.

30. Schlegel H. B. Optimisation of equilibrium geometries and transition structures. *J. Com. Chem. Soc.*, 1982. vol. 3. p. 214.

31. Goshkov V. S., Saveliev V. G., Fedorov N. F. *Physical Chemistry of Silicates and Other Refractory Compounds*. Higher School, Moscow, 1988. 400 p. (In Russian).

32. Mamedov Kh. S., Belov N. V. Crystal structure of wollastonite. *DAN SSSR.*, 1956. vol. 107, no. 3. pp. 463–466. (In Russian).
33. Bragg W., Claringbull G. *The Crystal Structure of Minerals.* Mir, Moscow, 1967. 325 p. (In Russian).
34. Little L. *Infrared Spectra of Adsorbed Molecules.* Mir, Moscow, 1969. 516 p. (In Russian).
35. Kiselev A. V., Lygin V. M. *Infrared Spectra of Surface Compounds and Adsorbed Substances.* Science, Moscow, 1972. 460 p. (In Russian).
36. Goshkov V. S., Saveliev V. G., Fedorov N. F. *Physical Chemistry of Silicates and Other Refractory Compounds.* Higher School, Moscow, 1988. 400 p. (In Russian).
37. Bragg U., Claringbull G. *The Crystal Structure of Minerals.* Mir, Moscow, 1967. 325 p. (In Russian).
38. Burshtein K. Ya., Shorygin P. P. *Quantum-Chemical Calculations in Organic Chemistry and Molecular Spectroscopy.* Science, Moscow, 1989. 104 p. (In Russian).

Chapter 9

Mechanochemical Treatment and Modification of Metal Systems

The ample possibilities of mechanochemical processing are illustrated by the fact that refractory substances and intermetallic compounds, inorganic and organic compounds were synthesized without dissolving or melting the reagents, and composite materials were created with the participation of a metal reagent [1–4]. During deformation and destruction of solids, various intermediate energy-rich intermediate states arise first, with the further decomposition of which releases a significant portion of heat. Between work and warmth, there is a so-called "black box"; Moreover, if energy is supplied to it faster than it will be converted into heat, then various and unexpected chemical transformations can be realized (Fig. 9.1) [5].

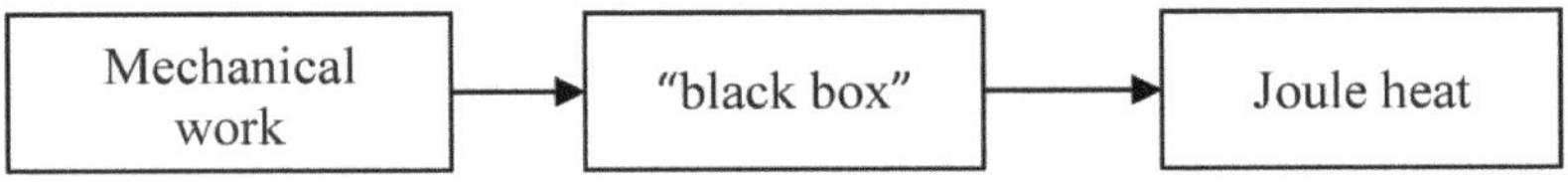

Figure 9.1 Scheme of implementation of mechanical work [5].

In the "black box" (Fig. 9.2), there can be vibrationally and electronically excited bonds, electrons and ions, metastable atomic structures, and other intermediate states, while its content is far from being exhausted.

Mechanochemical Synthesis of Composite Materials
Zulkhair A. Mansurov, Nina N. Mofa, Tlek A. Ketegenov, and Bakhtiyar S. Sadykov

ISBN 978-981-4800-88-4 (Hardcover), 978-1-003-12081-0 (eBook)
www.jennystanford.com

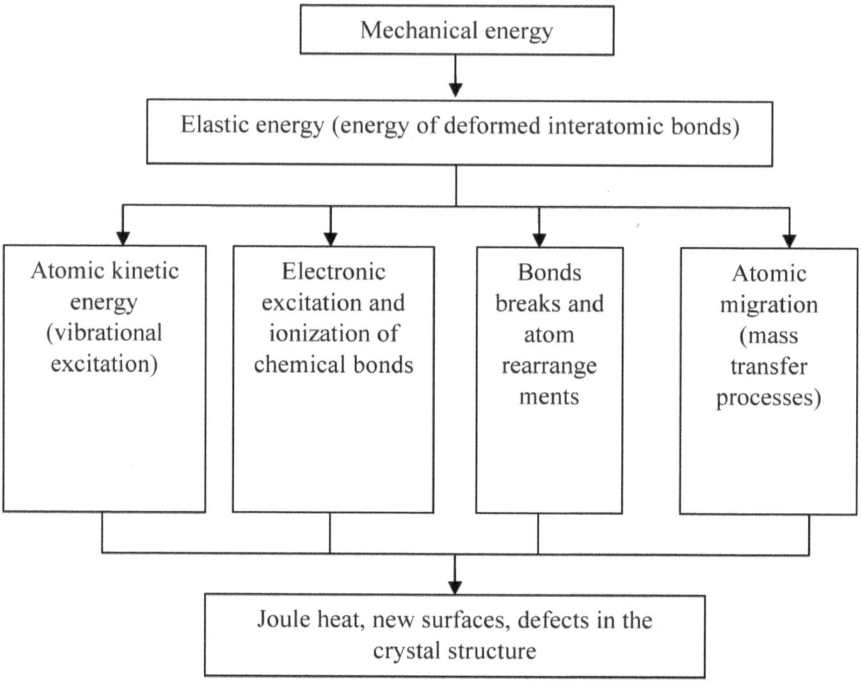

Figure 9.2 Diagram of energy transformation in solids [5].

The given scheme (Fig. 9.2) illustrates the main ways of the emergence and transformation of active nonequilibrium states arising from mechanical effects on solids. Each of the energy conversion channels has its own specific carriers of excess energy with a specific electronic structure, chemical structure, reactivity and relaxation mechanisms. Each state is characterized by a certain energy output and time [5].

Particular attention is paid to grinding of initial reagents to a nanosize scale [6–9]. When processing metal powders, most often, the average size of dispersed materials is not in the nanometric, but in the micron and submicron regions. In this case, the powder particles contain nanograins and can be considered as a non-textured polycrystalline material with a high density of grain boundaries [10]. Often crystalline materials crushed in ball mills have a grain size of particles up to 5–30 nm.

In the brittle-plastic systems, at the initial stage of grinding, particles of plastic metal powder are deformed as a result of a ball–powder–ball interaction, while brittle oxides or intermetallic phases are fragmented and crushed (Fig. 9.3). The course of the doping process in the "fragile-viscous" system depends on solubility in the solid state of the fragile component in the plastic matrix. For example, as a result of the dispersion of pure zirconium Zr (plastic) and $NiZr_2$ (brittle), a metal-oxide system or an intermetallic compound is formed. Starting from the moment of elastic deformation to brittle fracture, from the moment of accumulation of defects to a complete structural reorganization, a change in the energy state of a substance occurs in an ultra-small particle.

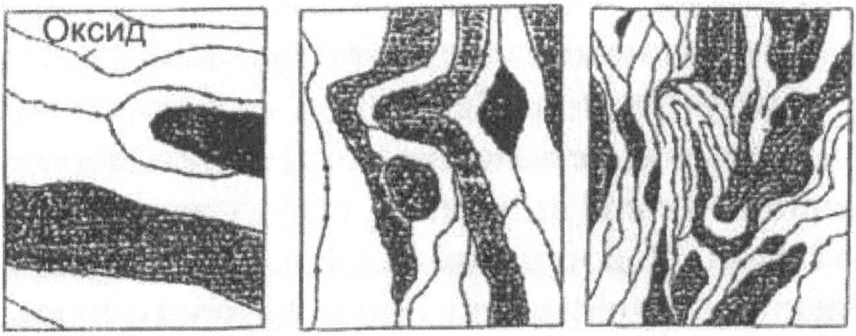

Figure 9.3 The evolution of the structure in the system "plastic-fragile" [11].

In recent decades, much attention has been paid to the use of MCT of metal powders, including that of aluminum [12, 13]. The interest in aluminum powders is due to their use in various energy systems. However, a dense oxide film on the surface of aluminum particles hinders the ignition of the compositions and causes a low burning rate. Various methods are used to remove or reduce the thickness of the oxide film. One of them is mechanochemical treatment [14]. During the mechanical treatment the particle size decreases, and the contact area between the reagents increases, the surface of the particles gets cleaned from the oxide film, defects in the bulk of the material get accumulated, this increasing the chemical activity of the powders. In case of MCT, it is important to select optimal conditions for processing of powders in a mill: time, the ratio of the mass of the balls and the powder, the type and mass fraction of the modifying agent.

9.1 Mechanochemical Treatment of Aluminum Powders with Organic Additives

Mechanical grinding of aluminum is difficult because of its plasticity; therefore, to facilitate the grinding process, surfactants, such as graphite, stearic acid, etc., are added to facilitate the grinding and modification of the surface of metal particles. A wide range of composition systems have been synthesized by the mechano-chemical method [1, 15]. Metal/organic energy composite systems have been developed, for example, such as aluminum/teflon, magnesium/teflon [16].

The choice of an organic additive during mechanochemical treatment (MCT) largely depends on both the degree of dispersion of aluminum particles and the formation of their structural features that determine the degree of activity of metallic reagents in subsequent combustion processes [17–20]. To study the structure and morphology of highly dispersed particles, electron microscopic analysis is usually used. This method allows us to study the change in the surface layers of the treated aluminum particles and visually follow the entire path of creating composite systems with the participation of organic and inorganic materials.

Studies on the MCT of aluminum were carried out on the powder grade PA4. Figure 9.4 presents the electron microscopic image, the energy dispersion spectrum, and the mass fraction of the elements of the original aluminum powder. From the presented picture, it follows that the particles of Al have a spherical shape. According to the particle size analysis, the average particle size is about 50 microns.

From the results of energy dispersive analysis of the initial powder, it follows that the amount of aluminum metal is 89.33 wt.%, and that of oxygen is 10.67 wt.%. The presence of oxygen atoms indicates the presence of a sufficiently dense layer of oxide film on the surface of particles. Powder particles are gray in color with metallic luster and do not contain foreign inclusions.

Mechanochemical treatment of aluminum was carried out with graphite (C), stearic acid ($C_{17}H_{35}COOH$) and polyvinyl alcohol ($(C_2H_3OH)_n$) with different amounts of the modifying additive (from 3% to 20%). The processing time was 20 min. The choice of the optimal time of MCT was determined by the results of previous studies [21]. In order to prevent aluminum particles from oxidizing with oxygen of the air after MCT and to estimate the changes actually associated with the mechanical action, the samples of the dispersed mixture were passivated with hexane (C_6H_{14}).

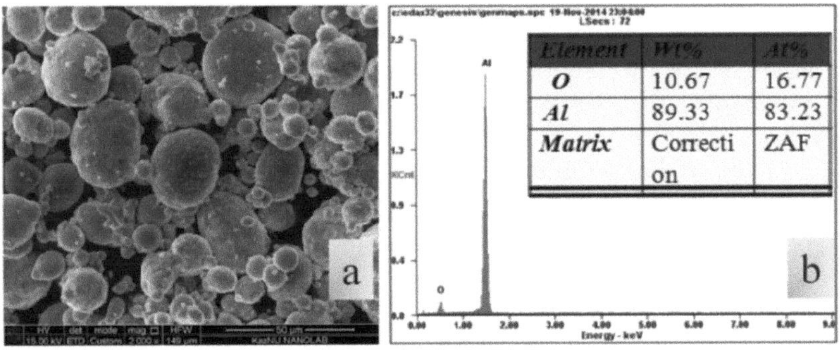

Figure 9.4 Characteristics of the original aluminum powder of grade PA4. (a) Electron microscopic image; (b) energy dispersive spectrum and mass fraction of elements.

After MCT of aluminum with graphite and polyvinyl alcohol, the particles have a lamellar (scaly) shape of various thicknesses. The outer surface is uneven and intersected by cracks. Aluminum particles treated in the presence of stearic acid retain a spherical shape. Figure 9.5 shows the images obtained on a scanning electron microscope of aluminum powder particles after MCT with 3% organic modifier, i.e., Al/C, $Al/(C_2H_3OH)_n$ and $Al/C_{17}H_{35}COOH$ composites. The micrograph shows large particles, crystal fragments, and aggregates of particles. In case of MCT of aluminum with graphite, the crystals of which are more sensitive to mechanical stress than aluminum crystals, during the grinding process, there takes place the change in the surface and formation of the Al/C composite layer structure. When processing with polyvinyl alcohol, aluminum particles agglomerate into granules, which retain the lamellar form and have well-defined cracks (Fig. 9.5b).

As a result of processing aluminum powder with stearic acid (Fig. 9.5c), aluminum particles are encapsulated in a more dense organic film, which prevents their oxidation by air oxygen. Particles of aluminum in this case have a higher degree of dispersion.

Figure 9.5 Electron-microscopic images of composites after the MCT. (a) Al/C; (b) Al/$(C_2H_3OH)_n$; (c) Al/$C_{17}H_{35}COOH$.

The use of aluminum organic additives in MCT substantially changes the morphology of the surface layer of particles. So, when processing with graphite, the sample presented in Fig. 9.6a has a translucent lamellar formation on the surface, which belongs to single crystals of graphite. Translucent masses of graphite-like carbon envelop the surface of the particles.

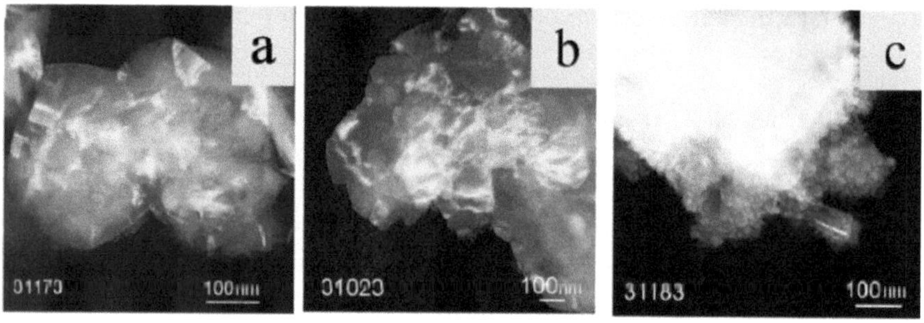

Figure 9.6 Transmission electron microscopy of composites after MCT. (a) Al+3%C; (b) A+3%$(C_2H_3OH)_n$; (c) Al+3%$C_{17}H_{35}COOH$.

After the MCT with the participation of 3% polyvinyl alcohol, the surface of the particles is also covered with a translucent coal-coated layer. Carbon, in the form of small crystals, enters the film on the surface of dense single-crystal particles and forms an independent friable formation of rounded particles (Fig. 9.6b). After MCT with 3% stearic acid, the particles of the sample with a size of about 2–3 microns are encapsulated into a fairly dense formation. Fragments in the form of flattened rounded particles (20–40 nm) are observed at the boundaries of the particles.

There are also elongated formations (about 30 nm in width and 100 nm in length). Carbonaceous formations are amorphized and are the "linking" link for small aluminum particles (Fig. 9.6c).

The size distribution of aluminum particles, both before and after MCT, was determined using the Malvern Mastersizer 2000 instrument. Figure 9.7 presents the results of the analysis of the distribution of the original PA4 aluminum particles. The average particle size of the powder is 50 μm, more than 30% of the main fraction refers to the range of 30–50 μm. The specific surface area is 3.692 m^2/g. The gravimetric density of the powder is not less than 0.96 g/cm^3.

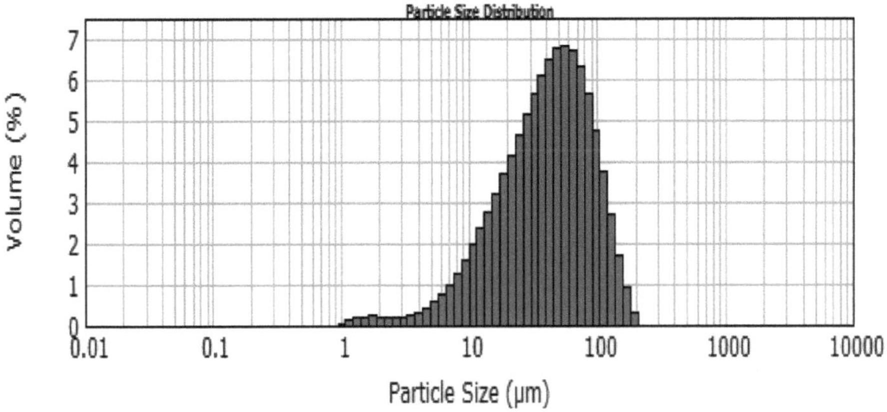

Figure 9.7 Distribution of the particles of the original aluminum powder in size.

After MCT, the results of the analysis of the particle size distribution of the powder show that for aluminum treated with 3% graphite, the average particle size is 36.0 μm and the surface area increases to 4.379 m^2/g (Fig. 9.8). With an increase in the amount of graphite to 20%, the average particle size of aluminum during MCT decreases to 17–18 μm with an increase in the specific surface area to 9.554 m^2/g. A significant part of the powder refers to the fraction of particles with a size of up to 10 microns.

When aluminum is activated and modified in the MCT process with polyvinyl alcohol (PVA) and stearic acid (SA), agglomeration of particles occurs simultaneously with grinding. The resulting formations have a layered structure. The average particle size in the case of 3% PVA is 16.1 μm, and from 20% PVA it increases to 30.5 μm (Fig. 9.9a). The specific surface after MCT increases to 4.976 and 14.648 m^2/g, respectively, for composites with 3% and 20% polyvinyl alcohol, which may be due to the high inconsistency of the surface layer of particles, namely the presence of cracks, pores and other types of structural heterogeneity. The effect of agglomeration of aluminum particles during MCT with stearic acid is enhanced and the average particle size with 3% stearic acid is 38 μm. As the amount of acid increases to 20%, aggregation increases and the particle size increases to 46 μm (Fig. 9.9b). Adsorbed organic reagents create an encapsulating layer on the surface of aluminum particles, more effective than the oxide layer, which prevents oxidation of the metal. According to the results of BET analysis, the machining process causes an increase in the specific surface area of aluminum particles from 5.31 (Al + 3% $C_{17}H_{35}COOH$) to 9.167 m^2/g (Al + 10% $C_{17}H_{35}COOH$). In the composite with 20% stearic acid, the specific surface area of aluminum particles, on the contrary, decreases to 2.568 m^2/g. This is due to the fact that during MCT of aluminum with a large amount of stearic acid, agglomeration occurs and the particles have a spherical shape.

Elemental analysis was carried out and electron microscopic images were taken for composites obtained after the MCT (Al–C, Al– $(C_2H_3OH)_n$ and Al–$C_{17}H_{35}COOH$). The energy dispersive spectra of the composite (Al + 3% C) are presented in Fig. 9.10.

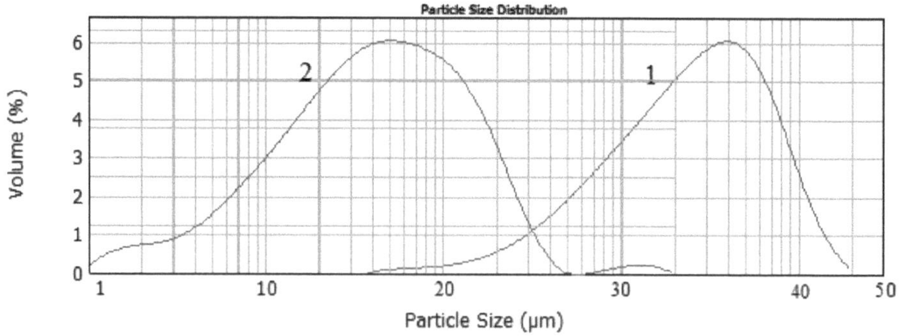

Figure 9.8 Distribution of aluminum particles after MCT with graphite. (1) (Al+3 % C); (2) (Al+20 % C).

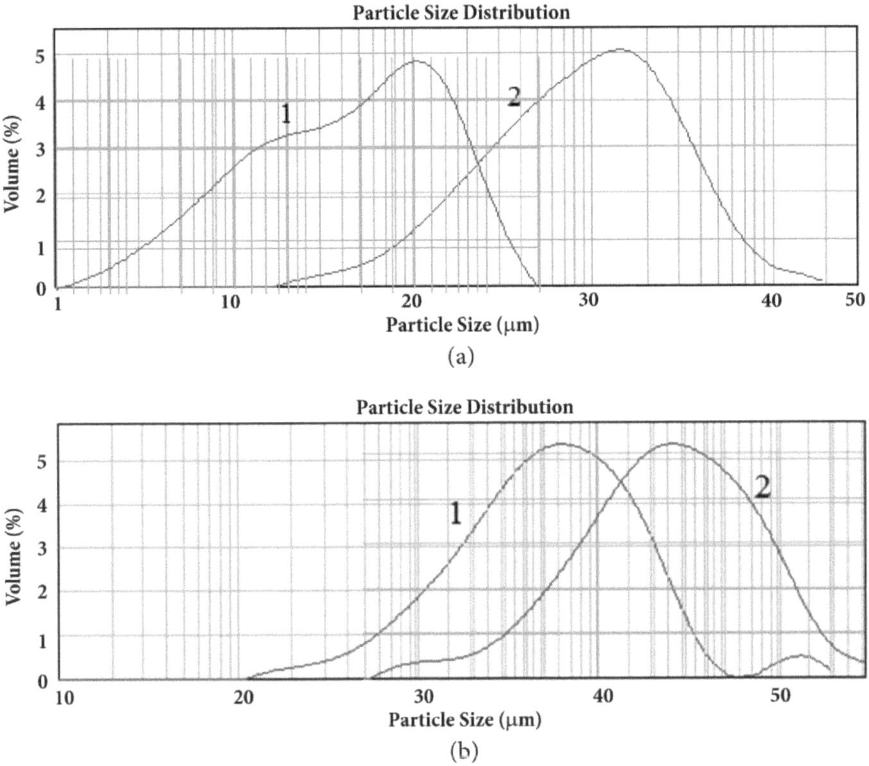

Figure 9.9 Distribution of particles of composites after MCT. (a) 1, (Al+3 % PVA); 2, (Al+20 % PVA); (b) 1, (Al+3 % S.A.); 2, (Al+20 % S.A.).

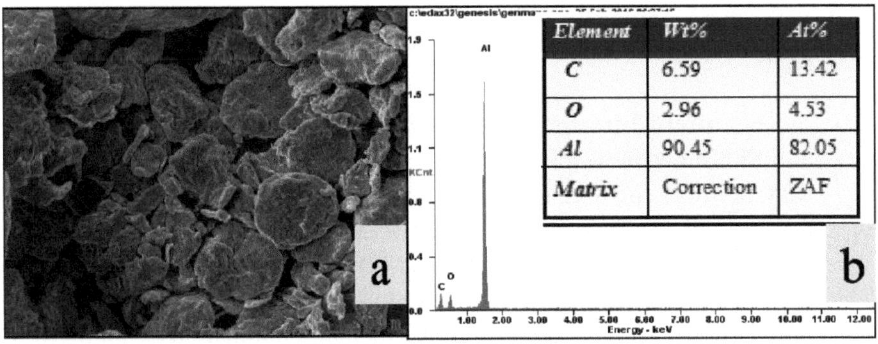

Figure 9.10 Structural characteristics of the composite (Al + 3% C) after MCT. (a) electron microscopic image; (b) energy dispersive spectrum and mass fraction of elements.

According to the results of the EDX analysis, the aluminum mass fraction in the Al + 3% C composite is 90.45%, and the carbon content is 6.59% of the total mass of the sample. The amount of oxygen does not exceed 2.96%. In the initial Al powder, the oxygen content is 10.67%. Thus, in the process of MCT of aluminum with 3% C and obtaining a composite Al–C, aluminum is partially reduced in the surface oxide layer of the particles and the oxygen content in the composite decreases.

The particles of the composite (Al + 3% C), after MCT, were studied by XRD (Fig. 9.11). The diffractogram shows the absence of other phases besides aluminum, which could reduce the activity of the powder particles.

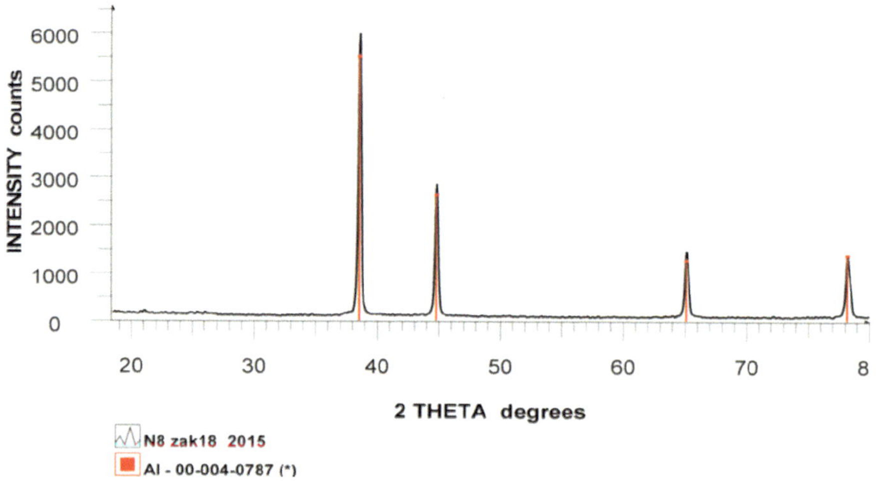

Figure 9.11 Diffraction pattern of Al + 3% C sample after MCT.

Figure 9.12 presents the results of the elemental composition and SEM images of the composite (Al + 20% C). The results of the EDX method show that the content of oxygen atoms is less than 6%, which is present in the oxide film on the surface of the particles. The aluminum content is 80.69%, i.e., as the amount of modifying additive increases, aluminum is reduced. This may be due to the fact that during the mechanochemical treatment of aluminum with graphite in the reactor an inert medium is formed and the aluminum metal is reduced by graphite from aluminum oxide.

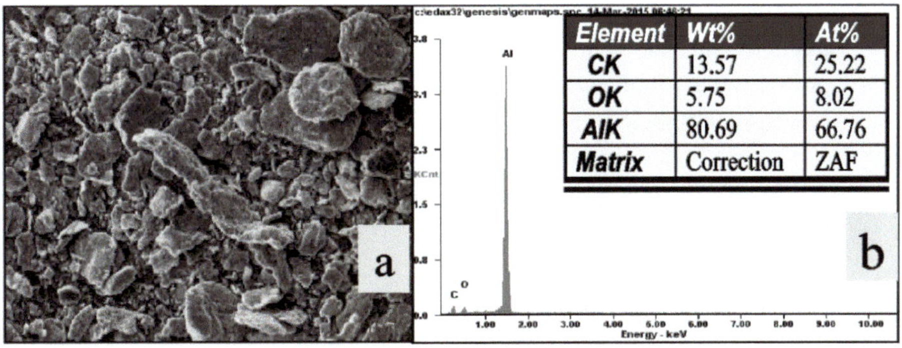

Figure 9.12 Structural characteristics of the composite (Al + 20% C) after the MCT. (a) Electron microscopic image; (b) energy dispersive spectrum and mass fraction of elements.

According to the results of X-ray phase analysis in the composite sample under study (Al + 20% C), after MCT, carbon in crystalline form is present in a very small amount, its main part as a result of dispersion is in X-ray amorphous state (Fig. 9.13).

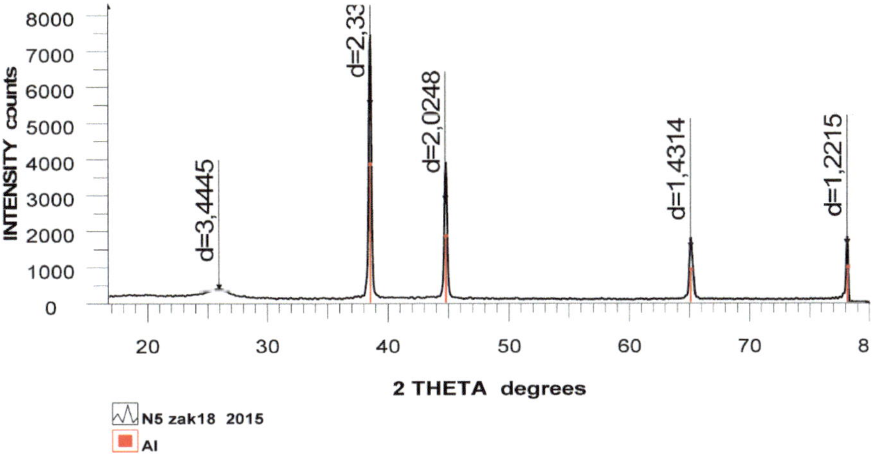

Figure 9.13 Diffraction pattern of the sample (Al + 20% C) after the MCT.

When machining aluminum with 3% polyvinyl alcohol, the main component of the composite is aluminum, which amounts to 92.72% (Fig. 9.14). The mass fraction of carbon is about 4%, and the amount of oxygen does not exceed 3.4%, i.e., there is a significant reduction of aluminum in the oxide layer of the particle surface. The aluminum particles after treatment with polyvinyl alcohol have a scaly form, and their surface has a rough structure (Fig. 9.14a).

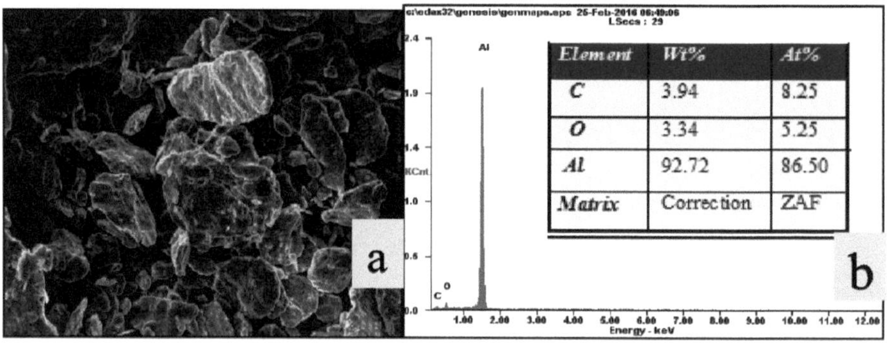

Figure 9.14 Structural characteristics of the composite (Al + 3% $(C_2H_3OH)_n$) after MCT. (a) Electron microscopic image; (b) energy dispersive spectrum and mass fraction of elements.

By increasing the amount of introduced organic additives to 20% (polyvinyl alcohol), the carbon content in the composite increases to almost 34% (Fig. 9.15).

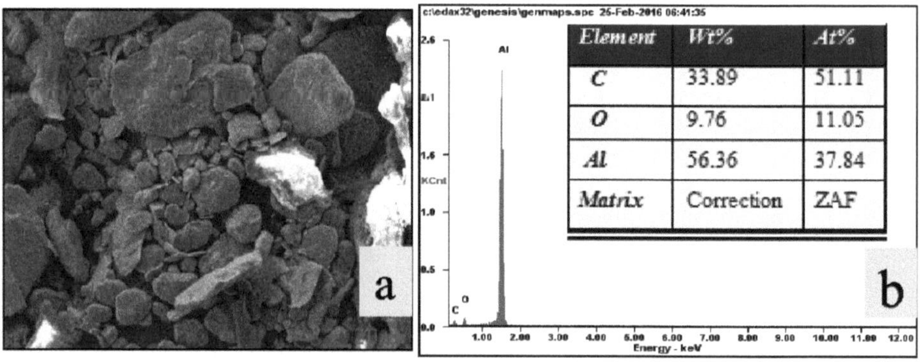

Figure 9.15 Structural characteristics of the composite (Al + 20% $(C_2H_3OH)_n$) after MCT. (a) Electron microscopic image; (b) Energy dispersive spectrum and mass fraction of elements.

The increase in the amount of carbon is due to the fact that during machining the polymer layer of the surface of aluminum particles is saturated with a passivating agent (hexane—C_6H_{14}). The oxygen content is reduced to 9.8% compared with the initial state (in the initial state it was 10.67%). Consequently, aluminum is partially reduced in the oxide film of the surface layer. SEM images show that the aluminum powder particles in the composite after MCT are plate-shaped.

Figure 9.16 shows electron microscopic photographs, the energy dispersion spectrum, and the mass fraction of the composite elements (Al + 3% $C_{17}H_{35}COOH$) obtained during MCT of aluminum with stearic acid.

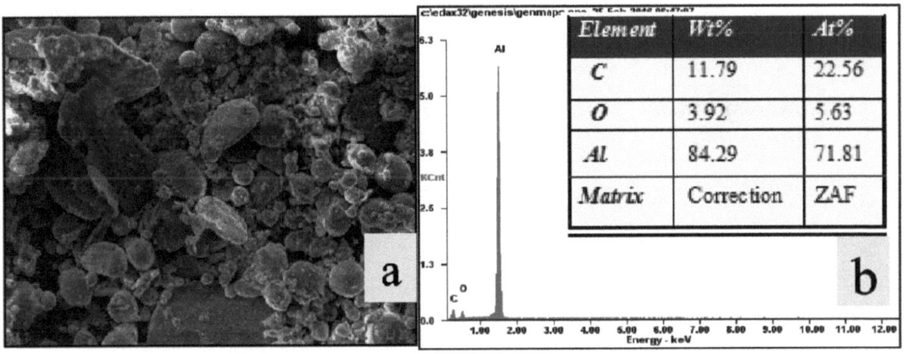

Figure 9.16 Structural characteristics of the composite (Al + 3% $C_{17}H_{35}$ COOH) after MCT. (a) Electron microscopic image; (b) energy dispersive spectrum and mass fraction of elements.

The figure shows spherical particles of aluminum encapsulated in carboxylic acid. The mass fraction of the elements is: Al, 84.29%; C, 11.79%; O, 3.92%. The increase in carbon content is due to the fact that stearic acid, as in the previous case, for samples with 20% polyvinyl alcohol absorbs a passivating agent (hexane—C_6H_{14}). The main proportion of particles of the composite has a particle size of several microns.

To assess the substructural features of aluminum particles after MCT, crystallite sizes were studied by XRD in the resulting Al/C, Al/$(C_2H_3OH)_n$ and Al/$C_{17}H_{35}COOH$ composites. According to the results of the analysis, in the process of mechanochemical treatment, the size of crystallites varies with the type and amount of the modifier used (Table 9.1). During mechanical action, there take place both accumulation and redistribution of defects throughout the particle volume. As a result of MCT of aluminum oxide with graphite, the growth of crystallites is observed with the increase of carbon content in the Al–C composite. This may be due to the fact that during MCT carbon atoms penetrate into the grain of the aluminum particle and, together with defects, diffuse throughout its volume under the action of mechanical stresses. Probably, this process promotes the removal of defects to the grain boundary of a particle and, as a consequence, an increase in the size of crystallites [22].

Table 9.1 Specific surface area and size of aluminum crystallites after MCT of Al composites + modifier

Composition of composites	Specific surface area, m^2/g	Crystallite size L, Å
Al initial	3.692	560
Al+3 % C	4.379	410
Al+5 % C	4.843	490
Al+10 % C	7.955	540
Al+20 % C	9.554	690
Al+3% $(C_2H_3OH)_n$	4.979	374
Al+5% $(C_2H_3OH)_n$	1.485	408
Al+10% $(C_2H_3OH)_n$	1.303	415
Al+20% $(C_2H_3OH)_n$	14.648	343
Al+3 % $C_{17}H_{35}COOH$	5.31	500
Al+5 % $C_{17}H_{35}COOH$	4.371	700
Al+10 % $C_{17}H_{35}COOH$	9.167	500
Al+20 % $C_{17}H_{35}COOH$	2.568	700

Treatment of aluminum with polyvinyl alcohol under similar conditions reduces the crystallite size to 374 and 343 Å with the addition of 3% and 20% modifier to the composite, respectively. In the presence of stearic acid, changes in the structure of particles during MCT do not occur unambiguously, the crystallite size increases markedly to 700 Å in composites Al + 5% $C_{17}H_{35}COOH$

and Al + 20% $C_{17}H_{35}COOH$. The reason for the growth of crystallites may be the sink of dislocations generated by mechanical (force) impact on the sample surface in the areas of chemical interaction of the particle with organic compounds-modifier.

The specific surface area of composites after MCT was investigated by the BET method. According to the BET analysis, the process of machining aluminum with graphite causes an increase in the specific surface area of aluminum particles from 3.692 to 9.554 m^2/g. Composite particles [Al + 20% $(C_2H_3OH)_n$] have a high specific surface area which makes up 14.648 m^2/g.

Thus, the use of organic modifiers (graphite, polyvinyl alcohol, stearic acid) during MCT of aluminum according to all analyzed characteristics contributes to a change in the morphology and structure of particles when composites are formed. The observed analogy in changing the particle size of aluminum modified with organic additives is a consequence of the fact that carbon resulting from the destruction of organic compounds plays a decisive role in formation of the surface layer of particles in all cases. The specific surface increases from 3.7 for the original aluminum to 14.6 m^2/g for the composite with 20% polyvinyl alcohol and this correlates with a low crystallite size (343 Å). Stearic acid creates a dense encapsulating layer on the surface of aluminum protecting from oxidation.

9.2 Mechanochemical Activation of Aluminum Powders with Organic Modifiers in the Presence of Quartz

As shown in Chapter 7, quartz is characterized by good dispersibility, i.e., obtaining fine powder as a result of grinding in a mechanical reactor. Since it has high hardness values, the ground particles can serve as a good abrasive material when simultaneously treating quartz with soft and ductile aluminum. In addition, quartz is a piezoelectric. Lattice deformation caused by mechanical stress leads to a redistribution of electric charges in the presence of polar directions (negative electric charges occur on the negative edges of the hexagonal prism zone, positive ones on the positive edges). There is also a reverse piezoelectric effect, i.e., internal stresses arise in an electric field in crystals that are proportional to the stressed fields [23], which certainly plays its role during mechanochemical treatment (activation and modification) of aluminum in the presence of quartz and organic modifiers.

For the compositions, we have taken systems with graphite, polyvinyl alcohol and stearic acid, which showed the best results in changing the structural characteristics, namely the size of the crystallites and the specific surface of aluminum powder as a result of MCT. Quartz was introduced into these compounds from 5% to 20%. As a result of processing in the mill, the obtained powders showed that the degree of grinding and the distribution of aluminum particles in size greatly depend on the amount of graphite and quartz sand. With an increase in the content of graphite from 5% to 20%, and SiO_2 from 5 to 10% there took place the most significant grinding of aluminum particles (Table 9.2). In the compositions Al + 5% C + 5% SiO_2 and Al + 20% C + 10% SiO_2, the average particle size decreases from 23 to 7.6 microns. In this case, they also have a minimum crystallite size.

Table 9.2 Changes in specific surface area, average particle size and size of aluminum crystallites after MCT of composites (Al + modifier + SiO_2)

Composition of the systems	Specific surface, m^2/g	Average particle size, μm	Crystallite size, L, Å
Al+5% C+5% SiO_2	2.447	23	451
Al+5% C+20% SiO_2	4.981	15	365
Al+20% C+10% SiO_2	5.991	7.6	343
Al+20% C+20% SiO_2	6.756	10	354
Al+20% $(C_2H_3OH)_n$ + 5%SiO_2	9.937	29	455
Al+20% $(C_2H_3OH)_n$ + 10% SiO_2	2.191	36	360
Al+20% $(C_2H_3OH)_n$ + 20%SiO_2	2.975	41	343
Al + 3% $C_{17}H_{35}COOH$ + 5% SiO_2	6.493	11.2	374
Al + 3% $C_{17}H_{35}COOH$ + 10% SiO_2	5.116	9.5	408
Al + 3% $C_{17}H_{35}COOH$ + 20% SiO_2	3.988	8.2	415

When aluminum is processed with polyvinyl alcohol in the presence of quartz, on the contrary, an increase in the particle size and, accordingly, a decrease in the specific surface area at sufficiently low crystallite sizes are observed. In all likelihood, in this case, the analyzed particles are a composite formation of aluminum and quartz particles, bound by a polymer, obtained on the basis of polyvinyl alcohol.

During MCT of the system with stearic acid, with an increase in the content of quartz in the composite, the particle size decreases, but the crystallite size increases, while the specific surface area decreases. The average particle size in the case of 3% stearic acid and 5% and 20% silica is 11.2 and 8.2 μm, respectively. In this case, most likely, aggregates of aluminum particles with quartz particles embedded in a metal surface and encapsulated in a polymer shell based on stearic acid are also formed. The characteristics obtained indicate complex structural changes in the compositions obtained after MCT of aluminum with graphite, polyvinyl alcohol and stearic acid both in the bulk of the particle and on the surface.

The change in the surface morphology of the particles as a result of MCT is clearly demonstrated by images obtained using a scanning electron microscope. Particles of aluminum composite powder with 20% graphite and 10% SiO_2 after MCT are presented in Fig. 9.17. It follows from the figure that the particles have a lamellar form and a pronounced surface heterogeneity. Each grain consists of a set of flattened subgrains, the average size of which is 7–6 microns. According to the results of energy dispersive analysis, it follows that the amount of aluminum in the composite is almost 65%, carbon is 25.57%, oxygen is 7.45% and silicon is 2.16%. Oxygen is a component of both silicon dioxide (quartz) and the oxidized surface layer of aluminum particles. Ground particles of graphite are embedded in the surface of deformed aluminum particles contributing to their destruction. The particles of quartz in the course of processing also abrade particles of aluminum, which ultimately leads to surface distortion and destruction of the oxide surface layer of particles.

Electron microscopic images of the composite (Al + 20% $(C_2H_3OH)_n$ + 10% SiO_2) showed that it consists of particles of different dispersity. The particles have a scaly shape and a fairly smooth surface (Fig. 9.18).

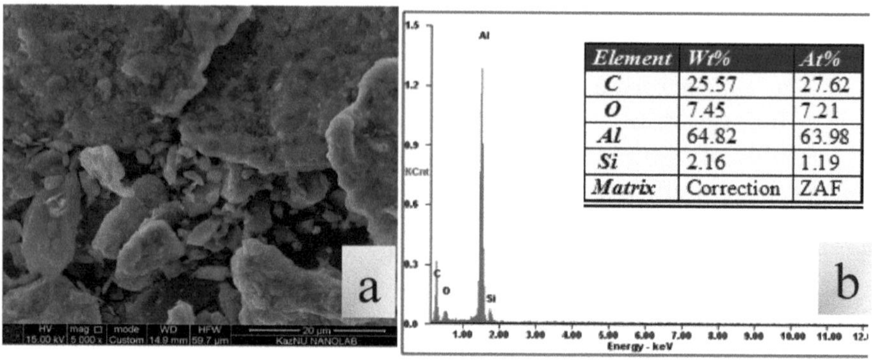

Figure 9.17 Structural characteristics of the composite [Al + 20% C + 10% SiO_2] after MCT. (a) Electron microscopic image; (b) energy dispersive spectrum and mass fraction of elements.

Figure 9.18 Structural characteristics of the composite [Al + 20% $(C_2H_3OH)_n$ + 10% SiO_2] after MCT. (a) Electron microscopic image; (b) energy dispersive spectrum and mass fraction of elements.

The latter indicates generation of composite formations of a polymer layer with dissolved carbon-containing inclusions on the surface. The specific surface of the particles is 2.191 m^2/g. According to the results of the energy dispersive spectrum, the aluminum content in the system is 52.38%. An increase in oxygen up to 15.21% is observed due to the fact that it is contained not only in oxides of silicon and aluminum, but is also present in large quantities in the modifier (polyvinyl alcohol).

Figure 9.19 shows electron microscopic photographs, the energy dispersion spectrum, and the mass fraction of the composite elements (Al + 3% $C_{17}H_{35}COOH$ + 10% SiO_2).

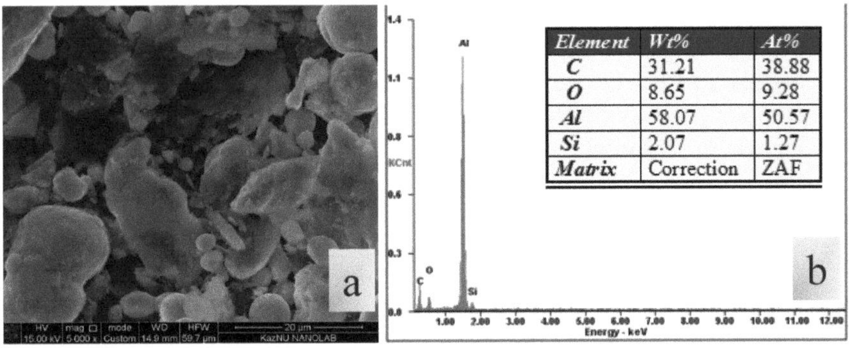

Figure 9.19 Structural characteristics of the composite [Al + 3% $C_{17}H_{35}COOH$ + 10% SiO_2] after MCT. (a) Electron microscopic image; (b) energy dispersive spectrum and mass fraction of elements.

Most particles are spherical. The aluminum content is 58.07%, i.e., less than in the case of graphite (65%), but more than when treated with polyvinyl alcohol. With the introduction of only 3% of stearic acid, the carbon content in the composite exceeds 31%, which may be due to saturation of the composite with a passivating substance (hexane), which is also used during MCT to prevent ignition of the system. The specific surface of the composite particles is 5.116 m^2/g.

Thus, as a result of MCT of aluminum in the presence of quartz and organic modifying additives, there takes place both the dispersion of the powder and transformation of the morphology and structure of the surface layer of particles. When using polyvinyl alcohol and stearic acid as modifiers, conglomerates are encapsulated into polymer films. All observed structural changes in the complex determine the qualitative changes in the state of aluminum as a result of MCT. More detailed information on the change in the surface structure of aluminum particles can be obtained by spectroscopic methods of analysis.

9.3 Study of Aluminum-Based Powders after MCT

Spectroscopic methods of analysis make it possible to see the relationship between the elements that represent the various functional groups from which the surface structure of the material particles is formed. After MCT of aluminum with various organic additives (graphite, polyvinyl alcohol, stearic acid) the composites were investigated by the method of IR-spectroscopy. Figure 9.20 shows the IR spectra of the Al + 3% C composite. After co-machining of aluminum powders with graphite, vibration lines of various C–H groups appear, this is probably due to the fact that hexane (C_6H_{14}) was added for passivation.

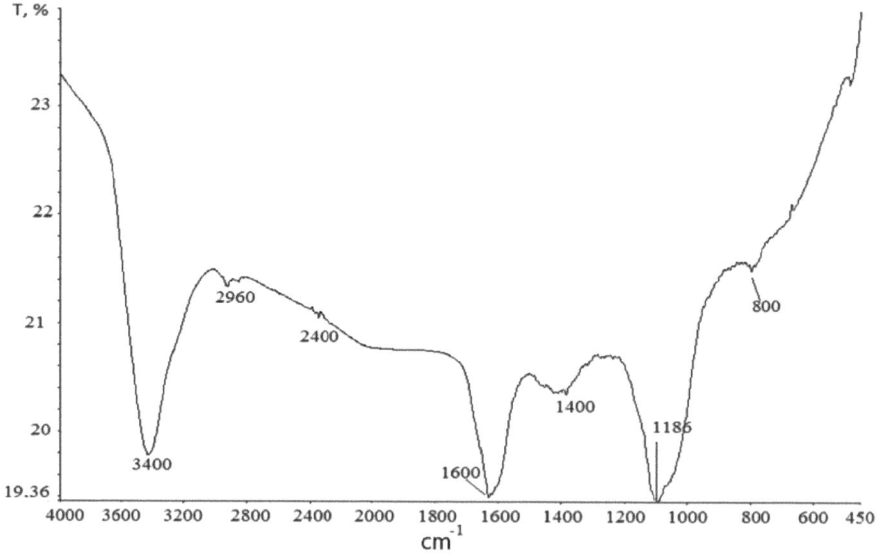

Figure 9.20 IR spectra of the composition [Al + 3% C] after MCT.

From Fig. 9.20 it follows that in the IR spectrum there are lines of stretching and deformation vibrations of the hydroxyl group –OH in the region of 3400 and 1600 cm^{-1}, and absorption bands of 2960 cm^{-1} with the type of $-CH_2$ and $-CH_3$ vibrations are observed. In the frequency range of 1400–1390 cm^{-1} there are signals of an average intensity of the deformation oscillation –CH groups, as well as a weak signal of the same group at a frequency of 800 cm^{-1}. At a frequency of 1186 cm^{-1}, an average signal of the deformation vibrations of the compound δ_{as}Al–OH is observed.

Thus, as a result of MCT of aluminum oxide with graphite, as well as in the presence of hexane, on the surface of crushed and deformed aluminum particles, in addition to hydroxyl groups, which are already present in the initial state of the powder, –CH groups of different valences are formed.

After MCT of aluminum with polyvinyl alcohol, significant changes are observed in the IR spectra (Fig. 9.21). Polyvinyl alcohol (PVA) has the following molecular structure:

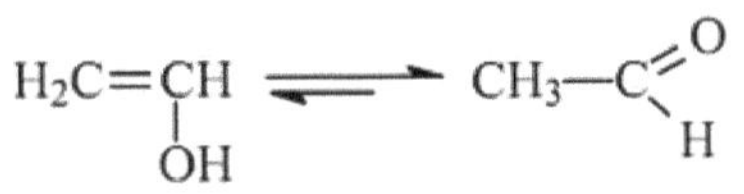

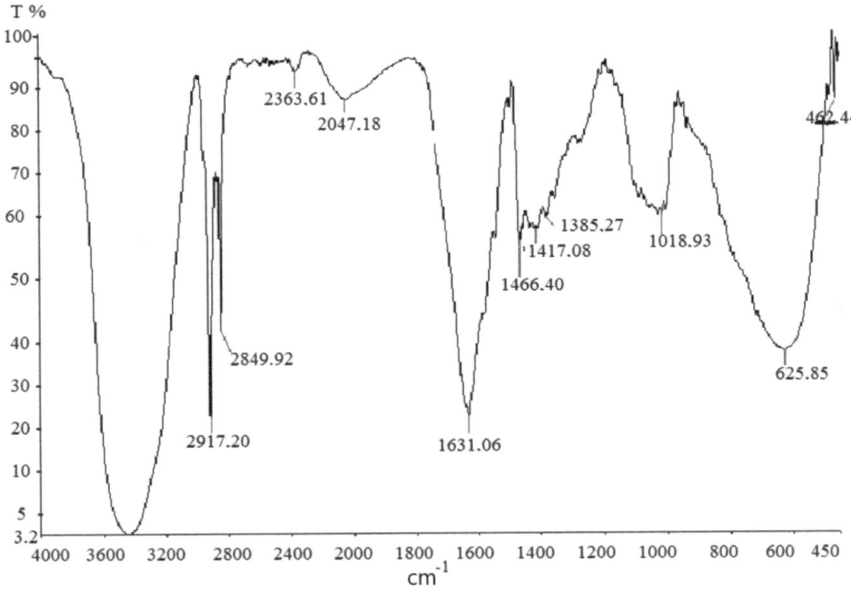

Figure 9.21 IR spectra of the composition [Al + 3% $(C_2H_3OH)_n$] after MCT.

As a result of MCT of Al + 3% PVA, on the surface of aluminum powders, one can observe absorption bands referring to the vibrations of $-CH_3$ and $-CH_2$ groups (2917.20 cm^{-1}, 2849.92 cm^{-1}), clear bonds of ν(OH) of a water molecule (1681 cm^{-1}) and carboxylate ion 1466.40 cm^{-1}, as well as wave numbers of primary, secondary and tertiary alcohols in the range of 1385 and 1417 cm^{-1}. Alcohols form polymer associates on the surface due to hydrogen bonds, which, being destroyed during dispersion, are restored after the load is removed. The intensity of the main lines, which were observed in the spectrum of the composition (Al + 3% C), significantly increased. This particularly applies to the hydroxyl group –OH (3400 cm^{-1}). As a result of MCT of aluminum oxide with polyvinyl alcohol, an intense absorption band is observed at a frequency of 1018.93 cm^{-1}, which belongs to the Al–O group. The observed spectra are the result of destruction of polyvinyl alcohol by mechanical action and grafting of radical formations to the surface of the particle.

Similar structural changes in the surface are observed when aluminum is combined with stearic acid in an amount of 3% (Fig. 9.22). During MCT, surface centers of the main character are formed on newly opened surfaces, clear bands ν(OH), δ(OH) of molecules of a bound OH group appear at frequencies of 3400 and 1632 cm^{-1}. In the frequency range of 2954, 2917 and 2849 cm^{-1}, strong intense signals of carboxylate ions with the type of vibrations ν^{as} C–O and ν^{s} C–O are observed. As with the composition (Al + 3% PVA), a signal is observed from the Al–O group, but shifted from 1018.93 cm^{-1} to a frequency of 1022.41 cm^{-1}. There are some more low-intensity signals on the right-hand slope of the main CH_2 band (pendulum oscillation), in the frequency range of 770–720 cm^{-1}.

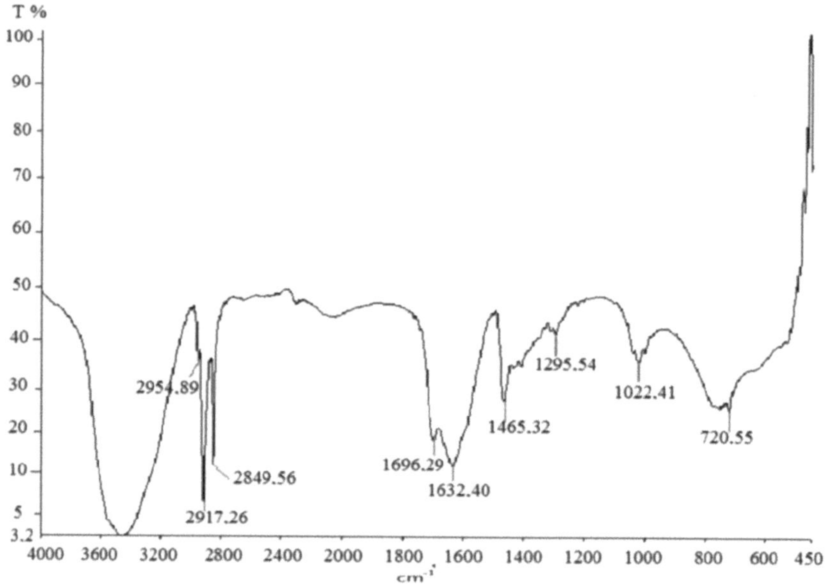

Figure 9.22 IR spectra of the composition [Al + 3% $C_{17}H_{35}COOH$] after the MCT.

With an increase in the amount of stearic acid in the system being processed to 20%, the intensity of the absorption bands from hydroxyl groups increases, especially at frequencies of 3464 and 1701 cm^{-1} (Fig. 9.23). In addition, signals from stretching asymmetric vibrations of aliphatic $-CH_3$ groups (2955 cm^{-1}) and stretching symmetric vibrations of aliphatic $-CH_2$ groups (2849 cm^{-1}) are observed in large quantities. The absorption frequencies of 1346, 1431 cm^{-1} belong to carbonyl groups, the 1701 cm^{-1} band belongs to the stretching vibrations of the C = O groups. Compared with the sample containing 3% stearic acid in the composite (Al + 20% $C_{17}H_{35}COOH$), intense bands of low-frequency vibrations of –CH groups in the frequency range (720–540 cm^{-1}) are pronounced. However, instead of Al–O, absorption bands of deformational vibrations of Al–O–Al appear in the frequency range 990–715 cm^{-1}. As a result, 20% of MCT of aluminum with stearic acid in the amount of 20%, the signal intensity of almost all groups, including –OH, $-CH_2$, CH_3, –C – O, and –C = O increases.

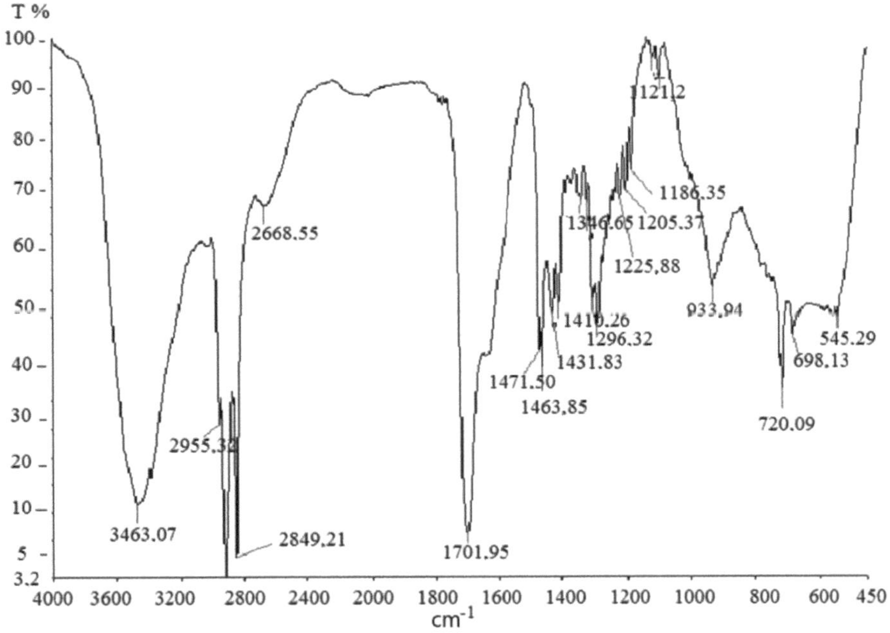

Figure 9.23 IR spectra of the composition [Al + 20% $C_{17}H_{35}COOH$] after MCT.

Thus, the results of infrared spectroscopy of mechanically treated and modified aluminum showed that during mechanochemical processing, changes occur in the structure of the surface layer of particles, which is manifested in changes in the IR spectra (the appearance of new absorption bands, frequency shift and band deformation). During MCT of aluminum with organic additives, several types of active centers are formed that can enter into chemical reactions when using the obtained compositions as part of mixtures with solid-phase combustion and as part of mixed solid rocket fuel (MSRF).

For a more detailed analysis of the surface state of modified aluminum particles, the methods of thermogravimetric and electron paramagnetic resonance were used. The method of electron paramagnetic resonance (EPR) allows you to capture the absorption of electromagnetic radiation in the range of radio frequencies by substances placed in a constant magnetic field associated with the presence of magnetic moment in electronic systems. The EPR phenomenon can be observed: on atoms and molecules that have an odd number of electrons in their orbitals; free radicals—methyl radical, nitroxyl radicals, etc.; electronic and hole defects.

The EPR spectrum recorded for aluminum in the initial state shows a relatively small amount of some impurities with a g-factor of 2.04992 and a line width of $\Delta H = 25.313$ (Fig. 9.24).

After MCT of aluminum oxide with graphite and passivation with hexane, the EPR spectra change dramatically (Fig. 9.24b). A strong spectrum signal with a g-factor of 2.00202 and a line width of $\Delta H = 44.586$, which can be attributed to –CH groups recorded in the IR spectra, is recorded between the third and fourth lines of Mn^{2+} in the MgO reference standard embedded in the GEOL spectrometer.

When 10% of silicon dioxide is introduced into the composite, the parameters of the EPR spectra change (Fig. 9.24c). Between the third and fourth lines, a stronger signal is recorded with a g-factor of 2.01068 and a line width of $\Delta H = 17.835$. This suggests that the number of radical metastable groups on the surface of modified aluminum particles has increased. In the EPR spectra of composites obtained by treating aluminum with polyvinyl alcohol and stearic acid, similar changes are observed.

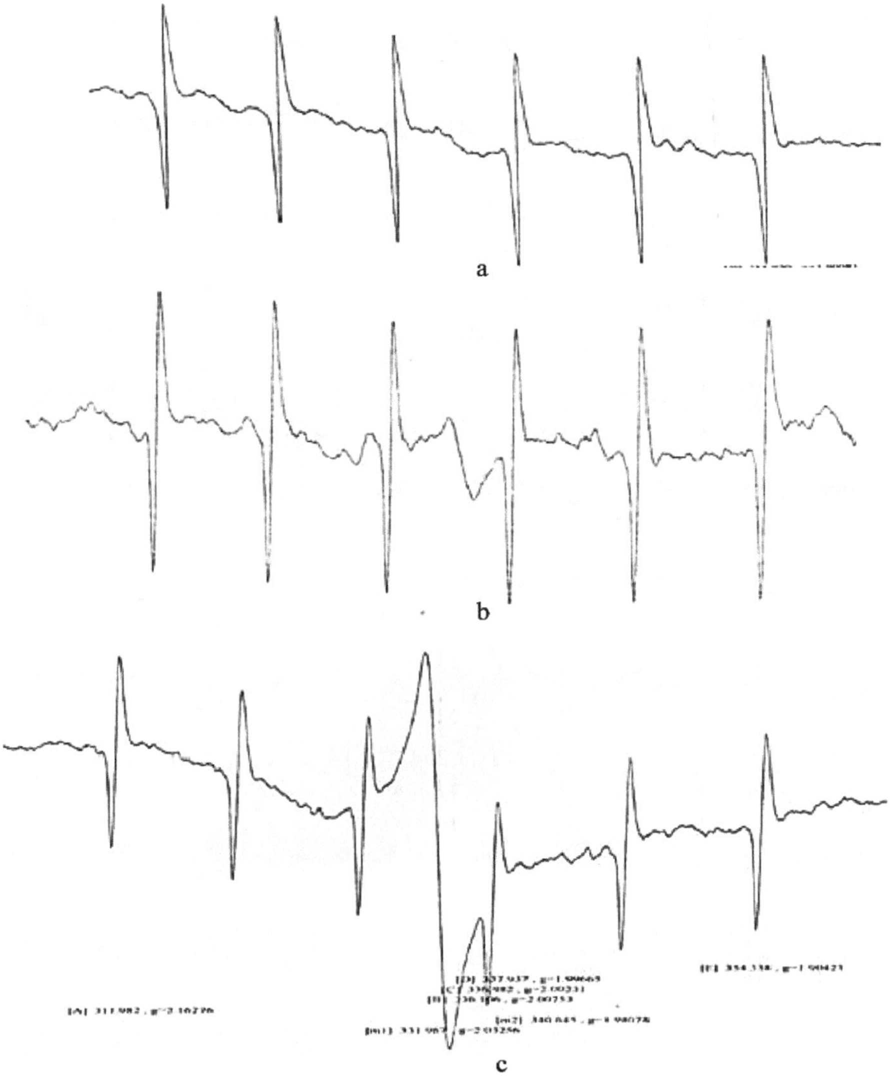

Figure 9.24 EPR spectra of the original aluminum and composites after the MCT. (a) Al_{ini}; (b) $[Al+3\%\ C]_{MCT}$; (c) $[Al+20\%C+10\%SiO_2]_{MCT}$.

Thermal analysis (thermogravimetry (TG)) of the samples was carried out using a NETZSCH 449F3A-0372-M instrument, in a nitrogen atmosphere of 99.99% purity and in the temperature range of 30–1000°C, at a heating rate of 10 K/min. Figure 9.25

shows a thermal analysis of the original aluminum. The mass loss in the sample of initial Al when heated to 600°C is 3.52%, which may be due to separation of the hydroxide layer from the surface of the powders. Aluminum melting and complete transformation into a liquid form occurs at a temperature of 665.6°C. Prior to melting, the exothermic effect is fixed at 647.4°C, which is associated with oxidation of aluminum. As the temperature rises to 909°C, the sample mass increases by 2.59%, which may be due to saturation of liquid aluminum with nitrogen and formation of aluminum nitride.

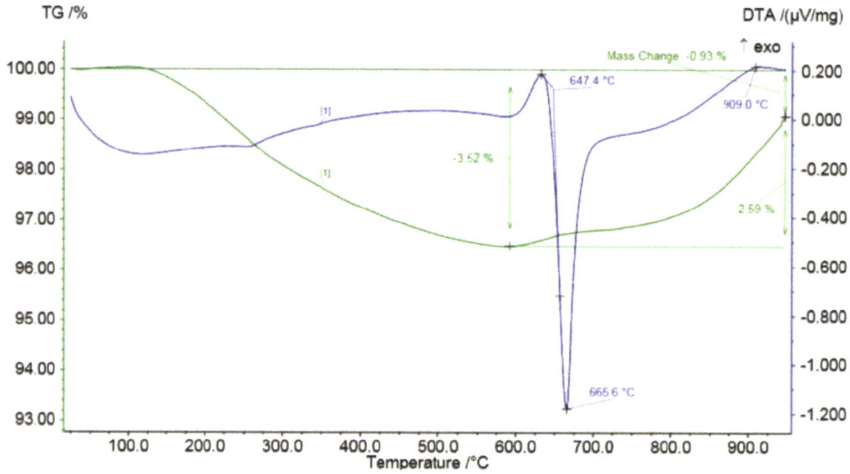

Figure 9.25 Thermogravimetric analysis of the original aluminum.

Figure 9.26 presents the results of thermal analysis of aluminum composites with 3% and 20% graphite after MCT. In the Al + 3% C composite, aluminum oxidation is not observed during the heating process, but after melting at 660°C, formation of aluminum carbide is possible:

$$4Al + 3C \rightarrow Al_4C_3 \qquad (9.1)$$

This process is exothermic and as a result, the temperature rises up to 700°C and the mass of the sample increases (Fig. 9.26a). With further heating from 800 to 1000°C, a change in mass by 5.27% is observed, which is possibly due to formation of aluminum nitride:

$$N_2 + 2\mathrm{Al} \xrightarrow{800\text{–}1200^\circ} 2\mathrm{AlN} \tag{9.2}$$

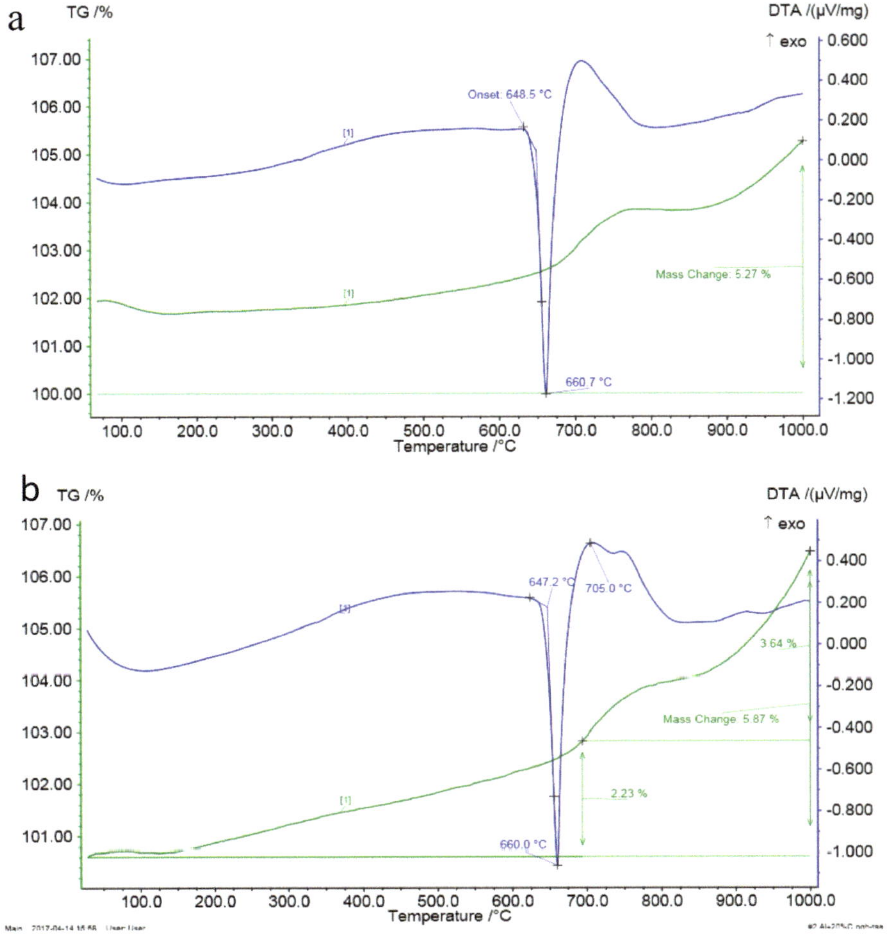

Figure 9.26 Thermogravimetric analysis of composites after MCT. (a) [Al + 3% C]$_{\mathrm{MCT}}$; (b) [Al + 20% C]$_{\mathrm{MCT}}$.

With an increase in the amount of graphite in the composite up to 20% (Al + 20% C) when the sample is heated, the processes proceed similarly as in the sample with 3% graphite. At a temperature of 705°C, the reaction of carbidization of aluminum begins and the process proceeds in the temperature range up to 770°C. Then the processes associated with the interaction of aluminum with nitrogen proceed. As a result, the weight gain is 5.87% (Fig. 9.26b).

Figure 9.27 presents the results of the presence of silicon dioxide in the sample on its state during MCT.

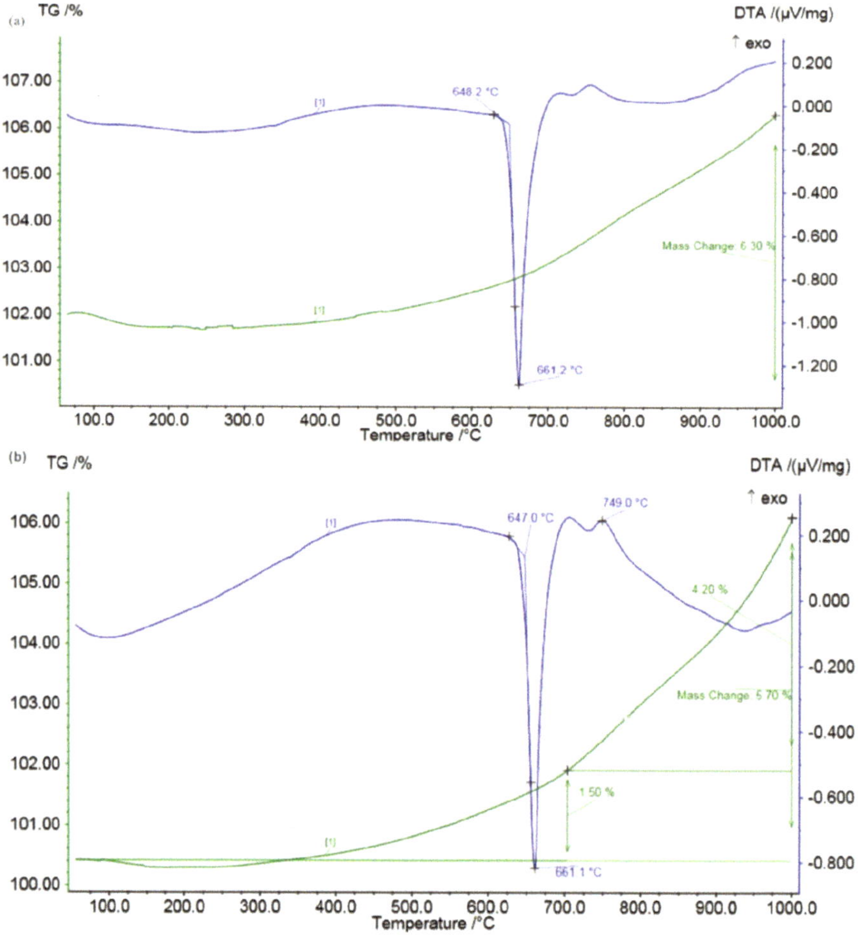

Figure 9.27 Thermogravimetric analysis of composites after MCT. (a) [Al + 3% C + 10% SiO_2]$_{MCT}$; (b) [Al + 20% C + 10% SiO_2]$_{MCT}$.

Oxidation of the sample to the melting of aluminum with 3% graphite is not observed. The increase in temperature and mass of the sample from 20% C to melting may be due to oxidation (burning) of carbon. After melting, there takes place an exo-effect from a redox reaction between the components of the composite:

$$Al + SiO_2 = Al_2O_3 + Si \tag{9.3}$$

The increase in mass in the sample with 3% C is 6.3%. With an increase in the amount of graphite to 20% [Al + 20% C + 10% SiO_2] at a temperature of 749°C and higher, the reaction (2) proceeds. Since the carbon content in the composite is 20%, an endothermic reaction is possible with formation of mullite and aluminum carbide [24]:

$$12Al + 13SiO_2 + 9C = 2[3Al_2O_3 \cdot 2SiO_2] + 9SiC \tag{9.4}$$

The total change in sample mass is 5.70%.

Thermographic analysis of aluminum after MCT with polyvinyl alcohol showed that at temperatures above 329°C alcohol decomposes and the weight reduction at 500°C is 16.93%, i.e., almost complete decomposition of the organic additive occurs (Fig. 9.28a). However, after melting of aluminum powder, an increase in the sample mass by 7.51% is observed. Above 900°C there is a rise in temperature and a decrease in the mass of the sample as a result of the reaction:

$$Al_2O_3 + 9C \rightarrow Al_4C_3 + 6CO \tag{9.5}$$

In the composition [Al + 20% $(C_2H_3OH)_n$ + 10% SiO_2] at 400°C, destruction of polyvinyl alcohol is also observed and the mass loss is 13.65% (Fig. 9.28b). At a temperature of 900°C and above, an exothermic reaction occurs with the release of a large amount of heat. The increase in sample mass is 17.36%. It is assumed that in this case the products of the reaction may be mullite and aluminum carbide [25].

Thus, adding quartz to the composition of the system results in formation of a material containing corundum, mullite and silicon carbide, and also the formation of small amounts of silicon, aluminum carbide, the ratio of which depends on the amount of carbon in the composition.

Melting of pure massive aluminum is characterized by the following parameters at a pressure of P = 101325 Pa: melting point is t_{melt} = 660.2°C, melting enthalpy ΔH_{sp} = 10.8 kJ/mol (specific melting heat ΔH^{sp}_{melt} = 400 J/g) [26].

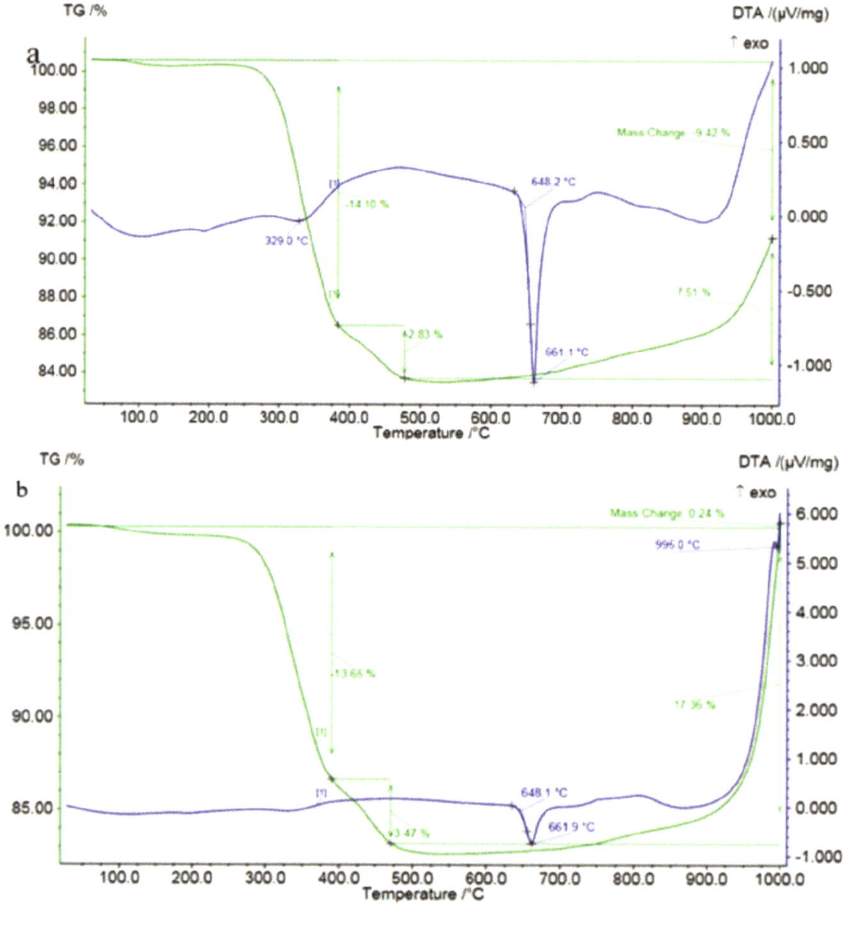

Figure 9.28 Thermogravimetric analysis of composites after MCT. (a) [Al+20% $(C_2H_3OH)_n$]$_{MCT}$; (b) [Al+20% $(C_2H_3OH)_n$+ 10% SiO_2]$_{MCT}$.

Figure 9.29 presents the TG curves of the samples [Al + 3% $C_{17}H_{35}COOH$] and [Al + 3% $C_{17}H_{35}COOH$ + 10% SiO_2]. There is a mass loss in the composite in accordance with the content of the organic additive in the sample. The exoeffect before melting and the weight gain can be related oxidation of powders. The weight gain after melting is related the dissolution of the gas phase in liquid aluminum, in particular nitrogen, and is 5.35%.

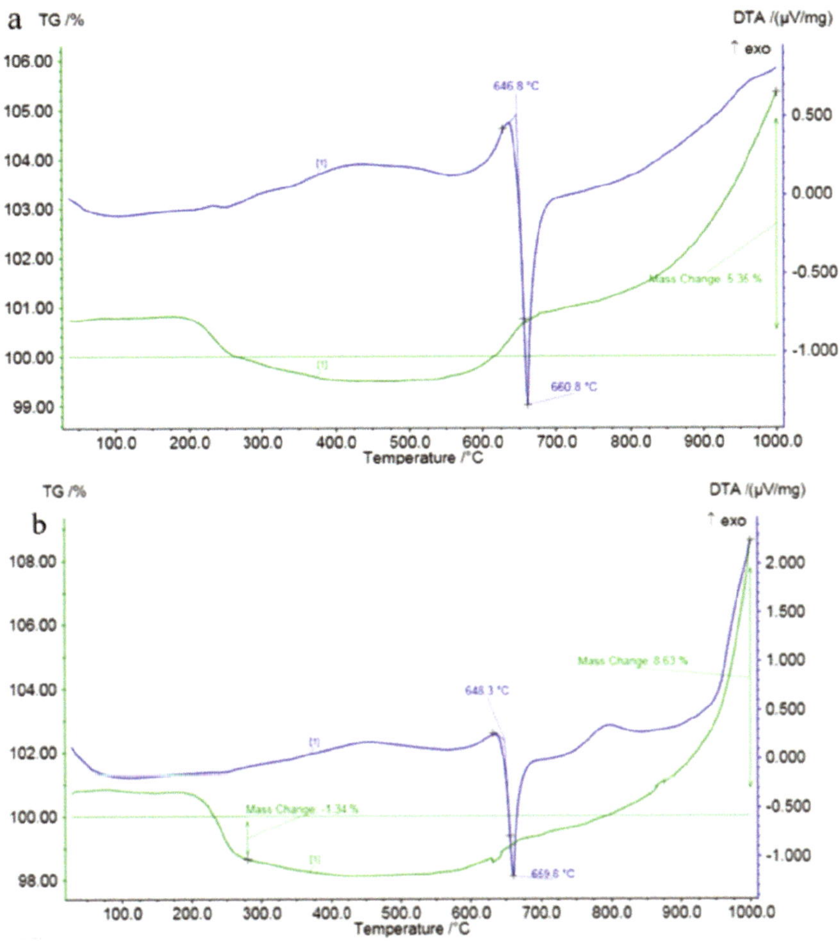

Figure 9.29 Thermogravimetric analysis of composites after MCT. (a) $[Al+3\%\ C_{17}H_{35}COOH]_{MCT}$; (b) $[Al+3\%\ C_{17}H_{35}COOH\ +10\%\ SiO_2]_{MCT}$.

For the sample $[Al + 3\%\ C_{17}H_{35}COOH + 10\%\ SiO_2]$ above 900°C, a significant heat release and a weight gain of the sample by 8.63% are observed as a result of the intense reaction between silica and aluminum. The presented results of the thermogravimetric analysis of aluminum powders show that mechanochemical treatment leads to a change in their activity in the interaction with gases and solid components of the system.

Thus, the results of IR spectroscopy, EPR spectroscopy and thermogravimetry of activated and modified aluminum powder showed that as a result of mechanochemical treatment, there takes place changes in the main spectra (new lines, frequency shifts and deformation bands are observed) and the intensification of interaction processes with various components of the system. Changes in the intensity of the absorption bands and the degeneracy of a number of spectra are observed, starting with deformation vibrations and formation of a destructible modifier on the surface of particles of groups of elements and radicals. All this together should affect the activity of the processed powder.

9.4 Determination of the Activity of Modified Aluminum Powders

The activity of crushed and modified with organic additives aluminum in a mechanical reactor was evaluated by the volumetric method for the evolution of hydrogen in the interaction of aluminum with a 20% aqueous solution of sodium hydroxide. The reaction with water proceeds according to the following equation:

$$2Al + 4H_2O \rightarrow 2AlOOH + 3H_2 \quad (9.6)$$

This method is based on measuring the volume of gas released during a chemical reaction between reactants. The solid reaction product is pseudo-boehmite AlOOH.

In the initial state, the activity of the used aluminum powder is 97.7%. For the studied compositions with different modifiers, the activity of aluminum varies. The content of active aluminum in the studied compositions is presented in Table 9.3. In this case, it should be taken into account that the mass fraction of aluminum in the mixture of compositions decreases from 5% to 20%, i.e., in all cases there is an increase in the proportion of active aluminum in the composite relative to the original aluminum. This is primarily due to the fact that the oxide film on the surface of the aluminum particle is partially reduced by carbon during the MCT process, thereby increasing the content of active aluminum in the composite. The lack of a stable pattern in the change in the content of active aluminum in the composite from the content of the modifying additive when processed with polyvinyl alcohol and stearic acid may be due to formation of a dense encapsulating film on the surface of the particles.

Table 9.3 Content of active aluminum in the composition of the systems of Al + modifier after MCT

No composition	Composition of the systems	The content of active aluminum in the composite, %
1	Al initial	97.7
2	Al+3% C*	96.9
3	Al+5% C	95.9
4	Al+10% C	90.35
5	Al+20% C	88.06
6	Al+3% S.A.**	98.9
7	Al+5% S.A.	96.75
8	Al+10% S.A.	85.8
9	Al+20% S.A.	79.95
10	Al+3% PVA***	91
11	Al+5% PVA	95.5
12	Al+10% PVA	86.05
13	Al+20% PVA	89.05

C*, graphite; S.A.**, stearic acid; PVA***, polyvinyl alcohol.

The relative rate of growth of active aluminum in the composite Al + modifier can be estimated by the formula [27]:

$$\Delta J = J_x \frac{J_0^* w_{\text{Al}}\%}{100\%}, \tag{9.7}$$

where ΔJ is the relative growth rate of hydrogen released in composites; J_x the experimental data on the content of active aluminum in the composite Al + modifier, %; J_0 the experimental data on the content of active aluminum for the original aluminum powder, %; and ω_{Al} the mass fraction of aluminum in the composite Al + modifier, %.

Figure 9.30 shows the dependence of the increase in aluminum activity after MCT of the blend Al + modifier on the content of the organic additive in the composite. The highest relative increase in aluminum activity relates to the Al + 20% C composite and amounts to 11.34%. For the Al + 20% PVA composite, there is also a relatively high growth rate of aluminum activity, which is 10.89%, whereas in the presence of 20% stearic acid, the aluminum activity gain is only 1.79%. The greatest increase in aluminum activity in MCT with stearic acid is observed only when its content is not more than 3%.

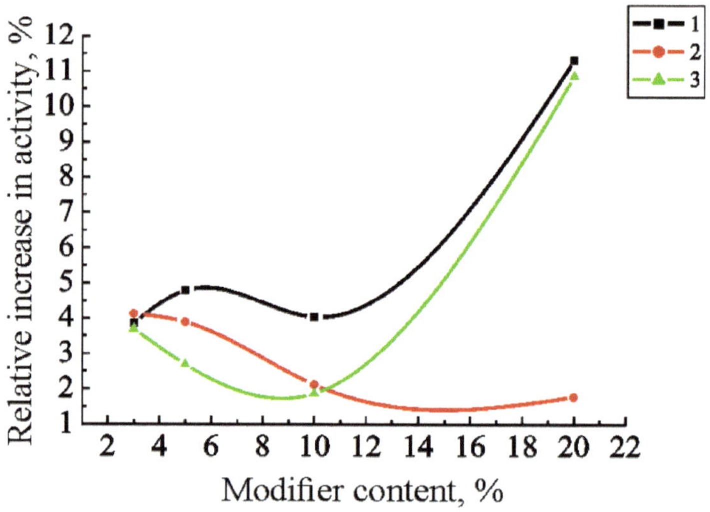

Figure 9.30 Relative growth rate of active aluminum in the composite (Al + modifier) after MCT with different content of organic additives. (1) (Al+C); (2) (Al+S.A.); (3) (Al+PVA).

From the presented data, it follows that with an increase in the content of carbon and polyvinyl alcohol in the composite, the activity of aluminum increases. However, with an increase in the content of stearic acid in the composite, the increase in the activity of the composition decreases. This may be due to the fact that in the process of grinding the mixture (Al + S.A.) with a large amount of stearic acid (more than 5%) on the surface of aluminum particles a dense encapsulating layer is formed, poorly soluble with alkali.

The change in the content of active aluminum was also evaluated for compositions with silica (Table 9.4).

Table 9.4 Content of active aluminum in the composition of the composite (Al + modifier + silicon dioxide) after MCT

The composition of the charge	The content of active aluminum in the composite, %
Al initial	97.7
Al + 5%C+5% SiO_2	91.7
Al + 5%C+10% SiO_2	87.3
Al + 5%C+20% SiO_2	80.9
Al + 20%C+5% SiO_2	69.2
Al + 20%C+10% SiO_2	75.1
Al + 20%C+20% SiO_2	61.1
Al + 3% S.A. + 5% SiO_2	114.4
Al + 3% S.A. + 10% SiO_2	96.9
Al + 3% S.A. + 20% SiO_2	85.1
Al + 20% PVA + 5% SiO_2	85.6
Al + 20% PVA + 10% SiO_2	91.55
Al + 20% PVA + 20% SiO_2	79.55

According to the results of volumetric measurements, calculations were made of the relative increase in aluminum activity in systems (Al + modifier + silicon dioxide) with different SiO_2 contents. From Fig. 9.31 it follows that for the composite (Al + 3% SA + 5% SiO_2) the highest growth rate of activity is observed (up to 24.5%). This is due to the fact that not only aluminum but also stearic acid reacts with alkali to form stearates [28].

With an increase in the amount of silicon dioxide in the composition of the aluminum composite with 20% polyvinyl alcohol, a steady trend of increasing activity from 12.3% to 20.93% is observed. Mechanical treatment of aluminum with graphite as a modifying additive in the presence of silicon dioxide leads to an increase in the relative activity of aluminum within 7% when the system contains 20% C and, conversely, reduces the activity of aluminum in the presence of 5% C. This is probably due to the fact that under mechanical action, a large part of graphite interacts with quartz, and not with aluminum particles, as a result of which more rapid oxidation of aluminum metal occurs.

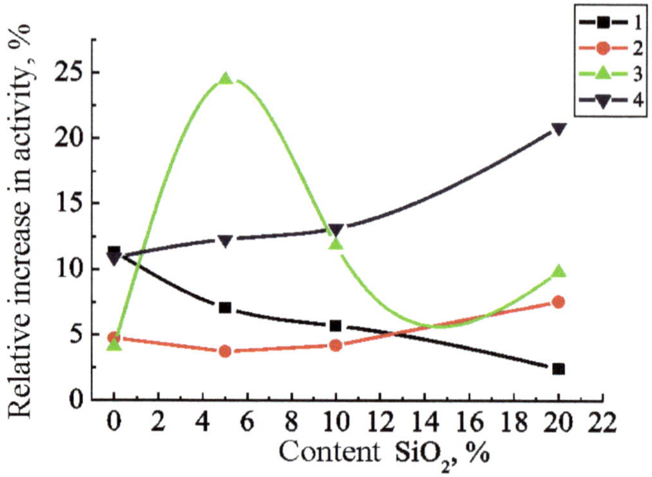

Figure 9.31 Relative indicator of the increase in the activity of aluminum in the composite (Al + modifier + SiO_2) depending on the content of SiO_2 after MCT with different organic additives. (1) (Al + 5% C), (2) (Al + 20% C), (3) (Al + 3% S.A.), (4) (Al + 20% PVA).

From the presented results, it follows that regardless of the type of modifier, the activity of aluminum increases with an increase in the content of organic additives in the composite. However, at a content of 10% graphite in the Al–C composite, the intensity of the increase in activity decreases. This can be explained by the fact that in the process of grinding the mixture of Al + 10% C for 20 min, a more significant temperature increase in the reactor was established, which contributes to the annealing of defects in dispersible aluminum particles, and consequently to a certain decrease in its activity [29].

During mechanical grinding of aluminum with various modifiers, the following processes occur: in the first stage, the oxide film covering aluminum is crushed in areas of increased stress, and fragments of crushed particles accumulate in the gaps between the particles (along the boundaries of subgrains). Here, the oxide impedes the movement of the boundaries and keeps the formed subgrains in the sizes of the initial powder particles. In the second stage of the accumulation of fragments, particles are dispersed by distributing oxide plates in the direction of plastic deformation. The consequence of this is the further fragmentation of subgrains.

The conducted studies allow us to conclude that MCT of aluminum with various modifiers and in the presence of quartz in a mechanical reactor leads to a change in the surface structure and morphology of the particles. The optimal conditions of the MCT for obtaining the maximum activity of aluminum are determined. The metal particles in the Al/C composite can be a solid solution of carbon in aluminum. They have a minimum crystallite size, which in a complex determines their high level of reactivity when used as part of combustible condensed systems.

Such an approach will make it possible to obtain a material in a highly active state, both due to a high degree of dispersion, high specific surface, and also the accumulated imperfection of the structure in the particle volume. Surface modification will prevent the discharge of defects from the particle volume onto its surface, i.e., to fix the active state of metal particles and the energy intensity of the combustible system as a whole.

Thus, a comprehensive analysis of activated and modified aluminum and aluminum powders with quartz showed that during the mechanochemical treatment the surface of the particles is saturated with solid solutions of carbon-containing compounds. The use of such powder materials is effective, in particular, in the preparation of SHS composites with enhanced chemical and mechanical stability.

Control Questions

1. What happens during mechanochemical processing of materials?
2. What types of organic modifiers have been used in the MCT of aluminum powder?
3. What functional groups appear on the surface of aluminum particles during mechanochemical processing with organic modifiers?
4. What is the effect of mechanical processing on the chemical activity of the material obtained?

References

1. Mansurov Z. A., Mofa N. N. *Mechanochemical Synthesis of Composite Materials*, Kazak University, Almaty, 2016. 376 p. (In Russian).

2. Koch C. C., Whittenberger J. D. Mechanical milling/alloying of intermetallics. Intermetallics, 1996. vol. 4, no. 5. pp. 339–355.
3. Suryanarayana C. Mechanical alloying and milling. *Progress in Mater. Sci.*, 2001. vol. 46, no. 1–2. pp. 1–184.
4. Gaffet E., Bernard F., Nierce J.-C., Charlot F., Gras C., Le Caer G. Some recent developments in mechanical activation and mechanosynthesis. *J. Mater. Chem.*, 1999. vol. 9. pp. 305–314.
5. Butyagin P. Yu. Problems and prospects of development in mechanochemistry. *Chemistry*, 1994. vol. 63, no. 12, pp. 1031–1043. (In Russian).
6. Suryanarayana C. Mechanical alloying and milling. *Prog. Mater. Sci.*, 2001. vol. 46, no. 1–2. pp. 1–184.
7. Yelsukov E. P., Dorofeev G. A., Barinov V. A., Grigorieva T. F., Boldyrev V. V. Solid state reactions in the Fe–Sn system under mechanical alloying. *Mater. Sci. Forum.*, 1998. vol. 269–272, no. 1. pp. 151–156.
8. Grigorieva T. F. *Mechanical Fusion in Two-Component Metal Systems with Participation of a Low-Melting Metal: Dis. ...* Dr. chemical sciences., IHTTMSORAN, Novosibirsk, 2005. 285 p. (In Russian).
9. Eckert J., Holzer J. C., Krill C. E., Johnson W. L. Structural and thermodynamic properties of nanocrystalline fcc metals prepared by mechanical attrition. *J. Mater. Res.*, 1992. vol. 7, no. 7. pp. 1751–1761.
10. Gaffet E., Bernard F., Nierce J.-C., Charlot F., Gras C., Le Caer G. Some recent developments in mechanical activation and mechanosynthesis. *J. Mater. Chem.*, 1999. vol. 9. pp. 305–314.
11. Novikov I. I. *Defects in the Crystal Structure of Metals*. Metallurgy, Moscow, 1983. 232 p. (In Russian).
12. Dossi S., Paravan C., Maggi F., Galfetti L. Enhancing micrometric aluminum reactivity by mechanical activation. *AIAA*, 2015. https://doi.org/10.2514/6. pp. 4206–4221.
13. Streletsky A. H., Kolbanev KV., Troshin K. Ya., Borisov A. A., Leonov A. V., Mudretsov S. N., Artyomov V. V., Dolgoborodov A. Yu. Structure and reactivity of mechanically activated Mg(Al)MoO_3 nanocomposites. *Chem. Phys.*, 2016. vol. 35, no. 7. pp. 79–91. (In Russian).
14. Dolgoborodov A. Yu., Streletsky A. H., Makhov M. N., Kolbanev K. V., Fortov V. E. Explosive compositions based on mechanically activated metal-oxide mixtures. *Chem. Phys.*, 2007. vol. 26, no. 12. pp. 40–45. (In Russian).
15. Grigorieva T. F., Barinova A. P., Lyakhov N. Z. Mechanosynthesis of nanocomposites. *J. Nanoparticle Res.*, 2003. no. 5. pp. 439–453.

16. Lyakhov N. Z., Grigoriev T. F., Barinov A. P., Vorsina I. A. Mechanochemical synthesis of organic compounds and composites with their participation. *Successes Chem.*, 2010. vol. 79, no. 3. pp. 218–233. (In Russian).
17. Sadykov B. S., Mofa N. N., Galfetti L., Mansurov Z. A. *The Effect of Mechanochemical Activation and Modification of Aluminum on Technological Combustion of Condensed Systems Application. XII International Conference "HEMs-2016" "High-energy materials: demilitarization, anti-terrorism and civilian use".* Tomsk, 2016. pp. 55–58. (In Russian).
18. Mofa N. N., Sadykov B. S., Mansurov Z. A. Production of nanodispersed modified metal powders for high-energy condensed systems. *IX International Symposium "Physics and Chemistry of Carbon Materials/ Nanoengineering" and the International Conference "Nanoenergy Materials and Nanoenergy".* Almaty, 2016. pp. 2–8. (In Russian).
19. Mansurov Z. A., Mofa N. N., Sadykov B. S., Sabaev Zh. Zh., Bakkara A. E. Mechanochemical treatment, structural features, properties and reactivity of SHS systems based on natural materials. Part 3: The influence of mechanochemical processing and modification of oxide materials on technological combustion. *Eng. Phys. J.*, 2014. vol. 87, no. 5. pp. 1000–1008. (In Russian).
20. Mansurov Z. A., Mofa N. N., Sadykov B. S., Sabaev ZH. ZH., Bakkara A. E. Mechanochemical treatment, structural features, properties and reactivity of SHS systems based on natural materials. Part 4: Production of SHS-ceramics based on mechanically activated materials. *Eng. Phys. J.*, 2016. vol. 89, no. 1. pp. 221–228. (In Russian).
21. Avakumova E. G. *Fundamental Principles of Mechanical Activation, Mechanosynthesis and Mechanochemical Technologies.* Science, Novosibirsk, 2009. 342 p. (In Russian).
22. Mikhailov A. A., Krasnov A. A., Dyuryagin B. S., Gaidomako I. M. Open-work composite fibrous materials with nanostructured elements. *Struct. Made Composite Mater.*, 2003. no. 3. pp. 29–38. (In Russian).
23. Smagin A. G., Yaroslavsky M. I. *Quartz Piezoelectricity and Quartz Resonators. Energy*, Moscow, 1970. 488 p. (In Russian).
24. Volochko A. T., Podbolotov K. B., Dyatlova E. M. *Refractory and Refractory Ceramic Materials.* Belarusian. Navuka, Minsk, 2013. 385 p. (In Russian).
25. Watson E. S. A differential scanning calorimeter for quantitative differential thermal analysis. *Anal. Chem.*, 1964. vol. 36. pp. 1233–1238.

26. Borchardt B. J., Daniels F. The application of differential analysis to the study of reaction kinetics. *J. Am. Chem. Soc.*, 1957. vol. 79. pp. 41–46.

27. Bakkara, A. E., *Effect of Additives of Metal Nanoparticles on Combustion System Combustion: Dis. ...* Dr. Philosoph. (PhD): 6D073400. Almaty, 2017. 122 p. (In Russian).

28. Duranti E., Sossi A., Paravan C., Deluca L. T., Vorozhtsev A. B., Gromov A. A., Lerner M. I., Rodkevich N. G., Savin N. *Nano-sized Aluminum powders as energetic additives for hybrid propulsion: Physical analyses and performance tests. 21st AIDAA Congress*, Venice, Italy, 2011.

29. Bakkara A. E., Smagulova G. T., Sadykov B. S., Atamanov M. K., Lesbaev B. T., Mansurov Z. A. The effect of the addition of metal nanoparticles on the combustion of condensed systems. *VIII International Symposium "Physics and Chemistry of Carbon Materials/ Nanoengineering"*. Almaty, 2014. pp. 144–146. (In Russian).

Chapter 10

Fields of Application of Composite Materials Obtained Using MCT

10.1 Mechanochemical Synthesis of Multifunctional Sorbents Based on Nanostructured Composite Quartz-Containing Systems

One of the practically important applications of silicon dioxide in the form of its various modifications—crystalline, amorphous, polymeric—and in the form of compounds with other oxides (a wide class of silicate materials), as well as in compositions with other substances, including organic substances, is purification of water from various types of pollution. Silicate particles are well enveloped in an oil film and can either be mechanically removed from the surface of water, or sink to the bottom of the tank (reservoir), where oil, petroleum and petroleum products undergo natural decomposition [1].

Sewage of many industries, especially of the metallurgical industry, is polluted with compounds of chromium, manganese, nickel, copper, zinc, cadmium, lead, and other heavy metals. Depending on pH of the medium, the type of anion and the presence of organic impurities, these metals exist in water in the form

Mechanochemical Synthesis of Composite Materials
Zulkhair A. Mansurov, Nina N. Mofa, Tlek A. Ketegenov, and Bakhtiyar S. Sadykov

ISBN 978-981-4800-88-4 (Hardcover), 978-1-003-12081-0 (eBook)
www.jennystanford.com

of ions, complex compounds, or hydro complexes. To remove compounds of heavy metals in water in a dispersed state, clay collectors together with flocculants are used as a coagulant [2]. Clay minerals also provide water purification from ion-soluble impurities of heavy metals. On the basis of clays, complex sorbents with polyacrylamide and polyvinylbenzyltrimethylammonium are obtained, which provide high-quality water purification from heavy metal ions. More effective in this regard is vermiculite.

Zeolite, shabazite, and other natural aluminosilicate dispersed materials, as well as waste products of similar composition are widely used as sorbents in various types of production. Strengthening their sorbing ability is provided by various types of modifying treatment, in particular by carbonization. In this case, composite systems are obtained, which have shown themselves to be effective as sorbents of heavy metal ions. The content of carbon in the carbonized sorbent was 1–5%, and the adsorption activity with respect to heavy metals reached 20% and 60% for cadmium and lead, respectively.

Silicate materials of various structures and forms [3–5] were and remain the most frequently used for water purification from oils and petroleum products. Different methods have been developed for producing particles from silicon dioxide in the form of hollow microspheres, freely floating on the surface of water and absorbing oils [6, 7]. The materials are brought to a high degree of dispersion in mills and are treated with different organic compounds to form shells on the surface of particles, which provide high hydrophobicity and enhance their sorption properties [8, 9]. Porous silica (silica gel) is one of many forms of amorphous silica. Silica gel has found several different applications: in sorption, catalysis, as a carrier of catalysts and filler [10]. A new field of application of silica as a selective adsorbent appeared as a result of chemical modification of its surface with organic substances.

Among silicate materials, polymeric (or silicone) compounds occupy a special place. On the one hand, they have high sorption capacity and, on the other, chemical resistance to aggressive corrosive media [11, 12]. Much attention has been paid to the stability of siloxane bonds in various systems and conditions of exposure to alkaline and acidic media. This factor is extremely important in the manufacture of complex sorbing and filtering systems consisting of silicate and polymer components of the composition. The use of polymeric materials, in particular super-crosslinked polystyrene, in production of microfiltration systems, has recently become more and more common for the purification of water, process fluids, liquid food products, drugs, etc.

Various methods are being developed to enhance the sorption capacity of silicon-containing materials. As a rule, they are brought to a high degree of dispersion and are treated with various organic compounds to form shells on the surface of particles providing their high hydrophobicity and enhancing their sorption properties [13, 14]. Carbon, quartz, as well as ashes and slags from various industries, in particular those containing quartz and phosphogypsum, are treated with organic substances in order to ensure their maximum hydrophobicity [15]. Fuel oil, diesel, and liquid bitumen may serve as a water repellent [16, 17].

In addition to chemical treatment, mechanochemical modification of sorbents is used [18]. This process allows us to simultaneously bring the material to the desired degree of dispersion, modify the surface of dispersible particles and obtain plastic, well-moldable mass. The latter fact is very important in obtaining durable granules. Mechanochemical treatment extends the range of modifiers used. In particular, the use of carboxylic acids makes it possible to purposefully form active centers on the surface of various layered silicate systems [19]. The volume of scientific papers and publications in the press, as well as those presented at the international conferences on mechanochemical modification, has increased significantly in recent years, reflecting the prospects and effectiveness of work in this direction.

10.1.1 Modified Quartz as a Sorbent Material for Water Purification from Various Types of Pollution

Studies on the changes in the structure and morphology of quartz particles during MCT showed that when the material is crushed in the presence of organic additives, nanoscale compounds are formed on the surface of particles, which results in a significant change of their properties, in particular, electric, magnetic and thermal. IR spectroscopy of modified quartz particles showed that the characteristic feature of the surface is the presence of intense siloxane groups capable of forming radicals upon contact with various organic and inorganic substances. The appearance of hydroxyl (siloxane) and carboxyl groups, which are the active sorption sites of organic and inorganic atoms and molecules, clearly indicates a change in the chemical activity of quartz after its mechanochemical treatment. These results on the change in the structure of the material are similar to the widely used chemical modification of silica in the preparation of selective sorbents [20].

The first indicator of the increase in the sorption capacity of silica modified during MCT is the results of determining the specific pore volume by the absorption of benzene vapor (Table 10.1). This characteristic is an indicator of the adsorption capacity of the powder material.

Table 10.1 Specific pore volume of quartz modified in the course of MCT for 20 min, depending on the type and amount of organic and inorganic additives

Organic additives, %	**Specific volume of pores V, cm^3/g**						
	NaCl, %				C, %		
	0	1	3	5	1	3	5
AA* 3	0.27	0.42	0.45	0.40	0.47	0.59	0.67
5	0.23	0.39	0.40	0.45	0.51	0.67	0.43
7	0.21	0.29	0.35	0.37	0.35	0.55	0.42
10	0.17	0.25	0.32	0.47	0.31	0.40	0.49
PS** 3	0.39	0.45	0.52	0.67	0.49	0.62	0.75
5	0.30	0.49	0.62	0.69	0.47	0.67	0.77
7	0.32	0.5	0.60	0.69	0.47	0.69	0.71
10	0.36	0.52	0.67	0.71	0.39	0.37	0.37

*AA, acrylic acid; **PS, polystyrene.

For modified quartz, the specific pore volume increases by 30–40 times or more, especially when organic substances with a high polymerization and destruction activity are used as a modifier, followed by depolymerization. Such modifiers, in particular, are acrylic acid and polystyrene. Variation in the amount of organic additives (from 5% to 10%) in combination with carbon and sodium chloride, which affect the process and degree of polymerization, significantly changed the value of the adsorption capacity of the material. Moreover, these changes are ambiguous. Judging by the $V_{specific}$ values, the optimal amount of acrylic acid, providing an effective increase in the chemical activity of modified quartz, is 3%, with the additional introduction of 3% NaCl or 5% C. When used as a modifier of polystyrene, the variations in the choice of the optimal composition of modifying additives are broader. When modifying quartz using 3% and 5% polystyrene, it is effective to additionally introduce 5% carbon into the mixture. When using a larger amount of polystyrene during MCT of quartz, it is desirable to add up to 5% sodium chloride into the treated mixture.

When using powdered sorbent material by its functional purpose, as a rule, it is subjected to granulation. To obtain granules, it is important to choose a binder, which, first of all, must maintain the sorption capacity of the material being molded and optimally approach the structure of the sorbent. The molding method must also ensure the strength of the granules. After testing several types of ligaments, liquid glass diluted with water in the ratio of 1:2 was chosen as a binder. To obtain a water-insoluble binder, liquid glass was solidified with sodium silicofluoride in a stoichiometric ratio to liquid glass.

The choice of the binder was determined by the fact that, according to IR spectra, there are ≡Si–OH centers on the surface of the modified quartz, regardless of the type of modifier, i.e., silicic acid. After molding, granules with a size of 1–2 mm were subjected to heat treatment (at 70–80°C for 2 h), ensuring their strength and resistance to abrasion, water resistance, and high adsorption capacity. Numerical values for the specified characteristics in the averaged version for granules of quartz, modified by various additives, are given in Table 10.2.

The sorption capacity of the granular material was investigated in relation to organic matter and heavy metal ions. Pellets were used as a fixed bed in columns. According to the results of research on the quality of water purification from mineral oils, it can be seen [20] that the most effective is water purification with quartz, modified polystyrene in the presence of NaCl and carbon (Table 10.3). The optimal time of mechano-chemical treatment was 30 min.

Table 10.2 Characteristics of granulated modified quartz

Material	Material quality indicators			
Quartz + modifier	The strength of individual granules, g	The resistance of the granules to abrasion, %	Phenol adsorption capacity, mg/l	Water resistance, %
	19–36	85–93	138–217	93–97

Table 10.3 Purification of water from oil by filtration through a granular sorbent

	Purification degree, %		
	Mechanochemical treatment time, min		
Material	10	20	30
SiO_2 + 3% AA*	37	29	31
SiO_2 + 3% PS**	43	67	63
SiO_2 + 3% C	39	63	62
SiO_2 + 3% AA* + 5% C	45	44	60
SiO_2 + 3% AA* + 5% NaCl	54	39	50
SiO_2 + 3% PS** + 5% C	36	72	77
SiO_2 + 3% PS** + 5% NaCl	68	75	82

*AA, acrylic acid; **PS, polystyrene.

To determine the degree of purification depending on the volume of oil-containing water, granulated sorbent modified with polystyrene was used. It was stated that for the analyzed sample of the sorbent, a degree of purification is at least 50% takes place when the volume of polluted water is not more than 250 cm^3 (Fig. 10.1).

It follows from the calculations that the working capacity of the sorbent is 17.5 cm^3 of polluted water per 1 g of sorbent. After spilling 500 cm^3 of polluted water, the sorbent underwent thermal regeneration (roasting at 200–300°C for 1 h). After regeneration, not only the properties of the sorbent are restored, but also the degree of water purification from oil increases to 98.8%. The quality of the synthesized sorbents is due to the fact that they are composites in two levels: first, the particle itself consists of a non-porous substrate (quartz) and a surface modified with polymeric compounds; secondly, composite particles are combined into conglomerates—the actual sorbent. By selecting specific modifiers, it is possible to provide selective wastewater treatment against organic compounds of various classes.

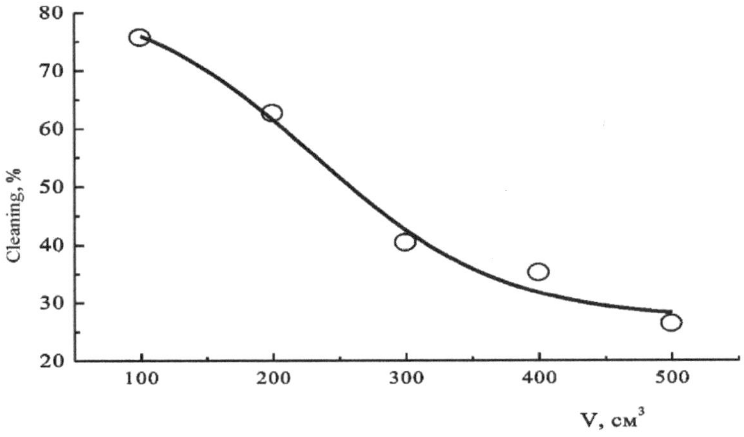

Figure 10.1 The degree of water purification from oil, depending on the volume of oily water.

Experimental samples of sorbents, the composition of which is indicated above, were also used to purify water from Fe, Pb, and Mn ions. Studies were carried out by filtering water in the columns through a fixed bed of granular sorbent. The results of removing of Fe ions through a chromatographic column with granules of the sorbent material are presented in Table 10.4. Acrylic acid is the most effective modifier, providing high sorption capacity of quartz sorbent.

Purification of water from lead ions was also carried out through granules in a chromatographic column. Quartz modified by all selected substances gives a degree of purification of up to 90% or more. X-ray phase analysis of the sorbent after purification showed the presence of lead compounds in the material and a decrease in crystallite size, this indicating the procedure of chemical sorption on the filters. X-ray analysis confirmed quantitatively the presence of lead in the spent sorbent. The most effective purification of water from manganese ions was also obtained by a material modified with acrylic acid in the presence of sodium chloride. X-ray analysis of the used sorbent confirmed the presence of manganese in the material in an amount corresponding to the results of chemical analysis. When water is purified from sulfates with a sorbent synthesized by MCT, the degree of purification is 70–80%, which may be of interest for power plants.

Table 10.4 The degree of water purification from iron, lead, manganese and sulfur ions, depending on the composition of the granulated sorbent after mechanochemical treatment

	The degree of water purification from ions of various elements, %			
Material	**Iron**	**Lead**	**Manganese**	**Sulfur**
SiO_2 + 3% AA*	87.0	92.0	87.0	77.0
SiO_2 + 3% PS**	82.0	94.0	67.0	52.0
SiO_2 + 3% AA + 5% NaCl	89.7	89.2	91.0	65.0
SiO_2 + 3% AA + 5% C	93.2	92.4	88.0	80.0
SiO_2 + 3% PS + 5% NaCl	86.7	91.0	75.0	23.0
SiO_2 + 3% PS + 5% C	32.7	88.5	62.0	68.0

*AA, acrylic acid; **PS, polystyrene.

As a result of a comprehensive analysis of all the above-considered experimental data on structural changes in silica modified by MCT and on its sorption capacity to various water pollutants, the sorbents were completed with modifiers and their activity to organic and other types of pollution. Modification of the quartz particle surface in the process of mechanochemical treatment leads to an increase in the specific surface, specific pore volume, enhancement of its structural characteristics, and ultimately to an increase in the sorption activity of the synthesized material [21–23]. Introduction of sodium chloride simultaneously with the organic additive affects the formation of the structure of the modified layer, changes its density, porosity and the overall value of the specific surface. The presence of sodium chloride also enhances grinding of quartz particles and the degree of dispersion of the material and, hence, its sorption characteristics.

10.1.2 Magnetic Sorbents Obtained by Mechanochemical Treatment of Quartz-Containing Systems to Collect Oil from the Water Surface

In Chapter 5, it was shown that the mechanochemical treatment of quartz leads to strong structural changes that ensure the appearance of induced ferromagnetism in dispersed and modified particles. The observed effect of magnetization of non-magnetic materials can be practically implemented to obtain sorbing magnetic materials, in particular for collecting oil products from the surface of water.

Sorbents with magnetic properties differ, as a rule, in increased absorption capacity with respect to organic and inorganic substances. Powder materials based on quartz with magnetic properties synthesized in our works [24–26] by a mechanochemical method also proved to be active sorbents in water purification from oil and oil products spills. For such purposes, polystyrene was the most effective modifier, and quartz was used in a mixture with ash and slag. Ash-and-slag wastes of the energy industrial complex (ashes and slags of heat and power plants) are ready-made mixtures of various oxides: SiO_2, Al_2O_3, Fe_2O_3. In addition to crystalline oxide compounds and mullite, ash and slag contain in large quantities the amorphous component—the glass phase and partially unburned coke (carbon) in an amount of 3–10%.

Optimal indicators of both the sorption capacity and the magnetic characteristics of the powder were obtained with a certain amount of quartz and polystyrene mixed with ash and slag. This clearly follows from the volume diagram of the dependence of the magnetic permeability μ_{rel} on the concentration of the additionally introduced polystyrene into the system of quartz and modifier at 20 min of mechanochemical processing (Fig. 10.2).

With an increase in the amount of quartz from 20% to 40%, the magnetic permeability of the system increases from μ_{rel} = 6–8 to μ_{rel} = 16–18, depending on the time of mechanochemical processing and the concentration of the modifying additive. Introduction of a modifying additive into the system renders an ambiguous action on the value of magnetic permeability. Thus, in a system containing 20% additional SiO_2, the highest magnetic permeability is observed after 20 min of treatment with the introduction of 10% polystyrene. For a system in which there is 30% of additionally introduced quartz, the magnetic permeability is μ_{rel} = 14.5 after 20 min of treatment with the introduction of 5% polystyrene. A similar picture is observed for the system with the content of 40% of the additionally introduced quartz, but the magnetic permeability of the material increased to 17.5. Therefore, to obtain high values of magnetic permeability of the mechanically treated ash-slag mixture, additional introduction of quartz into it is necessary, which is explained by formation of a special microstructure (or nanostructure) with formation of ferromagnetic compounds on the surface of quartz particles. In addition, since quartz during the mechanochemical processing acts as an electromagnetic emitter, the presence of its particles provides for the directed formation of the structure and properties of the mixture being processed.

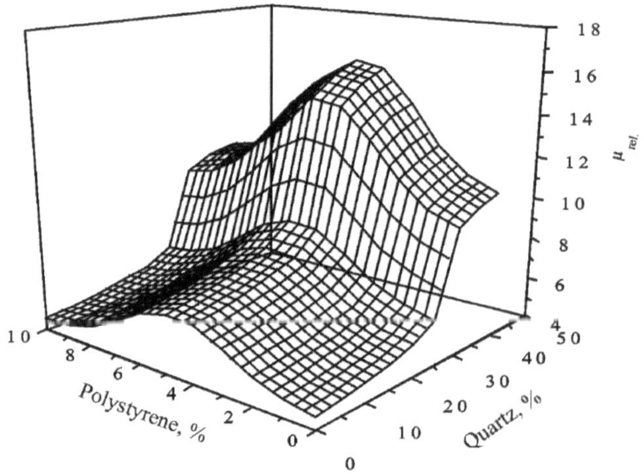

Figure 10.2 The dependence of the magnetic permeability of ash on the quartz content in the presence of a modifier.

The results obtained show how strongly the magnetic properties of mechanically treated ash and slag depend on the content of the optimum amount of quartz in it. This indicates the fact that the magnetic properties manifest themselves in the synthesized material not so much due to the number of iron-containing components in the system, but owing to the surface structure features of quartz particles with ferromagnetic compounds and a protective organometallic (polymer) film incorporated [35–37]. Quartz, in this case, is a kind of matrix, the basis for the synthesis of magnetic sorbent by mechanochemical processing, and ash and slag is the raw material containing the necessary components, both organic (carbon) and iron oxides for modifying the surface of quartz particles. The optimal amount of quartz, i.e., crystalline SiO_2, for the synthesis of an effective magnetic material from an ash and slag mixture is 45% magnetic permeability of such a mixture reaches μ_{rel} = 27.5–28 (Fig. 10.3).

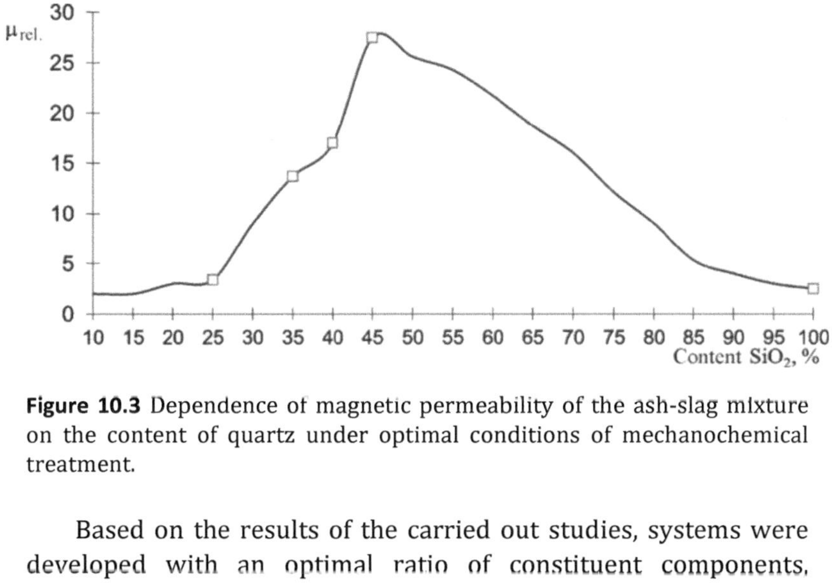

Figure 10.3 Dependence of magnetic permeability of the ash-slag mixture on the content of quartz under optimal conditions of mechanochemical treatment.

Based on the results of the carried out studies, systems were developed with an optimal ratio of constituent components, including ash and slag, quartz, oxides and iron salts, as well as organic modifying additives, which, after a short treatment in a mechanical reactor (centrifugal planetary mill), showed a rather high magnetic permeability, specific surface area and specific pore volume (Table 10.5). The choice of the composition and conditions of mechanochemical processing, determining the formation of the physicochemical properties of the material, depends on the specific purpose of the resulting magnetic powder material. As a result, the composition of a magnetic quartz-containing sorbent was developed for water purification from oil and oil refining products: ash (20–80 wt.%), quartz (20–80 wt.%), organic component (5–8 wt.%), iron oxide or chloride (2–5 wt.%). The sorbent, based on this mixture, was obtained with a small consumption of electricity (processing in a mechanical reactor for 10–20 min) [40].

Table 10.5 Specific surface area and specific pore volume of ash and slag systems after mechanochemical treatment

Material	T_{MCT}, min	$S_{specific}$, m^2/g	$V_{specific}$, cm^3/g
Fly-ash (initial)	—	6.24	0.01
Fly-ash (MCT)	10	8.08	0.01
Fly-ash+quartz (20%)	20	11.1	0.19
Fly-ash+quartz (30%)	20	75.0	0.29
Fly-ash+quartz (40%)	20	89.6	0.36
Fly-ash+quartz (20%)+Fe_xO_y(10%)	20	110.0	0.45
Fly-ash+quartz (20%)+polystyrene(10%)	20	145.0	0.42
Fly-ash+quartz (20%)+Fe_xO_y(10%)+ethylene glycol (5%)	20	270.0	0.50
Fly-ash+quartz (20%)+Fe_xO_y(10%)+stearin (5%)	20	340.0	0.60
Fly-ash+quartz (20%)+Fe_xO_y(10%)+oil(5%)	20	350.0	0.65

Fe_xO_y, mixture of iron oxides (Fe_2O_3/Fe_3O_4 – 3/1).

To determine the absorption capacity of the obtained sorbent, it was scattered in an amount of 5 g per 10 g oil spill, spread on the water surface, then the sorbent saturated with oil was collected with a magnet and the mass of the entire mixture (powder + oil) was determined. The absorption capacity of the sorbent oil was calculated as the ratio of the mass of the powder and oil mixture to the amount of sorbent used. Its value was 0.54–0.78 g/g, depending on the composition. Thus, joint processing of ash and slag in a mechanical reactor leads to a high degree of material dispersion and amorphization of particles. Corundum (Al_2O_3), which is present in ash and slag and has a high hardness, contributes to a finer grinding of the input components, and the glass phase, in turn, coagulates around solid quartz particles, increasing the amorphous layer. This contributes to an increase in the specific surface area and specific pore volume of the synthesized magnetic sorbent.

Table 10.6 shows the limiting values of the main physico-chemical characteristics with recalculation of the values according to the standard system for magnetic sorbents synthesized by mechanochemical treatment. The carbon component of the ash and slag during processing in the mill participates in the reduction of oxides, in particular iron oxides to ferromagnetic forms, envelops mineral oxide particles and forms carbide compounds. Introduced as modifying additives complex organic compounds (polystyrene, stearin and oil) are involved in the polymerization and the formation of nanostructured layers on the surface of the modified particles. The synthesized structures ensure stability of the magnetic properties of the resulting heterogeneous material, on the one hand. On the other hand, due to the chemical "affinity" for oil and oil products, the introduced modifiers help to improve the quality of hydrotreating with the synthesized material. Oxides and iron salts are introduced to ensure the required level of the magnetic properties of the powder due to the optimal amount of ferromagnetic particles formed by their reduction with carbon in the process of mechanochemical treatment.

Table 10.6 The main characteristics of the magnetic adsorbent

Composition	Specific surface area, m^2/g	Absorption capacity, g/g	Magnetic susceptibility, $\times 10^6$
Ashes, quartz, organic component, Fe_xO_y or $FeCl_3$	340–350	0.54–0.78	28.9–33.9

Thus, in a mechanical reactor a magnetic material is synthesized, which has all the characteristics of an adsorbent, such as a high specific surface area and specific pore volume sufficient for sorption of petroleum hydrocarbons, this being most important in water treatment processes. The magnetization ability of the material ensures the effectiveness of the material usage in purification of large volumes of water in natural and artificial reservoirs and the subsequent magnetic separation of the collected mixtures (powder + oil).

The problem of water purification from oil and oil products spillage is important for the Republic of Kazakhstan [27–30]. The synthesized magnetic adsorbent can provide an environmental safety solution at oil production, oil refining and transportation enterprises in the Kazakhstan region of the Caspian Sea, as well as at industrial enterprises for the treatment of wastewater and sedimentation tanks. The synthesized material was tested in laboratory conditions on a pilot plant [3–34], the scheme of which is presented in Fig. 10.4.

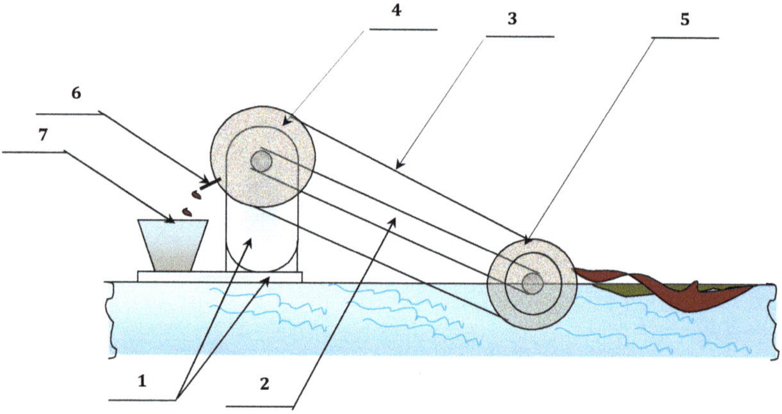

Figure 10.4 Scheme of a pilot plant for collecting oil from the surface of water. (1) supporting structure; (2) conveyor frame; (3) conveyor belt; (4) upper drive roller; (5) lower magnetic roller; (6) oil product remover; (7) tank for collecting oil products.

The quality of the material was estimated by the magnitude of the absorption capacity of petroleum products by sorbents, as the ratio of collected oil to spilled (in%). Such method of evaluation was chosen because purification of water from oil products with similar materials leads to the formation of colloidal solutions, which reduces the mobility of the adsorbed substance particles. The experiment was conducted on eight compositions, different in their adsorption, coagulation (degree of compressibility of the oil slick) and magnetic characteristics (Table 10.7).

The variation in the composition of the mixtures leads to a significant change in the indicators of properties, first of all, due to the changes in the structure of the material both in volume and on the surface of the particles, ensuring the process of oil sorption. The most important indicators for the synthesized adsorbents are the specific pore volume (i.e., sorption capacity) and magnetic permeability. The first one determines the possibilities of oil (or oil products) absorption by it, the second one is the efficiency of the material during collection and separation of the collected oil/powder mixture. According to the data presented in Table 10.7, in this case compositions IV, VI, and VIII should be the most effective.

Table 10.7 Composition and main characteristics of systems designed to collect oil from the surface of water

	Main components		Additives					
No	Quartz, %	FA*, %	IC**, %	Organic, %	ρ_{bulk}, g/sm^3	S_{spec}, m^2/g	V_{spec}, sm^3/g	M_{rel}
I	20	80	—	—	0.98	75.0	0.29	28
II	20	75	Fe_3O_4–5	—	0.88	145.0	0.42	30
III	20	70	$FeCl_3$–5	Oil–5	0.79	350.0	0.65	29
IV	25	60	$FeCl_3$–5	Oil–10%	0.83	250	0.48	33
V	20	65	Fe_3O_4–5 $FeCl_3$–5	Stearin–5	0.91	340.0	0.60	27
VI	25	60	$FeCl_3$–5	Stearic–10	0.73	395	0.73	22
VII	20	70	$FeCl_3$–5	Polystyrene–5	0.81	180	0.55	19.5
VIII	30	60	—	Polystyrene–10	0.89	310	0.65	27.5

*FA, fly-ash; **IC, iron containing.

The results of collecting oil from the surface of water with selected compositions showed that for composition III, collection of oil when using 20–30% magnetic powder to the amount of spilled oil amounts to 92–94%, which is associated with the introduction of coagulant $FeCl_3$ into this composition. When using organic additives in a mixture of mechanochemically synthesized magnetic adsorbent, the quality of oil recovery significantly increases (up to 98%) with a smaller amount of used adsorbent. For Composition IV, in spite of the highest values of magnetic permeability, water purification from oil slicks (spills) does not exceed 95% when using 60% or more powder per volume of spilled oil. The effectiveness of compositions V and VI in their practical use for collection of petroleum products is ~20%. Optimal indicators on the quality of collection were obtained using compositions VII and VIII, modified with polystyrene. Moreover, polystyrene itself, degraded by the MCT process, plays the role of a coagulant, providing constriction of oil spills by 50–60% with the ratio of the mass of the used adsorbent to spilled oil in the amount of 10–20%.

Table 10.8 Changes in magnetic permeability (μ_{rel}) and oil recovery quality (M, %) with repeated use of compositions

Test stages	Compositions							
	I	II	III	VI	V	VI	VII	VIII
				Magnetic permeability (μ_{rel})				
1	28	29	30	35	27	22	19.5	27.5
2	28	19.2	13.1	29	24.5	20	18	27.0
3	26.5	17	11	20	16.5	17	17	26.5
4	17	13	7	13	16	16	17	26.0
				The amount of oil collected M, %				
1	92	87	92.5	92.5	95	97	97	99
2	90	68	70	80	87	92	95	96
3	45	66	60	72	45	63	69	89
4	33	55	45	4252	6242	48	63	72

Oil from the collected mixture with magnetic powder was extracted by magnetic separation. Contamination of oil with magnetic powder is not more than 5%. In this case, the resulting oil can be used for its intended purpose, as a raw material for production of fuel, and a magnetic adsorbent with residual oil in the amount of 2–3 mass. % was decided to reuse to collect oil spills. Table 10.8 presents the results of repeated use of the magnetic adsorbent in collection of oil when the ratio of the mass of the adsorbent powder to the mass of oil is 0.5 g: 1.0 g. In addition to determining the quality of collection by the adsorbent, we measured the magnetic characteristics (μ_{rel}) after each collection stage. The first composition, having a high magnetic permeability, retains its rather high values when it is used in the collection of oil for the fourth time, but the indicators for the quality of oil collection decrease sharply in the third stage of using the powder. Compositions from the second to the seventh one are reinforced with coagulants, as a result, the percentage of oil recovery is significantly increases, up to fourfold use of the material. A similar picture is observed for composition VIII, which does not contain ferric chloride. The material modified with polystyrene (VII and VIII), depending on the amount of the added additive has a different value of magnetic permeability from 19.5 to 27.5. At the same time, throughout all the stages of collection, this indicator does not deteriorate, and the material has a high level of water purification from petroleum products.

Thus, analyzing the obtained results, it can be concluded that the magnetic adsorbent can be used repeatedly (up to three to four times) to collect oil spilled on the surface of water, using only magnetic separation to extract the magnetic sorbent from the collected oil. For the regeneration of the used magnetic adsorbent, it was placed in a drying apparatus at a temperature of 200–300°C and dried for 2–3 h. Then, this material was reused to collect oil from the surface of water. The sorption capacity and magnetic permeability values were determined. It is found that all indicators of adsorption activity and magnetic permeability are restored to almost the original values. Thus, the magnetic adsorbent after two-, threefold collection of oil from the surface of water can be regenerated by shot-time annealing at 200°C until the initial characteristics are almost completely restored. In this case, carbon of partially burnable oil plays an important role, i.e., there proceeds an additional carbonization of the magnetic adsorbent surface.

The quality of collection is determined by the process of interaction of oil with the surface of the powder material particles. The mechanism of molecular interaction, in particular, adsorption forces, is one of the important problems of surface physics and chemistry. Experimental study of various cases of adsorption reveals that the relationship of the adsorbed molecule with the adsorbent can vary from weak molecular potentials to high chemical energies [35, 36]. It is known [37] that the surface of any solid is covered by a double electric layer. Whereas oil molecules are dipoles, the interaction between the surface of the adsorbent and the oil film is quite obvious. The external surface of the synthesized magnetic adsorbent is inhomogeneous, which was clearly shown by the results of electron-microscopic analysis.

Figure 10.5 shows a phased sequence of interaction between a magnetic adsorbent and an oil film. In the places of contact of the adsorbent particles and the oil film, interaction occurs due to intramolecular forces, which leads to a decrease in the density of the oil film (Figs. 10.5b,c).

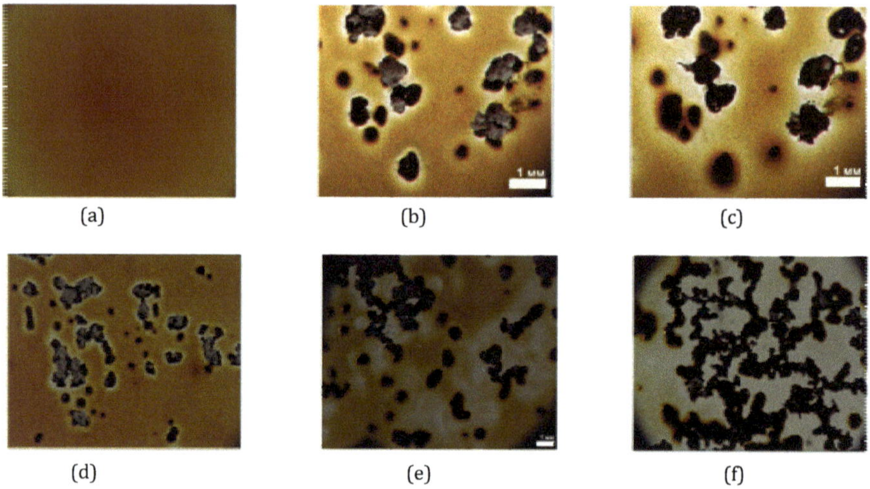

Figure 10.5 The process of interaction of the magnetic adsorbent and oil film on the surface of water.

Conglomerates powder + oil are characterized by some magnetic moment, the presence of which leads to attraction of particles to each other (Fig. 10.5d). An increase (summation) of the magnetic moment of such complexes enhances the effect of their attraction and the formation of dendride-like structures (Fig. 10.5e,f). Such a structure provides a rigid framework and stability of the position of an oil slick on the surface, at least in calm water conditions. Without external influence, such a system can be in equilibrium for an unlimited time. If it takes the entire process 3–6 min to proceed from the initial moment to the structured stable phase depending on the concentration of oil and sorbent, the framework itself is in the equilibrium state for 2–3 days. However, in natural conditions there can be no absolutely calm water due to hydrodynamic movements, as well as during operation of an electromagnetic device for collecting magnetized oil film. Therefore, in the process of structuring a second stable phase is formed—this is a complete coagulation of oil around the magnetic sorbent.

The hydrodynamic impact resistant and magnetically controlled clot of coagulated oil can be assembled into floating vessels using a magnetic conveyor [38]. Thus, the adsorbent developed on the basis of a quartz-containing ash-slag mixture using mechanochemical modifying treatment is characterized by a high level of basic characteristics: coagulating ability for oil and oil products: hydrophobicity, sorption capacity, and ferromagnetism. With an optimal ratio of components and conditions of mechanochemical treatment, providing a high specific surface area and organization of its active structural elements, the adsorbent performs almost complete cleaning of the water surface from oil spills.

The whole process of water purification from oil pollution by the adsorbent can be gradually represented, starting from the moment of immersion of individual particles (or their conglomerates) of the sorbent in the oil film on water and interaction with it according to the micelle formation scheme [39]. This process boils down to the fact that on the surface of a solid particle, due to its hydrophobicity, the process of adhesion of sorbate (oil) molecules and sticking into dense colloid formations enveloping the adsorbent particles as the core of the micelle begins. In addition, for the units of the adsorbent there starts the process of movement and introduction of oil into micelle complexes. The film of oil becomes thinner.

Due to the fact that the adsorbent particles have magnetic properties, the resulting micelles also exhibit magnetic properties. Primary complexes begin to contract between themselves, forming the structure of a mixture oil + sorbent. At the same time, the process of oil absorption by moving micellar complexes continues. The particles of the magnetic adsorbent get arranged so that to form an internal, fairly rigid "dendride-like" structure inside the remaining oil film on the surface of water. The number and branching of such colonies of micelles is determined by the ratio of the amount of spilled oil and the powder scattered over it—a magnetic adsorbent. With the complete exhaustion of the sorption capacity of the particles of the powder adsorbent, a stable structure of oil + powder is formed, ensuring stability of this mixture on water. The oil film does not spread, but is fixed on the surface of water and becomes magnetically controlled. To enhance the effect of restriction and complete purification from oil, it is necessary either to add an additional amount of magnetic adsorbent or to ensure the mobility of the remaining oil by "agitation" of the water surface. In this case, complete coagulation of the oil by the sorbent occurs. The resulting mixture exhibits magnetic properties and can be easily controlled when moving on water and being collected in a tank in various ways, including using a magnetic conveyor.

Thus, from the presented results, it follows that mechanochemical synthesis is a rather effective way of obtaining new materials—sorbents for various purposes. Selection of modifying additives, their quantitative ratio and the time of MCT determine the selectivity of the action of the obtained powder materials during water purification from various types of contamination (organic compounds, heavy metal ions and sulfates). Quartz-containing waste from the energy complex of the Republic of Kazakhstan (ash and slag) is a valuable raw material for mechanochemical synthesis of a magnetic adsorbent intended for hydrotreatment from oil spills and products of oil refining. During the mechanochemical treatment of ash and slag, the carbon contained in it acts as a water-repellent surface and modifier of the oxide part of the system, contributing to a change in the structure and manifestation of the magnetic properties of the synthesized material.

During the mechanochemical modification of quartz with organic carbon-containing compounds, nanostructural re-organization of the surface takes place depending on the type of organic modifiers. The presence of metallic iron nanoparticles in the organic film encapsulating the particle was found in the surface layer. The formation and growth of carbon films on the surface of quartz particles indicate the sequence of transformation processes of the structure of modified quartz during its mechanochemical processing. The heterogeneous distribution of reduced carbon on the surface of a modified quartz particle, i.e., the creation of carbon complexes, and the simultaneous presence of iron particles is the basis for the representation of the modified surface as a set of galvanic couples, determining the sorption activity of the synthesized material.

Questions

1. Describe the chemical modification as a method of enhancing the absorptive capacity of dispersed materials.
2. What are the advantages of mechanochemical modification of sorbents?
3. What are the mechanochemical modification of quartz as a method of changing its chemical activity in the preparation of selective sorbents?
4. What modifying organic and inorganic additives in the mechanochemical treatment of quartz provide a high adsorption capacity of the material?
5. Describe the role of sodium chloride in the formation of the structure and properties of the modified polymer layer on the surface of a quartz particle.
6. What are the results of water purification from Fe, Pb, Mn and S ions by filtering water in columns through a fixed bed of granulated sorbent based on modified quartz?
7. How is the magnetization of non-magnetic materials carried out during MCT and obtaining sorbing magnetic composite systems?
8. How is a magnetic sorbent obtained at mechanochemically treated mixture of quartz with ash and slag and polystyrene for collecting oil products from the water surface?
9. Describe the magnetic properties of the material synthesized in the course of MCT and their relation to the peculiarities of the surface structure of quartz particles with inclusions of ferromagnetic compounds and a protective organometallic (polymer) film.
10. What is the dependence of the absorption capacity of a quartz sorbent of petroleum products on the modifying organic additives involved in the polymerization and the formation of nanostructured layers on the surface of modified particles?
11. What are the conditions for the stabilization of the magnetic properties of the resulting heterogeneous material based on quartz?
12. What is the efficiency of collecting oil from the surface of water, depending on the composition?
13. How is recovery of oil from magnetic sorbent carried out by the method of magnetic separation and evaluation of changes in its sorption capacity and magnetic permeability?
14. Describe the methods of regeneration of the used magnetic sorbent.

10.2 Composite Systems with Fillers Modified by Mechanochemical and Ultrasonic Treatment

To increase the mechanical strength, thermal and electrophysical properties of ceramic, polymeric and other systems, fillers of concrete functional purposes are introduced into them. In order to provide the necessary "crosslinking" contact of the filler and the matrix, the surface of the filler material is subjected to treatment with special modifying components. As it was shown in Chapter 7, modification of particles of various minerals by mechanochemical treatment in mechanical reactors and ultrasonic treatment is an effective method of creating dispersed materials with the polymer structure of the surface layer of particles.

Quartz, calcite (marble) and wollastonite are widely used as fillers of various composite systems. Of these, wollastonite is given special attention. It has a number of valuable properties and advantages over other fillers [40, 41], especially when used in the polymer industry (low water absorption effect, low dielectric constant, low viscosity, high wear resistance and heat resistance). With an equal degree of filling, wollastonite ensures preservation of higher indicators of product strength than other fillers in polymer systems, while possessing an orientation effect in the feed direction (for example, during extrusion). Specificity of the properties of wollastonite $CaSiO_3$ is due to the elongated structure of crystals, the splitting of which gives off needle-shaped grains with clearly pronounced spatial-geometric anisotropy. The needle-shaped grain of wollastonite (Fig. 10.6) determines the main direction of its use as a micro-reinforcing filler [40–42].

It is used to reinforce coating films of paints and varnishes and to increase their durability and wear resistance. It is a substitute for harmful substances such as asbestos and talcum fiber, and is also used in the paper industry and dry construction mixtures. Currently, wollastonite has received the greatest application as a filler in the production of polymeric materials and plastics [42]. In the plastics industry, wollastonite is used in the manufacture of vulcanized and thermoplastic (molded and cast) rubber, bitumen and vinyl floor tiles, vinyl, polyester and epoxy resins and vinyl plasters and phenolic forms, as well as other materials.

Figure 10.6 The morphology of wollastonite particles [42].

Of particular interest is the use of wollastonite as a filler for epoxy resin sealants, as it reduces shrinkage, water absorption and improves the resistance to thermal shock, viscosity and stability. It also contributes to the improvement of heat resistance, high resistance to shock loads, chemical resistance. The elongated crystal of wollastonite with a chemically applied coating tends to be oriented in the direction of flow of the resin. Free carbon and hydrogen bonds coming from the coated particles are combined with free C–H chains of the resin and form chemical bonds between the filler and the resin.

In [43–45], the results of systematic studies of polymer compositions based on epoxy resin ED-20 and emulsion PVC, filled with natural wollastonite of the Sinyukhinsky deposit of the brands Voxil M 100, Vauxel M 300, Voxil M 1000 are presented. The introduction of wollastonite with l/d = 16–18 into a thermoplastic matrix of emulsion PVC in an amount of 30% leads to an increase in tensile strength. The introduction of wollastonite into an epoxy binder modified with acrylic copolymer (based on ED-20) leads to an increase in tensile strength by 80–100%, and shear strength by 15–20%. The possibility of grinding wollastonite by ultrasound is also shown. It was revealed that with the frequency of ultrasonic vibrations of 22.5 kHz, power 200 W, intensity 3.5 W/cm^2, exposure time from 30 to 90 min in an aqueous medium with 0.5–1.0% surfactant in wollastonite Voxil M 100 there takes place a change in the factor of anisotropy l/d from 5–6 to 7–8 with a exposure time of 60 min and up to 8–10 with a exposure time of 90 min.

The most important factor for the use of wollastonite in polymer systems is modification of its surface by compounds, which ensure compatibility of particles with the polymer matrix [46]. To solve this problem, the use of mechanochemical and ultrasonic treatment is effective, providing both dispersion and change in the structure and surface state of the particles to be ground. It is necessary to select the most effective conditions for modification of wollastonite for subsequent use as a filler and to obtain polymer-composite systems, in particular, based on epoxy resin.

It follows from the results presented in Chapters 6 and 7 that the pretreatment of wollastonites in a mechanical reactor and in an ultrasonic (US) mode with different frequencies has a significant effect on the structural changes in the particles, which affects their properties. The optimal time for the MCT and UST, as well as the frequency of ultrasonic exposure, ensuring the effectiveness of the structural and morphological changes of the processed wollastonite are determined.

After the MCT and UST wollastonite powder was used as a filler in the manufacture of a polymer system with an epoxy binder. First of all, the dielectric constant of the composition was measured. The values of dielectric constant allow us to estimate not only the electrical insulating characteristics of the polymer composition, but also to give a qualitative assessment of the change in the structure of the material depending on the state of wollastonite, which is used as a filler in an epoxy matrix.

In the presence of wollastonite, the polymerization process and the formation of the composite structure take place with participation of the filler, which, firstly, plays a reinforcing role, and secondly, the surface of the particles can participate in the formation and development of the cluster structure of epoxy polymer. The polymer based on epoxy resin, as is known [47, 48], has good electrical insulating properties. The dielectric constant values of the prepared epoxy samples correspond to 3.032. The dielectric constant of the wollastonite powder in the initial state is 2.27. With the introduction of up to 30% wollastonite into epoxy resin, the ε values for the composition slightly increase (up to 3.08), and with a further increase in the injected filler (up to 40–60%), the dielectric constant again decreases to 3.035. A further increase in the content of filler in the epoxy composite again leads to an increase in the values of ε (Fig. 10.7).

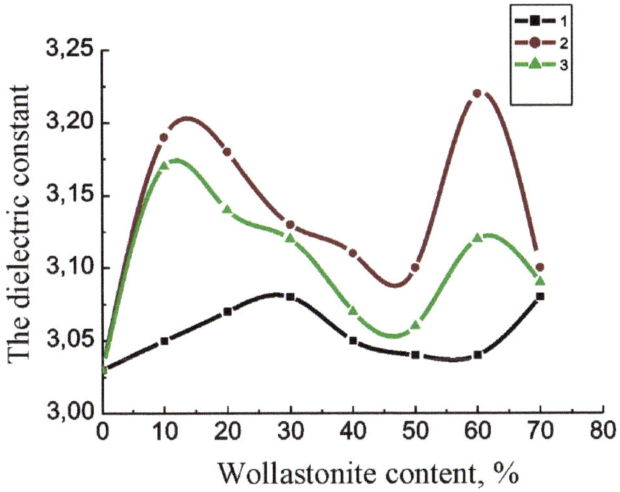

Figure 10.7 Dependence of the dielectric constant of the composition epoxy polymer + wollastonite on the amount of wollastonite in the initial state (1) and after MCT for 10 (2) and 20 min (3).

Thus, the dependence of the dielectric constant of an epoxy composition on the content of wollastonite is characterized by the presence of several areas with extreme changes in the values of ε—this is at 20–30%, 40–60% and 70% of wollastonite. The obtained patterns indicate both the fact of the presence of the filler in the composition and the peculiarities of formation of a cluster structure with a certain amount (40–50%) of the needle-shaped filler in the polymer composition when the orientational order in the structure of the bulk matrix changes [49].

After MCT of wollastonite and its introduction into the composition of the epoxy polymer, the dielectric constant of the composite increased slightly, but the patterns of the effect of the wollastonite amount on the change in ε remained. Two maxima are also observed: in the area of low concentrations (10–30%)

and high (60–70%). The highest values of ε showed samples with wollastonite after 10 min of the MCT. With an increase in the processing time of wollastonite, there is a tendency for a decrease in the dielectric constant values of the epoxy composition. According to the presented patterns, one can judge the structural changes in the synthesized polymer composite. The maximum values of the dielectric constant are due to the formation of structural elements with a high sensitivity to the electric field.

The use of wollastonite for epoxy composites after ultrasonic treatment leads to a wider range in changing the dielectric characteristics of the material. In this case, both the change in the processing time and the frequency characteristics of the ultrasonic action are important. Stabilization of the dielectric constant of the composite, regardless of the content of wollastonite in the system, is observed after 40 min of processing at 40 kHz (Fig. 10.8).

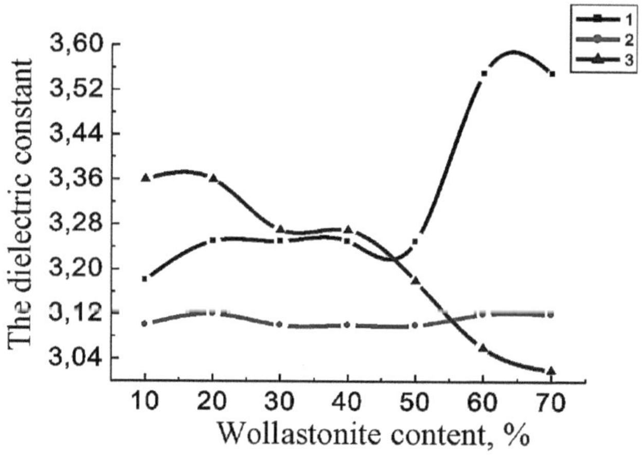

Figure 10.8 Dependence of the dielectric constant of the composition epoxy polymer + wollastonite on the amount of wollastonite after UST with a frequency of 40 kHz in water for 20 (1), 40 (2) and 60 min (3).

After shorter treatment, an increase in ε of compositions containing wollastonite 60% or more occurs. Conversely, after a longer UST (up to 60 min) of wollastonite, there is a steady decrease in ε of compositions containing more than 50% of the filler.

Thus, the dependences of the dielectric constant of composite systems obtained on the basis of epoxy resin and wollastonite filler indicate the structural features of polymerization (hardening) of samples depending on the conditions of pre-treatment of wollastonite (MCT or UST) and its amount in the composite. The resulting material is characterized by low dielectric constant values, i.e., is a high-quality electrical insulating material.

Another characteristic that is very sensitive to the structural characteristics of a composite material is the hardness index. The measurements were carried out according to the method of determining Brinell hardness. The introduction of wollastonite powder into the polymeric epoxy matrix in the initial state has already shown an increase in the hardness of the samples, especially with a content of 20 and 50–60% filler (Fig. 10.9). Samples containing 30–40 and more than 60% show a decrease in hardness, which is an indication of the differences in the formation of the network structure of the polymer composite and its fractal features [49–51].

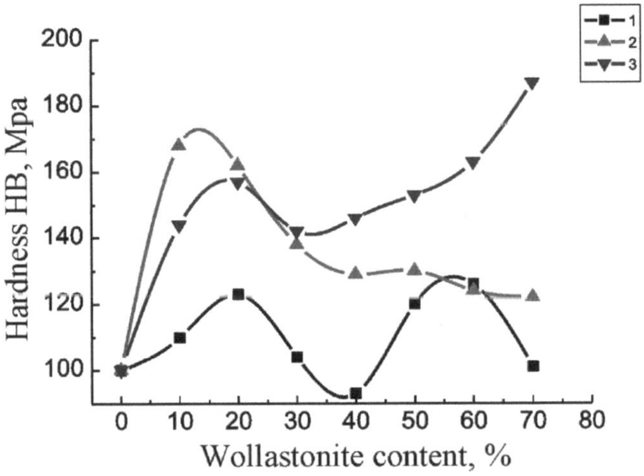

Figure 10.9 The dependence of the Brinell hardness of polymer compositions on the content of wollastonite in the initial state (1) and after the MCT for 10 min (2) and 20 min (3).

After the mechanochemical treatment (MCT) of wollastonite, the hardness of the composition increases significantly, especially at the points of extreme changes in the index, i.e., with a content of 10–20% and 50–60% wollastonite. And after 20 min of the MCT, an increase in hardness is observed with practically any degree of filling of the composition with wollastonite, and the more wollastonite, the higher the hardness, up to 190 MPa, i.e., compared to a non-filler polymer, the hardness has almost doubled.

After ultrasonic treatment of wollastonite in water, it also contributes to an increase in the hardness of the composite when introduced into epoxy resin (Fig. 10.10), but to a lesser extent than after MCT.

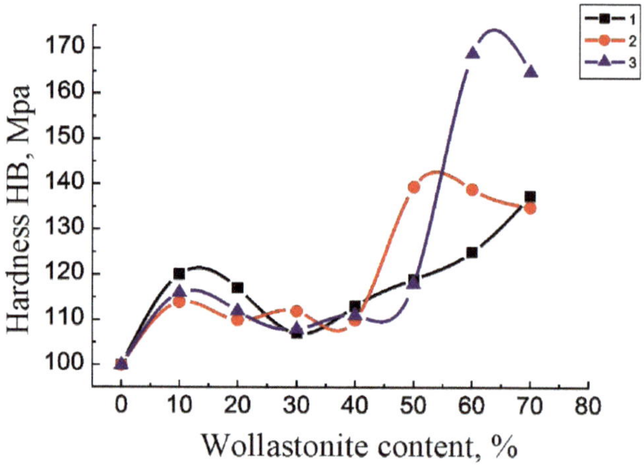

Figure 10.10 Dependence of Brinell hardness of polymer compositions on the content of wollastonite after UST at frequencies of 40 + 100 kHz in water for 20 (1), 40 (2) and 60 min (3).

Only with the introduction of 50–60% of the filler after 40–60 min of the UST, an increase in the hardness of the system to 140–170 MPa is recorded.

Thus, from the obtained data on the hardness of composite samples based on epoxy resin with a wollastonite filler, it follows that the processing conditions (MCT or UST) are important in changing the properties of the material. With an increase in the processing time, the hardening effect is more pronounced, both at low and high filler contents, up to 70%. However, the hardening effect is localized for systems with 10% wollastonite and from 50% or more.

Since epoxy composites are widely used in various industries and, especially, in construction, in assessing their manufacturability, it is first necessary to know the mechanical properties of strength and ductility. In the work, the plasticity was estimated by the bending angle and the bending strength index was determined. First of all, these two characteristics were evaluated for compositions with wollastonite in the initial state. The obtained regularities of the measured characteristics, as well as the ones considered above (dielectric constant and hardness) have a complex extreme character, indicating the specifics of the polymer structure formation depending on the amount of filler introduced. With the introduction of 10–20% wollastonite, the bending angle of the composition almost doubles, while simultaneously increasing the strength (Fig. 10.11).

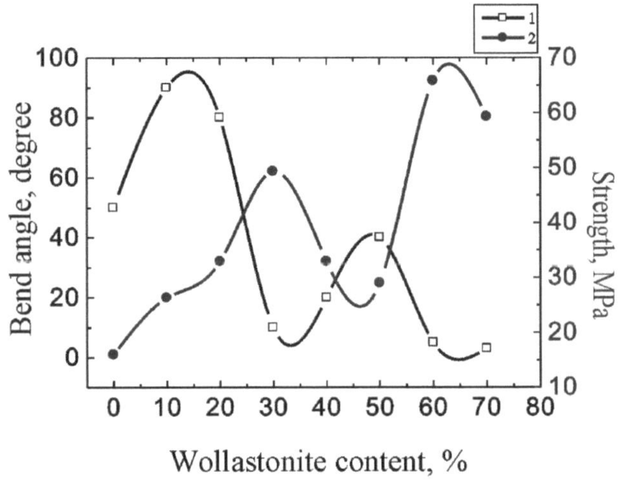

Figure 10.11 The change of the bending angle and strength of epoxy composition versus the content of wollastonite in the initial state.

With a further increase in the content of wollastonite in the composition, the bending angle begins to decrease, and the strength, on the contrary, increases. An exception is the composition containing 40–50% wollastonite, when there is a decrease in strength and an increase in the bending angle. Thus, in terms of strength and ductility, cyclicity is observed in changing the properties of epoxy compositions filled with wollastonite, which is undoubtedly related to cyclicity (or fractality) in the formation of the structure of the polymer composition.

After MCT of wollastonite, with an increase in the activation time, the mechanical characteristics of the composition decrease: the bending angle decreases to 20–10°, and the strength to 10–20 MPa, i.e., material becomes brittle with its high hardness. After ultrasonic treatment, a number of specific features are traced in the mechanical characteristics compared with the data obtained using wollastonite powder treated in a mill. First of all, there is a clear pattern in reducing plasticity and a significant increase in strength (up to 160 MPa) with an increase in the time of the UST in water (Fig. 10.12).

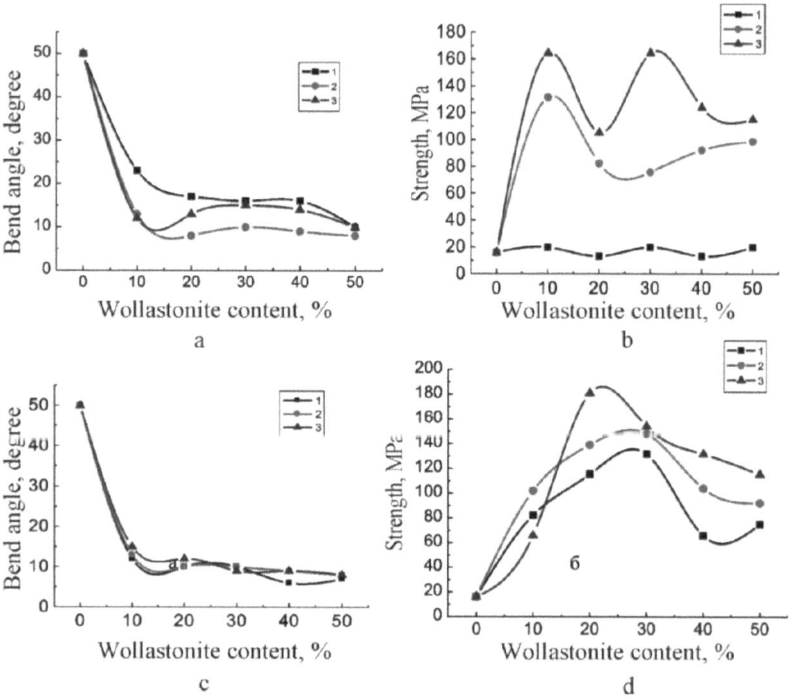

Figure 10.12 The change of the bending angle (a, c) and strength (b, d) of epoxy composition from wollastonite content after UST in water at 40 kHz (a, b) and 40 + 100 kHz (c, d) for 20 (1), 40 (2) and 60 (3) min.

With simultaneous exposure to frequencies of 40 and 100 kHz, structural changes in the surface of wollastonite particles, regardless of the processing time, provide an increase in the strength of epoxy compositions from 70 to 180 MPa, depending on the filler content. Composites with 20% treated wollastonite showed maximum strength.

Thus, the results of studies on the UST of wollastonite in the aquatic medium have shown the promise of this type of particle treatment for its subsequent use as a filler for epoxy compositions. Varying the conditions of ultrasonic treatment allows one to purposefully influence the mechanical properties of the synthesized composite systems with different wollastonite contents. Depending on the specific task, you can get a polymer-composite material with high performance properties with a relatively low content of wollastonite (10–20%) and a large degree of filling the composite (up to 50–60%) depending on the application area of the synthesized material and, therefore, taking into account other indicators of properties.

All the results of the research conducted in the complex characterize the synthesized polymer-composite material as promising, in particular, for its use in the manufacture of wear-resistant epoxy self-leveling floors and various potting compounds. This material can be effectively used for construction work in the repair and reconstruction of floors in the premises of various industries.

The indicators of the system properties qualitatively change during the modification of wollastonite in the process of the MCT and UST. Since it is known, alcohols are used as plasticizers for epoxy resin, and amino-containing compounds accelerate the hardening process of the resin, modifying additives of wollastonite particles in MCT and UST were butanol (monohydric alcohol—C_4H_9OH), glycerin (trihydric alcohol—$C_3H_5(OH)_3$), aqueous solution of ammonia (ammonia-NH_4OH) and urea ($CO(NH_2)_2$).

According to the results presented in Chapters 7 and 8, the optimal treatment regimes for wollastonite were determined. In case of MCT, the powder was modified for 10 min, and in the UST mode it was carried out for 40 min with simultaneous exposure to frequencies of 40 and 100 kHz. From 10% to 70% wollastonite was injected into the epoxy matrix. After hardening of the composition system, its indexes of Brinell hardness, bending strength and ductility, as well as the dielectric constant of the system were determined.

Introduction of wollastonite powder into the polymeric epoxy matrix in the initial state increases the hardness of the samples, especially at a content of 20 and 50–60% of filler (Fig. 10.13). Samples containing 30–40 and more than 60% show a decrease in hardness, which is an indication of differences in the formation of the network structure of the polymer composite and its fractal features [49, 52, 53]. After the mechanochemical treatment (MCT) of wollastonite, the hardness of the composition increases significantly, especially at the points of extreme changes in the index, i.e., with a content of 10–20% and 50–60% wollastonite. After processing wollastonite with butanol and introducing it into the epoxy resin, the hardness of the composition increased, but to a lesser extent than after the MCT without butanol (Fig. 10.13a).

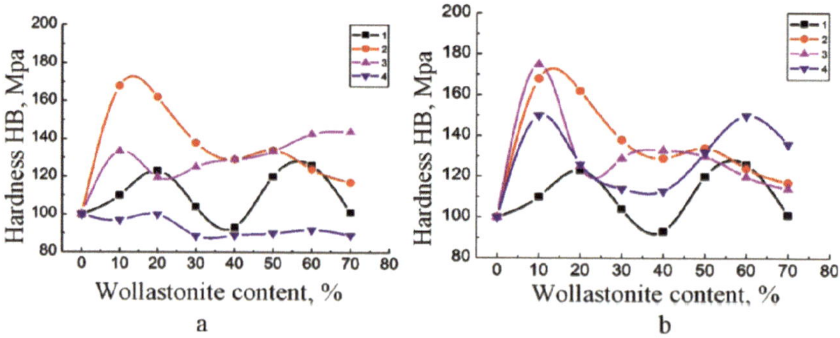

Figure 10.13 Dependence of Brinell hardness of polymer compositions on the wollastonite content in the initial state (1), after the MCT, for 10 min without the modifier (2) and with 10% modifier: (a) with butanol (3) and glycerin (4); (b) with ammonia (3) and urea (4). MCT time 10 min.

Butanol is known as one of the plasticizers of epoxy resin and the saturation of the surface of wollastonite particles with its radical groups after the MCT, on the one hand, accelerates the hardening process, and on the other, plasticizes the structure. The effect of plasticization is even more pronounced when using glycerin (triatomic alcohol) as a modifying additive. With the increase in the amount of wollastonite in the composition, the hardness of the polymer composition decreases to 75–90 MPa and becomes less than that of an epoxy polymer without filler (Fig. 10.13). The most significant decrease in hardness occurs when 30% or more of wollastonite modified with glycerin is introduced into the epoxy resin. The use of wollastonite as a modifier in MCT with amine-containing additives (liquid ammonia and urea) contributes to an increase in the hardness of the compositions, especially when the content of wollastonite in the epoxy matrix is 10% and 50–70% (Fig. 10.13b).

After ultrasonic treatment of wollastonite in water, it also contributes to an increase in the hardness of the composite (Fig. 10.14), but to a lesser extent than after MCT, when introduced into epoxy resin. Only with the introduction of 50–60% of the filler after 40 min of the UST, an increase in the hardness of the system to 140 MPa is recorded. Ultrasound processing of wollastonite in an aqueous 10% butanol solution slightly affects the strength of composites, especially those containing 50% or more filler (Fig. 10.14a). And after the UST in the glycerin solution, the hardness of the composite becomes lower than that of the unfilled polymer.

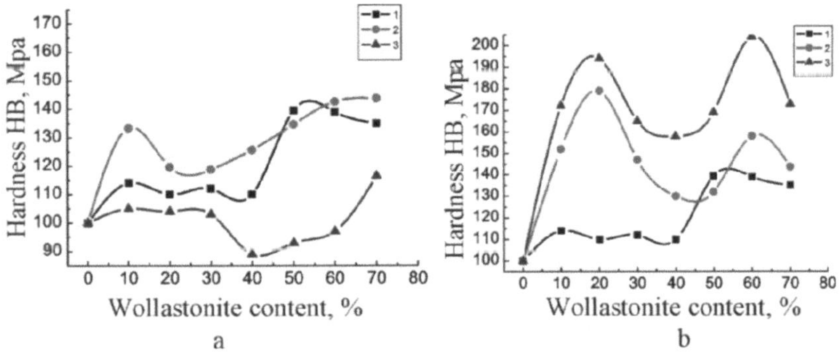

Figure 10.14 Dependence of Brinell hardness of polymer compositions on the content of wollastonite after UST at frequencies of 40 + 100 kHz in water (1) and in 10% aqueous solution: (a) butanol (2) and glycerin (3); (b) with ammonia (2) and urea (3). UST time 40 min.

With the introduction of amine-containing additives (ammonia or urea) into an aqueous solution and ultrasonic treatment in such a solution of wollastonite, the hardness of epoxy compositions with modified filler increases significantly (Fig. 10.14b). The hardness of composites containing 10–20% wollastonite, treated in the presence of ammonia, increases to 150–180 MPa. The average level of hardness of the samples, regardless of the content of wollastonite in them is 130–140 MPa. When used as a urea modifier, the hardness of the composite epoxy resin + wollastonite rises to 180–200 MPa with a filler content of 10% to 80%.

Thus, from the presented data on the hardness of composite samples based on epoxy resin with a filler in the form of modified wollastonite, it follows that a significant role in modifying the material properties, as a response to a change in the surface structure of the filler particles, is played by both the type of modifier and the processing conditions (MCT or UST). To increase the system hardness, it is more efficient to carry out modification of wollastonite in the UST mode using amine-containing modifying additives. With an increase in the duration of treatment, the effect of increasing hardness is more pronounced, both at low and high filler contents, up to 70%. An increase in the hardness of composites takes place with the use of wollastonite and after MCT with amine-containing additives. However, the effect of hardness growth is localized for systems with 10% wollastonite and 50% or more. For plasticization of the composite system, it is preferable to modify wollastonite with glycerol or butanol in the MCT mode.

Evaluation of flexural ductility and flexural strength of epoxy composites showed that if with the introduction of wollastonite in the initial state up to 10–20%, the bending angle of the composition doubles while simultaneously increasing the strength (Fig. 10.11) the bending angle begins to decrease, and the strength, on the contrary, increases. After MCT of wollastonite with butanol and glycerin, the mechanical characteristics of the composition decrease (Fig. 10.15).

A decrease in both the plasticity and strength of composite samples with an increase in the amount of wollastonite is also observed after MCT of wollastonite with ammonia. However, the strength level of these composites is significantly higher than when using glycerin as a modifier. With the increase of amine groups in the modifying additive, that is, when using urea, the indicators of both the strength and plasticity of the composite are reduced. In this case, the characteristic feature of dependencies is a more pronounced decrease in the bending angle and an increase in strength with an increase in the amount of filler in the composite.

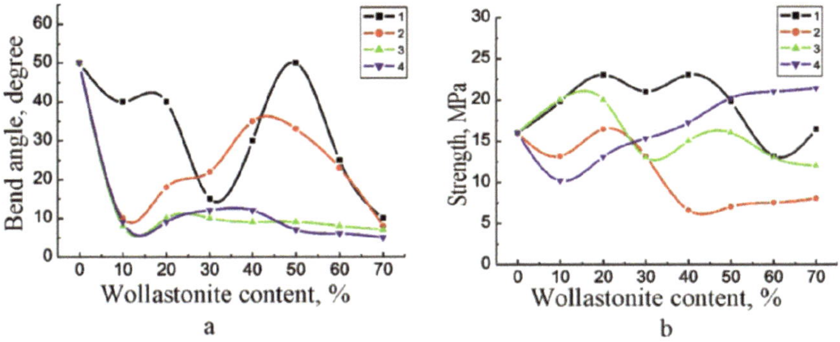

Figure 10.15 The change in the bending angle (a) and strength (b) of the epoxy composition versus the content of wollastonite after the MCT for 10 min with 10% modifier: butanol (1), glycerin (2), ammonia (3), and urea (4).

An increase in both the ductility and strength of the composite takes place after the MCT of wollastonite simultaneously with glycerol and urea (Fig. 10.16). A distinctive feature of these samples is the increase in both plasticity and strength of composites with an increase in the wollastonite content.

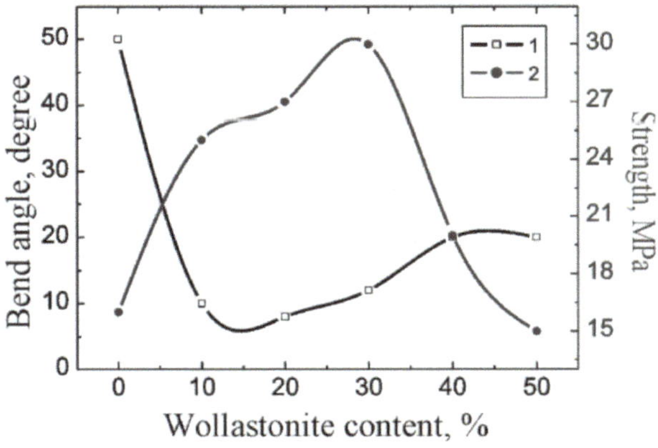

Figure 10.16 The change of the bend angle (1) and strength (2) of epoxy composition versus the content of wollastonite after MCT with 5% glycerin and 5% urea for 10 min.

Thus, the mechanical characteristics of a polymer composite filled with wollastonite do not change unambiguously with an increase in the concentration of the filler and depending on its pretreatment. The mechanochemical activation of wollastonite contributes to the improvement of the mechanical properties of the composite, especially at a content of 10–20% and 50% filler. The role of modifiers is different, alcohols lead to plasticization of the composite, and those containing amine groups contribute to increasing the strength of the material. A comprehensive selection of modifiers will provide a solution to the problem of increasing the mechanical characteristics of both the strength and plasticity of composites.

The results of measuring the hardness of samples of composites containing wollastonite after ultrasonic treatment, presented above, showed a number of fundamental differences in properties compared with samples filled with wollastonite after MCT. The results of the mechanical characteristics also traced a number of specific features in comparison with the data when using wollastonite powder treated in a mill.

First of all, there is a clear pattern in reduction of plasticity and a significant increase in strength (up to 160 MPa) with an increase in the time of the UST in water (Fig. 10.17). With simultaneous exposure to frequencies of 40 and 100 kHz, structural changes in the surface of wollastonite particles, regardless of the processing time, provide an increase in the strength of epoxy compositions from 70 to 180 MPa, depending on the filler content. Composites with 20% treated wollastonite showed maximum strength. Ultrasonic treatment of wollastonite in aqueous solution with 10% butyl alcohol contributes to the plasticization of the epoxy composition, especially that containing 20–30% (Fig. 10.17a). At the same time, there is a significant increase in the strength of samples from 200 to 400 MPa, depending on the content of wollastonite (Fig. 10.17b). After UST of wollastonite in an aqueous solution of glycerin, fundamental changes are observed in the dependences of the mechanical characteristics of the composite on the content of filler in it. First, there is a significant increase in the plasticity of samples containing from 20% and above wollastonite. The strength of samples containing from 10% to

20% of filler remains high (200–300 MPa). With a greater degree of filling of the composite, its strength decreases, but not less than 100 MPa.

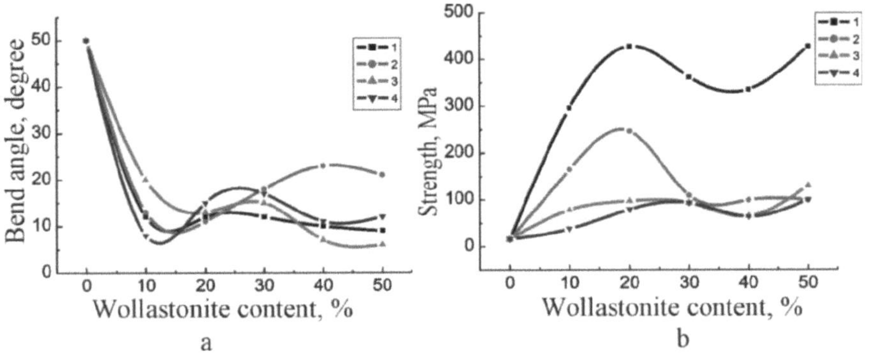

Figure 10.17 The change of the bending angle (a) and strength (b) of the epoxy composition versus the content of wollastonite after the UST in a 10% aqueous solution butanol (1), glycerin (2), ammonia (3) and urea (4). UST time 40 min, frequency 40 + 100 kHz.

After UST of wollastonite in solutions containing amine groups (ammonia), there is some tendency to increase plasticity of the samples with 10–30% of filler compared to that treated in pure water. The strength is somewhat reduced. When using a modifier with a large number of amine groups (urea) after UST, the tendency to increase the strength with some plasticization of the samples is even more pronounced. This pattern is most indicative of samples containing 30–40% wollastonite. Their strength is 140–160 MPa with ductility at bending of 15–17°.

From a comprehensive analysis of the results obtained on the formation of the mechanical properties of composites based on epoxy resin and filler (modified wollastonite), it clearly follows that UST in an aqueous solution of glycerin leads to an increase in the strength of compositions with 10–20% wollastonite and, conversely, to a significant increase in plasticity for those containing more than 20% filler. Modification of urea particles, under certain conditions of UST, contributes to a small increase in plasticity with a significant increase in the strength of the samples. It was of interest to conduct a joint ultrasound treatment of wollastonite in an aqueous solution containing both glycerol and urea. The introduction of thus treated wollastonite into epoxy resin showed (Fig. 10.18) the promise of its use in terms of simultaneously increasing the ductility and strength of composite samples containing from 20% to 40% wollastonite.

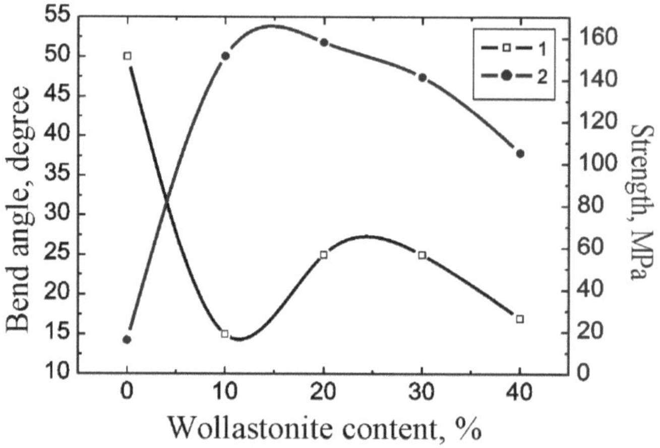

Figure 10.18 The change of the bending angle (1) and strength (2) of epoxy composition from the content of wollastonite after UST in an aqueous solution with 5% glycerin and 5% urea at 40 + 100 kHz for 40 min.

Thus, the results on UST of wollastonite in an aqueous medium with various modifying additives have shown the promise of this type of treatment and modification of particles for its subsequent use as a filler for epoxy compositions. Varying the conditions of ultrasonic treatment allows one to purposefully influence the mechanical properties of the synthesized composite systems with different wollastonite contents. Depending on the specific task, it is possible to produce polymer composition material with high indexes of properties at relatively low content of wollastonite (10–20%) and with a high degree of filling the composite (up to 50–60%) depending on the application area of the synthesized material and, therefore, taking into account other indicators of properties (electrical, chemical resistance and other indicators).

Thus, measurements of the dielectric constant of the composites obtained showed that its values vary from 3.00 to

3.60, depending on the amount of wollastonite in the system and the conditions of its preliminary preparation. With an increase in the filler content, there is a tendency to a slight increase in the values of the dielectric constant of the composite. After UST of wollastonite, as a rule, its values are slightly lower than after the MCT. As an example, Fig. 10.19 shows the results of measuring the dielectric constant of an epoxy composition with wollastonite after MCT and UST in the presence of glycerin and urea.

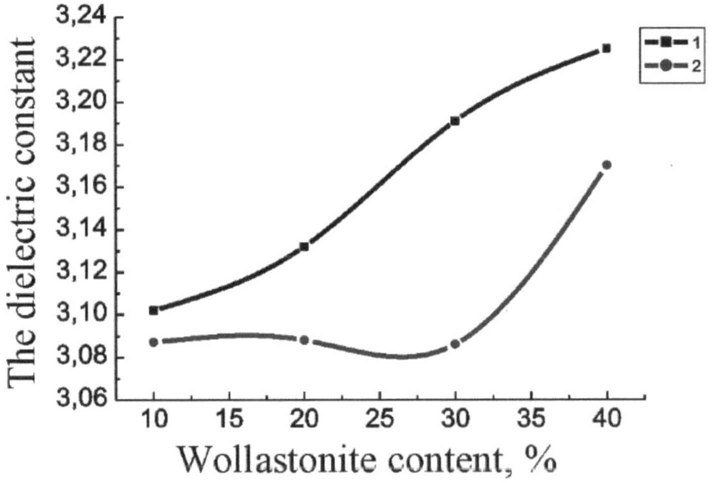

Figure 10.19 Dependence of dielectric permeability of epoxy composition on the amount of wollastonite modified with glycerol and urea at MCT for 10 min (1) and UST for 40 min (2) with combined exposure of 40 + 100 kHz.

The resulting material is characterized by low dielectric constant values, i.e., it is a high-quality electrical insulating material. The most effective use of UST with amine-containing modifiers is to obtain an electrical insulating composite with an optimal wollastonite content of 25–30%. All the results obtained in the complex characterize the synthesized polymer-composite material as promising, in particular, for using in the manufacture of wear-resistant epoxy self-leveling floors and various potting compounds. This material can be effectively used for construction work in the repair and reconstruction of floors in the premises of various industries.

Thus, the choice of processing modes and the type of modifying additive provides a directional change in the structure and properties of wollastonite powder. The modified wollastonite, as a filler in the polymer composition, provides a change in the mechanical and physical characteristics of the system epoxy resin + wollastonite, depending on the level of indicators of the properties of wollastonite particles modified by MCT or UST, which determines the quality of the composite on a polymer base with the required complex properties.

The hardness of composite samples based on epoxy resin with a filler in the form of a modified wollastonite, resulting from a change in the surface structure of the filler particles, largely depends on the type of modifier and the processing conditions (MCT or UST). To increase the system's hardness, it is more efficient to carry out modification of wollastonite in the UST mode using amine-containing modifying additives. With an increase in the processing time, the hardening effect is more pronounced, both at low and high filler contents, up to 70%. Hardening of composites takes place when using wollastonite and after MCT with amine-containing additives. However, the hardening effect is localized for systems with 10% wollastonite, as well as with 50% or more. For plasticization of the composite system, it is preferable to modify wollastonite with glycerin or butanol in the MCT mode. A comprehensive selection of modifiers will provide a solution to the problem of increasing the mechanical characteristics of both the strength and plasticity of composites.

The most promising is the use of ultrasonic treatment and modification of wollastonite particles for its subsequent use as a filler for epoxy compositions. Variation of the UST conditions allows targeted action on the mechanical properties of synthesized composite systems with different wollastonite contents. Depending on the specific task, you can get a polymer-composite material with high performance properties with a relatively low content of wollastonite (10–20%) and a large degree of filling the composite (up to 50–60%) depending on the application area of the synthesized material.

Questions

1. What mineral fillers are used for polymer-based composite systems?
2. What determines the need to modify the surface of the filler particles?
3. What is the specificity of the properties of wollastonite $CaSiO_3$, ensuring the effectiveness of its use when obtaining composite systems?
4. What is the dependence of the dielectric constant of the polymer composition on an epoxy base on the content of wollastonite filler?
5. Describe the effect of mechanochemical treatment of wollastonite on the change in dielectric constant of epoxy compositions.
6. What is the influence of wollastonite ultrasonic treatment on the change in dielectric constant of epoxy compositions?
7. How does the mechanochemical treatment of wollastonite on the change in hardness and mechanical properties of epoxy compositions?
8. How does the influence of ultrasonic treatment of wollastonite on the change in hardness and mechanical properties of epoxy compositions?
9. What modifying additives during MCT of wollastonite contribute to plasticity of the epoxy composition?
10. What modifying additives in MCT of wollastonite contribute to hardening and increase the hardness of the epoxy composition?
11. What is the perspective of UST of wollastonite in an aqueous medium with various modifying additives for its subsequent use as a filler for epoxy compositions?
12. What are the areas of use of polymer-composite systems based on epoxy binder and filler from modified wollastonite?

10.3 SH-Synthesis of Ceramic Materials Based on Pre-Activated and Modified Systems

10.3.1 The Main Macrokinetic Aspects of the Synthesis of SHS Systems and Methods of Process Control

One of the progressive ways to create materials both compact and with varying degrees of dispersion up to the formation of nanostructured systems today is self-propagating high-temperature synthesis (SHS), or in other words, technological combustion [54]. It was originated in 1967 by the works of A. G. Merzhanova, I. P. Borovinskaya and V. I. Shkiro. The advantages of SHS—insignificant energy costs and simplicity of the equipment, speed of the process, high productivity of the method, and purity of the products—gained attention of many materials science specialists [55]. Variations of the SHS method have appeared, combining the principle of carrying out a reaction in the combustion mode with energy and mechanical effects on the process using high-pressure techniques, deep-vacuum heat treatment, etc. [56].

The results of numerous experimental studies and theoretical calculations showed that the burning rate and temperature profile of the combustion wave during SHS are a function of many variables: the heating value of the mixture, the thermal conductivity of the combustion products, the volumetric heat release rate, the heat capacity of the products, the concentration of the initial substances and the reaction products, the diffusion coefficient of the reacting substance, activation energy of the reaction and its rate constant [57]. In the study of the SHS process, the quantitative characteristics of the process (linear combustion rates, maximum temperatures and temperature profiles, loss of stability of the front, thermal effects, degree of combustion, and composition of products) are used, as well as a qualitative description of the observed phenomena and effects [58, 59]. After initiation and some transition period, a certain reaction propagation mode is established.

With the help of various influences, it is possible to control the combustion process, ensuring its stationarity at certain values of speed [60]. During the course of SHS, a large amount of heat is released and a high temperature develops. The lowest temperatures are registered during the combustion of metals in hydrogen (1100–1400 K), the highest—during the formation of carbides and borides (3000–3500 K). The burning temperature (T_b) is most strongly influenced by the parameters related to the heat content of the compositions (ratio of reactants α, degree of dilution with inert products σ). With their help T_b can be changed to 500–700 K. The combustion temperature, as well as the velocity of propagation of the front, can be regulated within certain limits.

In SHS processes, T_b is most often used to reduce T_b by diluting the starting materials with combustion products and increasing T_r as a result of preheating the charge. A difficult but solvable problem is the regulation of the value of V at T_b = const and T_b at V = const for the same chemical system [61]. Short synthesis times (0.05–1 s) and high rates of heating a substance in a wave (up to a million degrees per second) indicate that SHS refers to the category of extreme chemical processes. A characteristic feature of SHS is the formation of condensed products that carry certain information about the process itself, and thus their analysis is an additional tool in studies of the SHS mechanism [62].

Methods of influencing the combustion process are techniques for changing the original SHS system and its components (various compositional variations, dispersion, preliminary fusion), and changing the conditions of the SHS process (temperature, pressure, impact on the ultrasonic field or centrifugal field system), as well as processing of final products in less or more time after the completion of the SHS process (pressing, grinding) [63, 64]. The results obtained so far indicate that the sensitivity of SHS processes to changes in the composition of the system and the parameters of the regime is the real basis for developing flexible technological processes for producing SHS products with the necessary properties. The use of ultrasonic action [65] and mechanochemical treatment [66], which allow changing not only the dispersion but also the energy state of the material, turned out to be effective.

In recent years, a fairly large number of papers have appeared in scientific literature in which SH-synthesis is preceded by mechanochemical treatment (MCT) of the powder materials being used. Preliminary MCT in mechanical reactors (dynamic action mills) allows to achieve a high degree of dispersion of particles (up to 100 nm), to change the structure, energy intensity and, consequently, the reactivity of the material. Mechanochemical activation facilitates diffusion processes in the material during subsequent thermal effects on it [67], as well as the change of combustion mechanisms [68]. In activated systems, combustion is more pronounced in the thermal explosion mode, i.e., in the entire volume of the sample and in the absence of temperature gradients, which favorably affects the structure of the final products and production of monophasic materials. In [66, 69], it was shown that the preliminary mechanical activation is an effective way to influence the reaction mixture in order to realize "true solid-phase combustion" in it.

Mechanical influence on the development of chemical reactions allows you to expand the possibilities of technological combustion. On the other hand, such an approach allows one to obtain new ideas on a number of phenomena that are observed in solid-phase combustion — these are autowave processes, conditions for violation of steady-state combustion, and patterns in the development of a thermal explosion. It is suggested that deformation plays a leading role in the mechanism of these phenomena [70], since deformation and structural transformations are thermodynamically interrelated. Consequently, by external mechanical actions, which ensure the internal instability of the system, both before and during chemical interaction, the kinetics and direction of the process can be changed. Structural defects arising from mechanical action facilitate the course of chemical reactions, change its rate and ultimately form the composition and structure of combustion products. All this is of great importance for SHS technology and the creation of new types of materials.

Mechanical treatment of a material in dynamic action mills leads to accumulation of various defects (vacancies, dislocations, electron-defect centers, broken bonds on the surface of particles) and stresses of various levels that occur in ground particles, which coordinately changes the reactivity of the substance [71, 72]. In addition to mechanical changes in the material being processed, there is a change in the electrical state of the system: emission of electrons at the moment of destruction, the electric charge on the surface of the deformed particle [71, 73]. No less important factor in the processing of substances in the mill is that the impact on the substance has a pulsed (cyclical) character. As a result of the periodic impact of the balls, changes in the structure of the material being processed occur in portions, including accumulation of defects, destruction of particles, annihilation of defects and their redistribution between the volume and the surface of the particle. The consequence of these periodic changes in the structure should be the periodicity in the change of electric charges on the surface of the particle.

When pre-activated systems are used in combustion processes, the task is to preserve the stored energy of the material for quite a long time, since it is well known that activated systems are subject to aging, that is, a gradual relaxation of the energy stored in them. The real way to solve this problem is to modify the surface of a particle with compounds that stabilize the energy state of the material, i.e., creating a capsule shell. The creation of such a shell can be facilitated by the resulting electrical charge during the grinding of particles of inorganic matter while simultaneously processing it in a mill with organic compounds, accelerating polymerization of the latter [74, 75]. As a result, formation of composite systems, including the polymer and the inorganic dispersed component is possible. Formation of the polymer in the process of mechanical action is also a cyclic process, since, along with polymerization, the chain segments of the formed polymer break as a result of the vibrationally excited state caused by the mechanical action of the grinding bodies.

Destruction of such a polymeric shell on the surface of an inorganic particle directly at the time of conducting SHS on a pre-activated material will allow the most complete realization of the energy of a substance stored during mechanical processing. It is important to choose the composition of the capsule in such a way that the products of its decomposition, interacting with the main reaction reagents, contribute to creation of SHS material of the required composition, structure and quality, which, in particular, was described in [76, 77].

Another effective way to influence the structure and condition of various materials (liquid, amorphous, crystalline) is ultrasonic treatment (UST) of a different range of frequency, power and time, depending on the material and processing medium [78]. When using ultrasound, metal, oxide, polymer and other systems are dispersed until nanomaterials are obtained. The main mechanisms for the ultrasonic dispersion of powder materials in the aqueous medium are cavitation and flows that occur in the working fluid during the collapse of cavitation cavities [79]. In a shock wave propagating from the place of collapse, the maximum pressure amplitude can reach several thousand kilogram-forces per square centimeter. The negative pressure next to the shock wave has a tensile effect, which leads to the destruction of the particles.

Erosion of the surface and destruction of particles under the influence of ultrasound leads to an increase in the chemical activity of their surface and accelerate the processes of interaction of particles of the material being processed with the processing medium. Thus, during UST, a mechanochemical effect is rendered on the material being processed. In both cases, there is a change in the structure, first of all, of the surface layers, the state and, consequently, the chemical activity of the material, which is then realized in various processes of synthesizing new materials for a specific purpose.

Numerous studies have shown that, depending on the parameters of ultrasonic vibrations (frequency, power, intensity) and the design of the emitter, substances can be destroyed or activated, promote synthesis, polymerization and dissociation, control the kinetics of a chemical or physico-chemical process, including processes or individual stages of mass transfer [78, 80, 81]. Acoustic radiation can significantly affect homogeneous redox and electrochemical reactions both with ion recharging and with the release of a new phase. In industry, as a rule, powerful ultrasonic fields are used in cavitation mode. The use of low-power ultrasonic fields can be important for evaluating the mechanism of sound action on the chemical and physico chemical characteristics of chemical processes and predicting the development of a certain type of reaction [82, 83].

10.3.2 Influence of Mechanochemical Treatment and Modification of Quartz, Calcite and Wollastonite on the Technological Combustion of Systems

In this section, it is shown how, when using, organic compounds containing bound water, as well as carbon and ammonia groups as modifying additives during MCT, one can influence the thermokinetic characteristics of the technological combustion process (self-propagating high-temperature synthesis (SHS)), ensuring the completeness of the reaction between the main components of the reaction mixture. SH-synthesis was carried out by the furnace method in the mode of spontaneous combustion of the sample (with the stoichiometric ratio of the components of the mixture) placed in a preheated furnace. The heating temperature of the furnace was equal to 900°C. The combustion temperature was measured with an IrconUltrimaxPlus UX10P pyrometric thermometer.

Thermograms of combustion obtained with the help of a pyrometer showed that, after the MCT of quartz, the maximum temperature of combustion, the induction period of ignition, the level and rate of temperature change at the postprocess stage change, when the phase composition of the synthesized material is actually formed (Fig. 10.20). Both the induction period of the system's self-ignition, and the maximum temperature of combustion vary with the time of the MCT of quartz are not linear. With an increase in processing time up to 20 min, an increase in the combustion temperature is observed. After a longer MCT of quartz, the combustion temperature of the system (SiO_2 + 37.5% Al) decreases slightly.

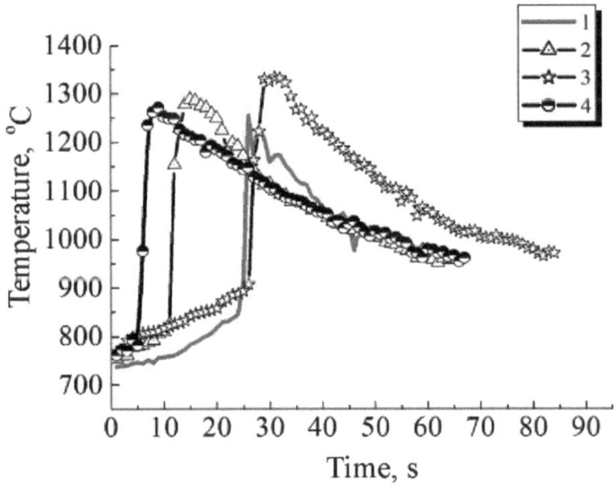

Figure 10.20 Combustion thermograms for the system (SiO_2 + 37.5% Al) with non-activated quartz (1) and after the MCT for 10 min (2), 20 min (3) and 30 min (4).

Processing in a quartz mill in the presence of ethyl alcohol leads to a greater degree of dispersion of particles (Fig. 10.21) and a more pronounced time dependence of the maximum combustion temperature. It was found that the MCT provides a rise in the combustion temperature to 1550–1600°C for 15–20 min, which corresponds to the maximum particle size reduction. With longer treatment, agglomeration of particles is observed and, as a consequence, a decrease in the combustion temperature.

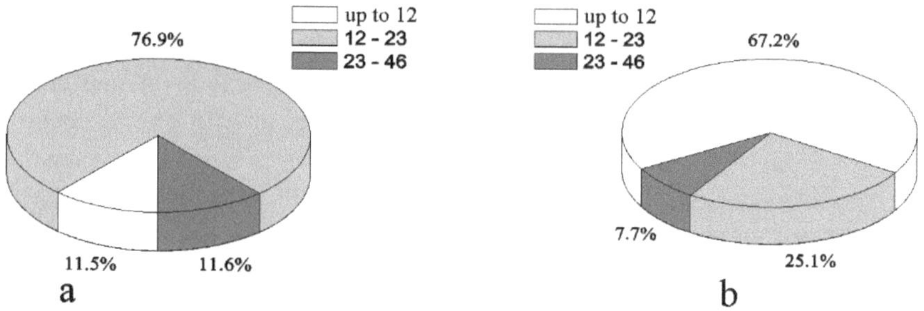

Figure 10.21 Particle size distribution after grinding quartz (a) and MCT of quartz with ethanol (b).

Thus, experiments with quartz crushed in the presence of monohydric alcohol have already shown a significant change in the combustion temperature depending on the time of the MCT, this indicating the change in the reactivity of the material depending on the conditions of its processing. According to the data obtained, the optimal treatment time is 15–20 min. Subsequent experiments with various modifying additives were carried out mainly at the specified time mode of processing.

The use of quartz with more complex modifying additives during MCT, as shown by the thermograms of the combustion system (SiO_2 + 37.5% Al), has a significant effect on all the thermokinetic parameters of the combustion of the samples (Fig. 10.22).

The presence of nitrogen-containing compounds has a significant impact on the development of the combustion process, reducing the induction period and increasing the temperature of combustion (Fig. 10.22a). The strongest effect, when using nitrogen-containing additives, showed the presence of an aqueous solution of ammonia in the sample. When using acids and alcohols as modifying additives, it was found that the kinetics of the process also significantly increased: the induction period decreases, the ignition rate increases and the combustion temperature rises (Fig. 10.22b).

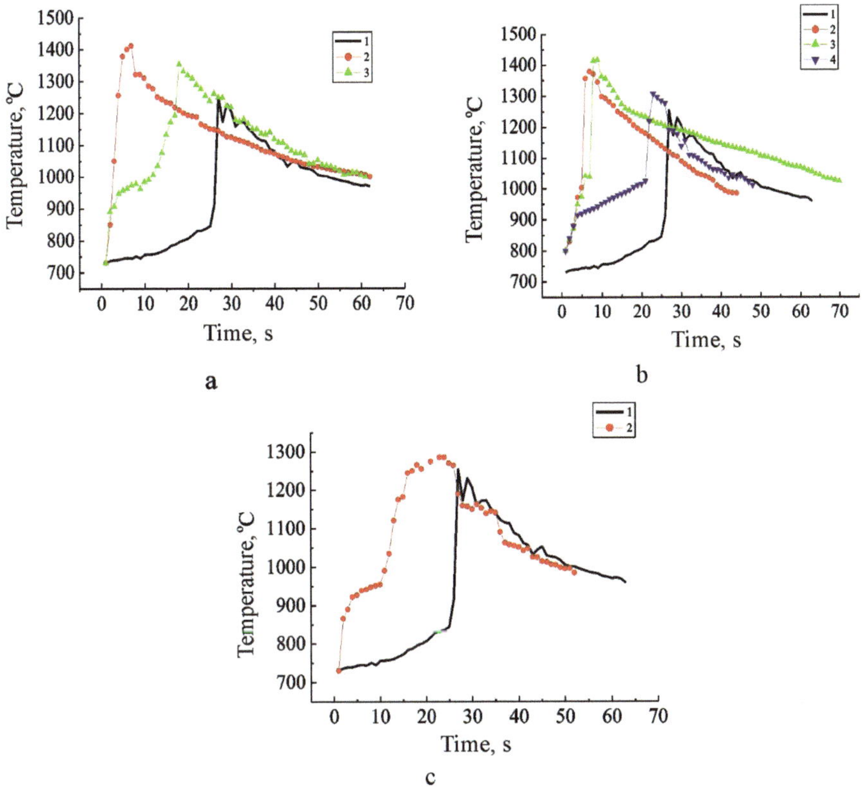

Figure 10.22 Thermograms of the combustion system (SiO_2 + 37.5% Al) with quartz in the initial state (1) and after the MCT in the presence of 5%: (a) NH_4OH (2) and $(NH_2)_2CO$ (3); (b) $H_2SiO_3{\cdot}n(H_2O)$ (2), $(C_2H_3OH)_n$ (3), and $C_4H_6O_4$ (4); in mixtures (5% $C_4H_6O_4$ + 5% NH_4OH) (2).

The greatest effect was obtained in the presence of aqueous silicic acid and polyvinyl alcohol, i.e., compounds containing a large amount of bound water, which when heated (above 200–400°C) is separated. If, when heating succinic acid, one molecule of water is separated and succinic acid anhydride is formed (in the form of a five-membered ring), then when polyvinyl alcohol and aqueous silicic acid is heated, n-amount of water is separated.

Accordingly, there takes place acceleration of the start of the mixture ignition and an increase in the combustion temperature. When succinic acid and ammonia solution are used together as modifiers, the processes become complicated. Multistep chemical reactions provide a higher level of temperature at the post-process stage (burn-out stage) during formation of the phase composition of the synthesis products (Fig. 10.22c).

Table 10.9 Phase composition of SHS samples based on quartz, activated and modified during MCT for 20 min

	Phase content, %							
	Samples							
Phases	**1**	**2**	**3**	**4**	**5**	**6**	**7**	**8**
Al_2O_3	48.4	59.0	67.1	67.4	64.0	56.9	58.2	61.7
Si	12.9	21.9	15.2	20.0	20.7	18.9	14.3	22.3
Al	12.2	4.2	3.5	2.5	2.6	2.1	2.0	0.3
SiO_2	24.7	2.6	0.7	1.2	0.9	2.8	4.1	1.6
$3Al_2O_3{\cdot}2SiO_2$ Mullite		7.9	7.1	7.1	4.8	7.8	9.0	5.7
$FeAl_3Si_2$	1.8	3.4	3.7		5.4	2.5	4.3	3.0
AlN						7.5	5.8	4.0
$FeSi_2$		1.0	2.7	1.8	1.6	1.5	2.3	1.2

(1) (SiO_2+Al); (2) [(SiO_2)$_{MCT}$ +Al]; (3) [(SiO_2+5% $H_2SiO_3{\cdot}n(H_2O)$)$_{MCT}$+ Al]; (4) [(SiO_2 + 5% $H_2C_4H_4O_4$)$_{MCT}$+Al]; (5) [(SiO_2+5%$(C_2H_3OH)_n$)$_{MCT}$+Al]; (6) [(SiO_2+5% NH_4OH)$_{MCT}$ +Al]; (7) [(SiO_2 +5%$(NH_2)_2CO$)$_{MCT}$ +Al]; (8) –[(SiO_2+5% $H_2C_4H_4O_4$+5%NH_4OH) $_{MCT}$ +Al].

Thus, mechanochemical activation and modification of the surface layers of a dispersible quartz particle can be used to purposefully influence the development of the combustion process, its thermokinetic characteristics, which ultimately must be realized in the phase composition of the synthesized material. Table 10.9 clearly shows that the preliminary activation of quartz contributes to a more significant use of reagents in the synthesis process, an increase in the amount of corundum, silicon in the reaction products and formation of mullite.

Modification of quartz intensifies this process even more. When using nitrogen-containing compounds as a modifier, aluminum nitride is formed in the synthesis products. The formation of $FeSi_2$ and $FeAl_3Si_2$ compounds is a consequence of the presence of iron in the mixture as a result of rubbing it from the surface of grinding steel balls during the grinding of quartz in the mill [84].

Thus, preliminary mechanochemical treatment of oxide compounds, especially in the presence of modifying additives, is an effective way to influence the development of the solid-flame combustion process, more complete realization of the initial components of the charge SHS mixture and formation of the necessary synthesis products [85–87]. The material synthesized in our case refers to refractories, the basis of which is corundum, and additional phases that provide a high level of refractoriness are mullite and aluminum nitride.

The role of the preliminary MCT for more complex systems is even more pronounced. Therefore, after the MCT of the mixture of quartz and calcite, the induction period significantly decreases and the longer the mixture is processed, the faster the sample ignites, but the burning temperature significantly decreases throughout the synthesis process (Fig. 10.23). The complex course of the temperature curves of combustion, namely, the jumps in the temperature change at various stages of the process, indicate a multi-stage phase formation both at the peak of the process and during the period of temperature decrease at the post-process stage. Variations in the change in the time of the MCT can significantly affect the development of combustion, both in terms of reducing the induction period and raising the temperature of combustion.

Introduction of wollastonite powder into the charge mixture (SiO_2 + 37.5% Al) first of all increases the induction period and reduces the burning rate, as well as the maximum process temperature (Fig. 10.24). Quartz for the charge mixture was used in the non-activated state. Wollastonite plays the role of a ballast, which consumes part of the heat of the exothermic reaction between quartz and aluminum, as a result of which the combustion process slows down (from 50°/s to 10°/s) and the maximum burning temperature somewhat decreases. The higher the content in the mixture of wollastonite, the lower these characteristics.

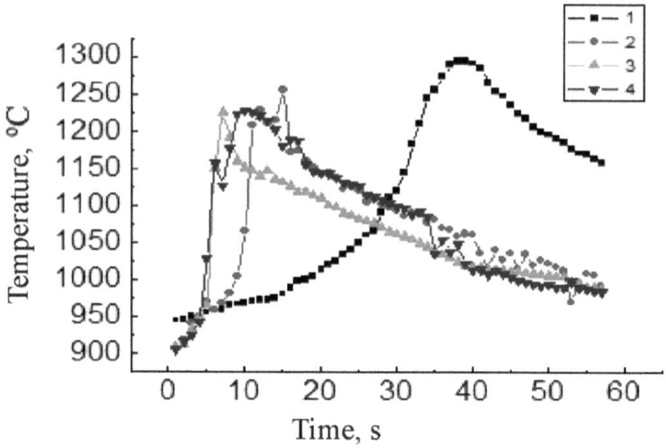

Figure 10.23 Thermograms of combustion of a mixture of quartz and calcite with a ratio of 50:50 components of the system: (1) non-activated mixture and after MCT for (2) 10 min, (3) 20 min, and (4) 30 min.

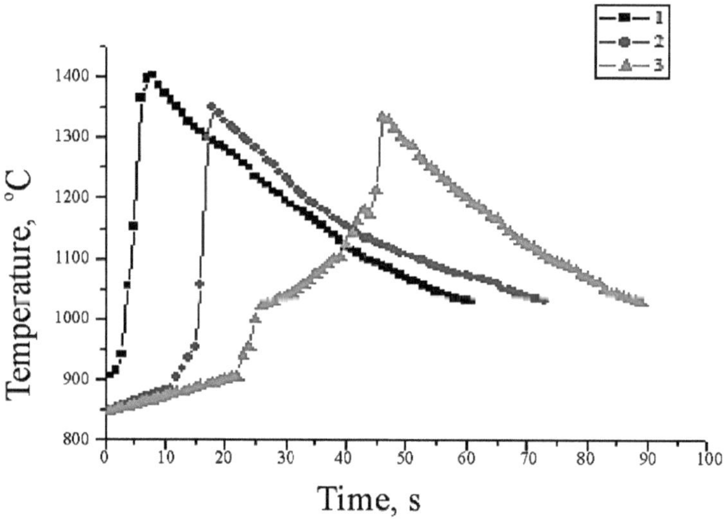

Figure 10.24 Thermograms of the combustion system [(SiO_2 + 37.5% Al] without (1) and with the addition of wollastonite in the amount of 5% (2) and 10% (3).

At the next stage, wollastonite processed in a mill with modifiers such as silicic acid, polyvinyl alcohol and succinic acid was introduced into the mixture (SiO_2 + Al). The use of modifiers makes it possible to qualitatively change the structure and composition of the surface layer of wollastonite particles as a result of partial decomposition of the organic additive and its interaction with the active surface of wollastonite particles.

From Fig. 10.25, it follows that after 10 min of the MCT of wollastonite with modifiers, its introduction in an amount of 5% to the charge leads to a decrease in the induction period and an increase in the burning rate. To the greatest extent, these indicators change, when using carbon-containing modifiers.

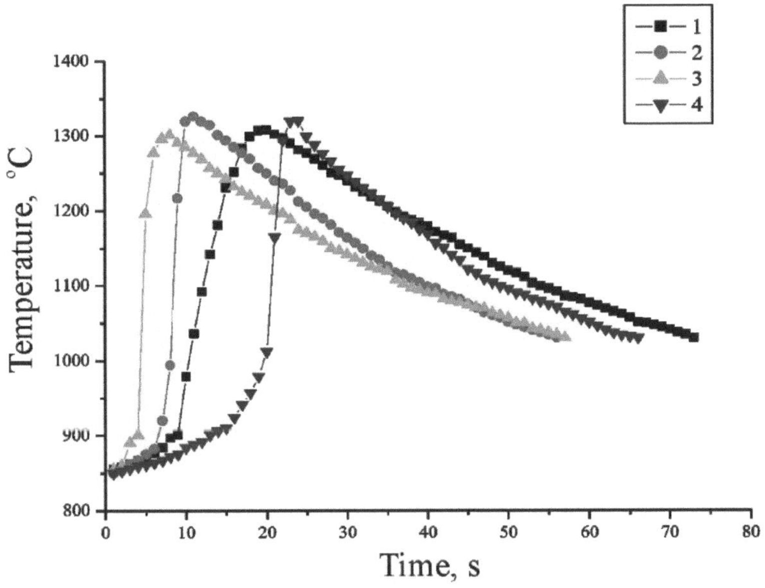

Figure 10.25 Thermograms of the combustion system [(SiO_2 + 37.3% Al] with 5% wollastonite FW100, activated for 10 min with silicic (1) and succinic acid (2), with polyvinyl alcohol (3) and without a modifier (4).

The more carbon in the modifier, the earlier the sample ignites and the burning process is more intense. However, the maximum burning temperature (1326°C) was recorded using wollastonite treated with succinic acid.

The dependence of the macrokinetic parameters of combustion on the processing time for each of the modifiers used is different. In a mixture of (SiO_2 + Al) with wollastonite modified with water-containing silicic acid, one can see that, after 20 min of treatment, the combustion process is accelerated, but the maximum temperature decreases from 1308 to 1225°C.

The main effect of silicic acid is associated with the presence of bound water in its composition, which, when heated (above 200–400°C), splits off and interacts with aluminum:

$$3H_2O + 2Al = Al_2O_3 + 2H_2 \tag{10.1}$$

The reaction proceeds with the release of a large amount of heat, and the formed hydrogen, being a strong reducing agent, reacts with silicon oxide and reduces silicon, according to the following equation:

$$SiO_2 + 2H_2 = Si + 2H_2O \tag{10.2}$$

The next step is the main reaction of reduction of silicon dioxide with aluminum:

$$3SiO_2 + 4Al = 2Al_2O_3 + 3Si \tag{10.3}$$

Thus, the presence of modified wollastonite in the mixture accelerates the beginning of interaction between the reactants and the reaction rate.

For the ternary system, which includes, in addition to quartz and wollastonite, calcium carbonate in the form of marble, prior activation and modification of the components lead to a significant increase in the thermokinetic characteristics of the combustion process (Fig. 10.26).

Thus, MCT using modifying additives, in particular mixtures of succinic acid with ammonia and polyvinyl alcohol, for both quartz with wollastonite [88], and complex systems (quartz + calcite + wollastonite) leads to a change in the kinetics of the process and temperature of combustion. The modifiers used and the presence of calcium carbide in the charge after MCT enhance the role of gas-phase reactions in the combustion process of the activated system. These results determine the optimal modes of modifying the multicomponent charge mixture. It was suggested that such conditions lead to the possibility of synthesis gas formation in the process of sample heating, which is actively involved in the combustion process, changing its thermokinetic characteristics. The result of these processes is a change in the strength of the SHS samples. The strength of SHS samples obtained on the basis of a pre-modified charge mixture increased from 10 to 50 MPa.

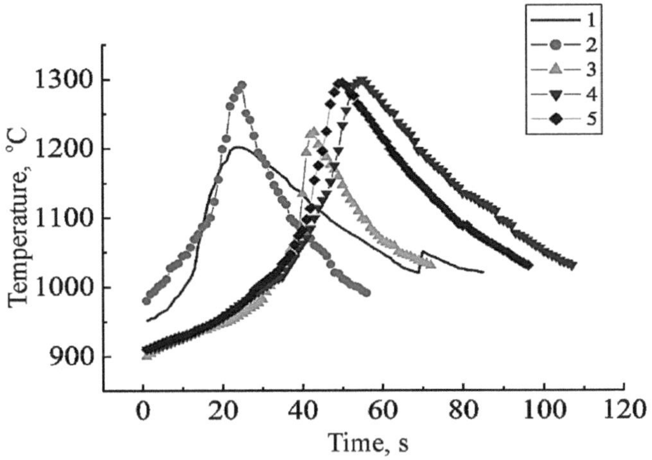

Figure 10.26 Thermograms of the combustion system [(30% $CaCO_3$ + 70% SiO_2) + 5% wollastonite] in the initial state (1), after MCT (2) and modification with compounds $C_4H_6O_4$ (3), with a mixture of 5% $C_4H_6O_4$ + 5% NH_4OH (4) and $(C_2H_3OH)n$ (5). MCT time 20 min.

The strength of the samples is determined by the phase composition of the synthesis products (Table 10.10). From the presented results, it follows that activation and modification of the components of the mixture contribute to the formation of anortite and helenite in the combustion process. After modifying wollastonite with succinic acid and activating a mixture of quartz and marble, silicon is more fully realized in combustion reactions, wollastonite decomposes during the synthesis process, and decay products are involved in the formation of anorthite, helenite and $CaAl_4O_7$.

The X-Ray results of the synthesized ceramic SHS-samples using activated quartz, activated mixture of it with calcium carbonate and when introduced into the mixture modified during MCT of wollastonite in a planetary centrifugal mill showed that the latter component plays an exceptional role in formation of the phase composition of the material obtained. A ceramic composite based on aluminum oxide (corundum), silicon and multicomponent compounds such as anorthite, helenite, etc., is synthesized. The increase in strength of such a ceramic composite is due to both the phase composition and the structural features of the material obtained.

Table 10.10 Phase composition of the combustion products of (SiO_2 + $CaCO_3$ + Al) system samples depending on the presence of wollastonite, modifier and conditions of the MCT

	Phase content, %						
	Samples						
Phases	**1**	**2**	**3**	**4**	**5**	**6**	**7**
Al_2O_3	42.7	20.8	32.8	30.7	32.8	37.3	30.4
Si	20.3	10.3	12.1	13.2	17.7	11.8	10.5
Al	7.8	11.1	5.7	11.6	8.2	3.5	9.6
SiO_2	9.2	3.4	1.9	1.5	10.6	1.1	4.5
$Ca(Al_2Si_2O_8)$	–	26.1	43.9	28.2	–	30.9	18.4
$Ca_2Al(AlSiO_7)$	10.5	8.8	3.5	6.5	11.0	5.8	14.4
CaO	5.8	–	–	–	–	–	1.0
$CaAl_4O_7$	–	13.7	–	–	16.2	–	–
$Ca_3Si_3O_9$	–	–	–	–	–	–	5.7
$FeAl_3Si_2$	–	5.7	–	6.7	3.5	7.2	5.3
$FeSi_2$	–	–	–	1.5	–	2.3	–

1. ($30\%CaCO_3+70\%SiO_2$) + Al.
2. $(30\%CaCO_3+70\%SiO_2)_{MCT\ 10\ min}$ + Al.
3. $[(30\%CaCO_3+70\%SiO_2)+5\%B^*]_{MCT\ 10\ min}$ + Al.
4. $[(30\%CaCO_3+70\%SiO_2)_{MCT\ 10\ min} + (5\%W+5\%SA^{**})_{MCT\ 20\ min}]$ + Al.
5. $[(30\%CaCO_3+70\%SiO_2)_{MCT\ 10\ min} + (5\%W+5\%SA+5\%Ammonia)_{MCT\ 20\ min}]$ + Al.
6. $[(30\%CaCO_3+70\%SiO_2)_{MCT\ 10\ min} + (5\%W+5\%PVA^{****})_{MCT\ 20\ min}]$ + Al.
7. $[(30\%CaCO_3+70\%SiO_2)_{MCT\ 10\ min} + (10\%W+5\%PVA^{****})_{MCT\ 20\ min}]$ + Al.

*W, wollastonite; **SA, succinic acid; ***PVA, polyvinyl alcohol.

Modified wollastonite is the center of focal synthesis, which accelerates the combustion process. Due to the increased reactivity of activated and modified wollastonite, its melting point and decomposition into constituent oxides CaO and SiO_2 decrease, and interaction with the components of the charge is accelerated with the formation of ultrafine complex compounds. It plays the role of a template, around which structural changes take place, and providing an ordered structure in the synthesized sample, this determining the increase in the strength of the material. At the same time, the modifier is a carrier of the components necessary for the synthesis and the source of the gas phase in the synthesis products, ensuring formation of a porous structure of various levels. The combination of these two factors must be effective in obtaining the material of a given structure and, consequently, of properties. First of all, this is indicated in increasing strength.

Thus, it is shown that MCT of natural minerals used as components of the charge mixture for the SH-synthesis of composite systems contributes to a change in the kinetic characteristics of the combustion process: a decrease in the induction period of ignition and an increase in the rate of combustion. The change in combustion temperature is complex, due to activation of both exothermic reactions (redox) and endothermic (decomposition of carbonate and hydrate compounds, formation of complex compounds such as anorthite, helenite, wollastonite, etc.).

Mechanical treatment of quartz, calcite and wollastonite in the presence of various carbon and nitrogen-containing modifying agents reduces the induction period, increases the burning rate and temperature during subsequent SHS due to the participation of carbon and nitrogen in reactions and contributes to the maximum formation of corundum, aluminosilicate and calcium silicate compounds. The particles modified in the course of MCT are composite formations and contain elements of modifying additives in the surface layer, which provide them with high activity and are an additional component that takes an active part in the subsequent SHS process. The change in the strength characteristics of the synthesized SHS composites depending on MCT conditions of the initial charge components is correlated with formation of various forms of wollastonite, anorthite, helenite, aluminum nitride and silicon carbide in the synthesis products.

10.3.3 Ultrasonic Treatment and SH-Synthesis of Composite Systems

In Chapter 8 it is shown that ultrasonic treatment of powders in an aqueous medium is a rather effective way of changing both the dispersity and the structure of the surface layer of particles. The main mechanisms of ultrasonic dispersion of powder materials are cavitation and flows that occur in the working fluid during the collapse of cavitation cavities [79]. Quartz, calcite and wollastonite were subjected to ultrasonic treatment. These natural minerals are used in the development of technologies and practical recommendations for the manufacture and use of heat-resistant wollastonite-containing technical ceramics for various purposes. SH-synthesis is one of the technological approaches in solving the problem of obtaining a new generation of heat-resistant, as well as heat-insulating and refractory materials. Ultrasonic treatment was carried out both in distilled water and in aqueous solutions of various organic compounds, as a result of which polymer films encapsulating the particles of the processed powder are formed on the surface of the particles.

After treating quartz with purity up to 99% by ultrasound in water and using it in a SHS-charge mixture (SiO_2 + 37.5% Al), an increase in the maximum burning temperature, temperature at the burning stage and duration of the entire burning process is observed (Fig. 10.27).

After UST of quartz in aqueous solutions of various organic compounds (butanol, glycerin, urea), the efficiency of changing the thermokinetic characteristics of the combustion process is even more pronounced. As an example, combustion thermograms are shown using quartz after UST in a 10% aqueous solution of urea (Fig. 10.27b). With an increase in the duration of treatment, the induction period of ignition decreases, the rate and temperature of combustion increase.

Thus, we have visual evidence of the effect of preliminary ultrasonic processing on the state in this case of quartz, which is realized in the subsequent process of the technological combustion of the SHS system. After the UST of quartz, in all cases the temperature is quite high and the combustion process is very long. This suggests the most possible completion of phase formation processes.

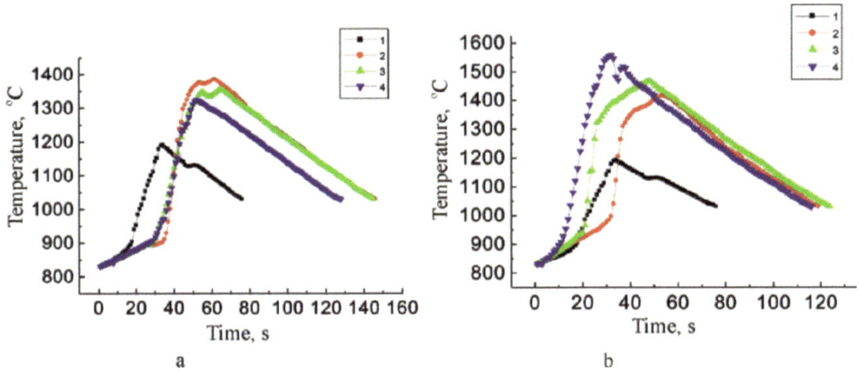

Figure 10.27 Thermograms of the combustion system (SiO_2 + 37.5% Al) with non-activated quartz (1) and after the UST in water (a) and 10% aqueous urea solution for 10 (2), 20 (3) and 40 min (4).

During UST of calcite, there takes place grinding and transformation of the particle surface (Chapter 8). With joint ultrasonic treatment of quartz and calcite (50:50 ratio), it is found that the system being processed is agglomerates, since there proceeds a partial interaction of quartz and calcite particles with each other. When a system is used for the SHS mixture, which contains, in addition to quartz, calcite in the ratio of 70:30, including also aluminum in the stoichiometric ratio, both in terms of quartz and calcite, the development of the synthesis process changes substantially. Calcium carbonate decomposes to CaO and CO when heated: CO creates a reducing atmosphere, and CaO can interact with aluminum. The presence of calcite leads to an increase in the induction period of ignition of the system, reducing the burning rate. The temperature profiles of combustion of a system containing both quartz and calcite indicate the fact that both at the stage of ignition of the mixture and at the stage of completion of the process (beyond the maximum temperature) of combustion in the sample, many phase-formation processes, both exothermic and endothermic, occur.

Figure 10.28 shows the temperature profiles of combustion systems containing SiO_2:$CaCO_3$ in the ratio of 70:30, both in the initial state and after the UST. The figure shows that after 10 min of the UST, the combustion temperature of the system has become significantly higher compared with the sample in the initial state of the components used, this indicating their high degree of activity. After 20 and 40 min of UST, there is a slight decrease in the combustion temperature, which is not associated with a decrease in system activity, and with the development, in addition to exothermic reactions, of endothermic processes, which take part of the heat to themselves, so-called endothermic reactions.

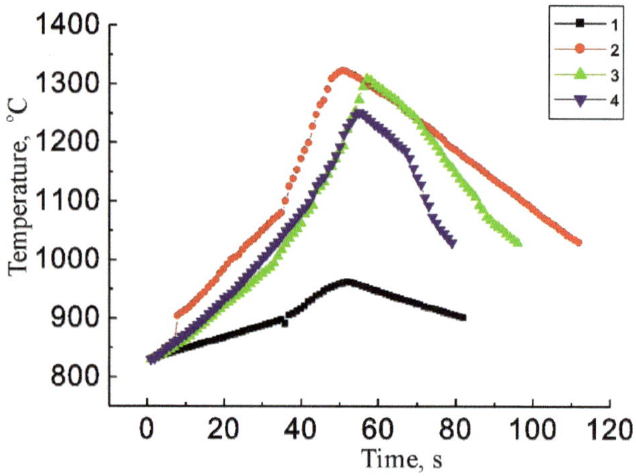

Figure 10.28 Thermogram of burning a mixture of quartz and calcite with a ratio of components 70:30. Non-activated system (1) and after UST in water for 10 (2), 20 (3) and 40 min (4).

Considering the combustion thermograms of quartz with calcite in the ratio of 70:30 after the UST in 10% aqueous solutions of butanol (Fig. 10.29a), it is clear that the combustion profile is also distinguished after 10 min of the UST. The burning rate is 7.11°/s, and after 20 and 40 min, of UST it decreases to 4.45°/s. In this case, the burning temperature and the duration of the period up to the maximum burning temperature, i.e., period of the active part of the burning increase. Probably, in this case, a more complete decomposition of $CaCO_3$ occurs. As a result of the SH-process, the formation of compounds between SiO_2, CaO and Al_2O_3 is also possible with the formation of mullite, pseudowollastonite, helenite, and other compounds. These reactions are endothermic and therefore a reduction in the rate and temperature of combustion is possible.

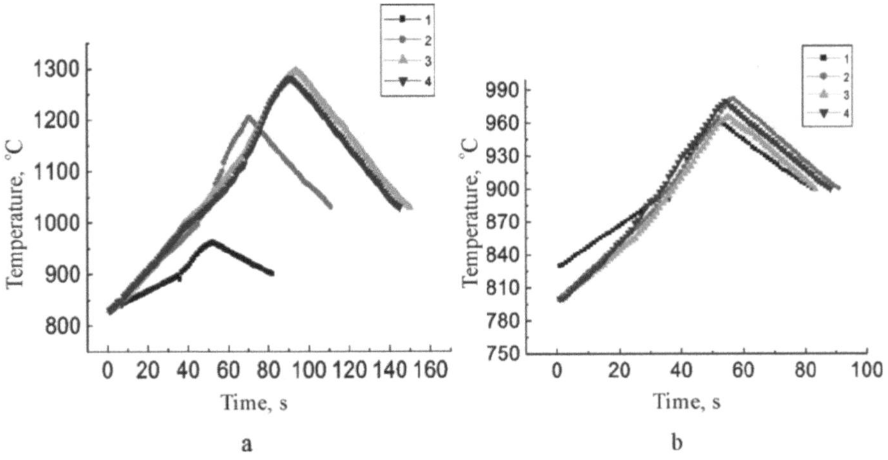

Figure 10.29 Combustion thermogram of a quartz mixture with calcite at a ratio of components 70:30. Non-activated system (1), after UST in 10% aqueous solution of butanol (a) and urea (b) for 10 (2), 20 (3) and 40 min (4).

Thermograms of combustion of quartz with calcite in a ratio of 70:30 after a UST in a 10% aqueous solution of urea showed (Fig. 10.29b) that the use of this modifying additive in UST reduces the speed and temperature of combustion of SHS samples. The temperature and rate of combustion are practically independent of the time of the UST of quartz with calcite in this case. And the maximum burning temperature does not exceed 990°C. Perhaps this is due to the decomposition of not only calcium carbonate, but also organic modifying additives. Urea decomposes to biuret, and in the presence of water vapor, complete hydrolysis with the formation of NH_3 and CO_2 is possible. Then ammonia is oxidized, which also affects the temperature and rate of reactions in the system.

Thus, the investigation results on the regularities of technological combustion of systems containing a mixture (SiO_2 + $CaCO_3$) with a component ratio of 70:30 showed that the presence of calcite leads to an increase in the induction period of the system ignition, reducing the burning rate. This is primarily due to the fact that, when heated, calcium carbonate decomposes with formation of calcium oxide, carbon oxide or carbon dioxide. The gas component can serve as a reducing agent for oxides, or at least be a medium that prevents oxidation of aluminum. After UST of the mixture (SiO_2 + $CaCO_3$), there is a significant change in the thermo-kinetic characteristics of the combustion process, the values of which depend on both the processing time and the modifier used in the UST. Thus, the presence of butanol in UST in an aqueous medium contributes to an increase in the combustion temperature, while urea, on the contrary, leads to its decrease.

The main result of changing the thermo-kinetic characteristics of the combustion process is the phase composition of the synthesis products. When the quartz + calcite + aluminum system burns, a lot of unreacted quartz and aluminum remain in the reaction products (Table 10.11). After UST of quartz with calcite for 40 min in a 10% aqueous solution of butanol, its use in the SHS charge leads to almost complete realization of quartz and significant consumption of aluminum (aluminum residue 10.4%) in the combustion reactions. In the process of synthesis, ternary compounds are formed, such as, helenite-$Ca_2Al((AlSi)O_7)$ and grassite, $CaAl_4O_7$. In the combustion products of the system under study, after processing a mixture of quartz and calcite in 10% aqueous urea solution, parawollastonite (4.8%) and more than 10% $Al_{2.144}O_{3.2}$, i.e., non-symmetric aluminum can be detected.

For the synthesized samples obtained on the basis of the binary system, including quartz and calcite (SiO_2:$CaCO_3$ = 70:30), after UST, the strength increases from 20 to 40 MPa, except for samples with quartz + calcite after UST in an aqueous solution of urea. The strength of such samples after 10 min of UST decreases sharply to 2 MPa, and after a longer treatment, it rises to 16 and 22 MPa. After UST for 40 min in the presence of alcohols, the strength of SHS samples increases (up to 30 MPa), and when using urea as a modifier, on the contrary, such a long ultrasonic treatment of the minerals used leads to a decrease in the strength characteristics of SHS samples (to 18 MPa). Low strength values are associated with formation of a porous structure of the samples (Fig. 10.30). The increase in strength at high porosity of the system samples (SiO_2 + $CaCO_3$) after prolonged ultrasonic processing of the powder is due to formation in the synthesis products of such phases as helenite-$Ca_2Al((AlSi)O_7)$ and grossite ($CaAl_4O_7$) and parawollastonite ($CaSiO_3$), which provide hardening of the ceramic parts of the sample.

Table 10.11 Composition of the combustion products of the samples (SiO_2 + $CaCO_3$ + Al) depending on the conditions of the UST and the type of modifier

	Phase content, %		
	Samples		
Phases	**1**	**2**	**3**
SiO_2-Quartz	48.1	0.6	33.6
Al_2O_3	5.1	39.1	15.4
Si	4.4	198	6.7
Al	34.3	10.4	22.6
$CaSiO_3$-Parawollastonite	8.0		4.8
$Ca_2Al((AlSi)O_7)$		12.8	6.0
$Al_{2.144}O_{3.2}$			10.8
$CaAl_4O_7$		17.3	

1. (SiO_2 + $CaCO_3$) 70:30 initial.
2. (SiO_2 + $CaCO_3$)70:30, UST (H_2O+10% C_4H_9OH) after 40 min of UST.
3. (SiO_2 + $CaCO_3$) 70:30, (H_2O+10%$(NH_2)_2CO$) after 40 min of UST.

(a) (b)

Figure 10.30 SHS-samples with a porous structure obtained on a system of 70% quartz + 30% calcite modified during UST in 10% aqueous solutions of butanol (a) and urea (b).

Thus, it is found that by varying the preliminary regimes of the UST of quartz and a mixture of quartz-calcite, it is possible to regulate the process of phase formation during the SH-synthesis process and the formation of a porous structure, which ultimately determines the strength properties of the synthesized composite material, and, consequently, the possibility of its practical use.

From the presented results, it follows that ultrasonic processing of minerals (quartz and calcite) is favorable for obtaining SHS-samples of the required quality from the standpoint of their phase composition. In addition, it also contributes to obtaining a granular structure, which ultimately can characterize the synthesized material as heat resistant with good heat-shielding properties.

A comprehensive study of the technological combustion of systems containing wollastonite after ultrasonic treatment (UST) was also carried out. As shown in Chapter 8, during UST in the presence of various organic modifying additives, there takes place not only grinding of wollastonite particles, but also erosion of their surface, as well as various complex structural changes on their surface, and encapsulation into colloidal conglomerates. After ultrasonic treatment in aqueous solution in the presence of various modifying additives, highly dispersed particles and their aggregates are encapsulated in sufficiently dense polymer films.

Depending on the conditions of UST (time, frequency), the nature and level of structural changes are somewhat different, this being indicated in thermokinetic characteristics of the combustion process. Introduction of wollastonite modified during UST into the charge mixture (SiO_2 + Al) in an amount of 10% contributes to a decrease in the induction period of ignition of the mixture, an increase in the rate and temperature of combustion, as well as a change in temperature at the post-process stage (Fig. 10.31).

The level of these characteristics varies with both the processing time and the frequency of ultrasound exposure. Intensification of the combustion process with introduction of wollastonite into the mixture after ultrasonic treatment in water can be associated with both an increase in the defectiveness of the powder particles and with the hydration of the surface layer of particles. As it was shown earlier, when using mechanochemical treatment to modify various minerals [89], modification of wollastonite with compounds containing bound water contributes to an increase in the thermokinetic characteristics of combustion. The changes in the maximum burning temperature are the most indicative (Fig. 10.32).

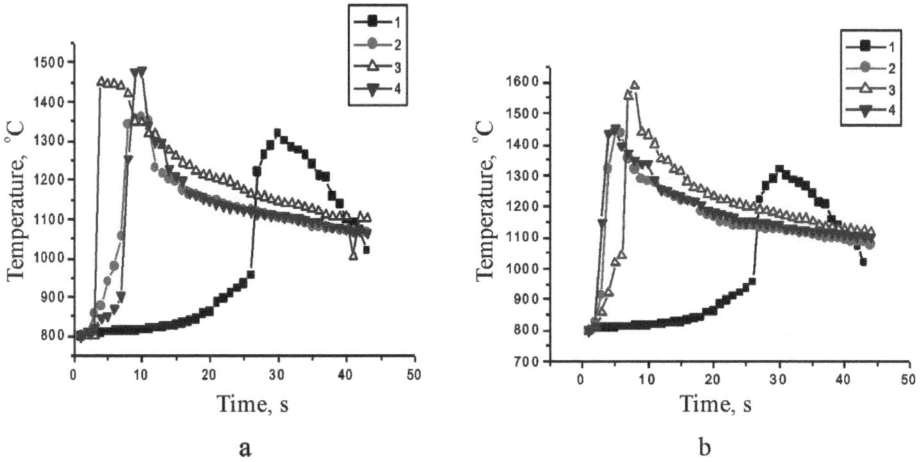

Figure 10.31 Thermograms of the combustion system [(SiO_2 + 37.5% Al]) with 10% FW100 wollastonite in the initial state (1) and after ultrasonic treatment in water for 20 (2), 40 (3) and 60 min (4) at a frequency of 40 kHz (a) and simultaneous exposure to 40 and 100 kHz (b).

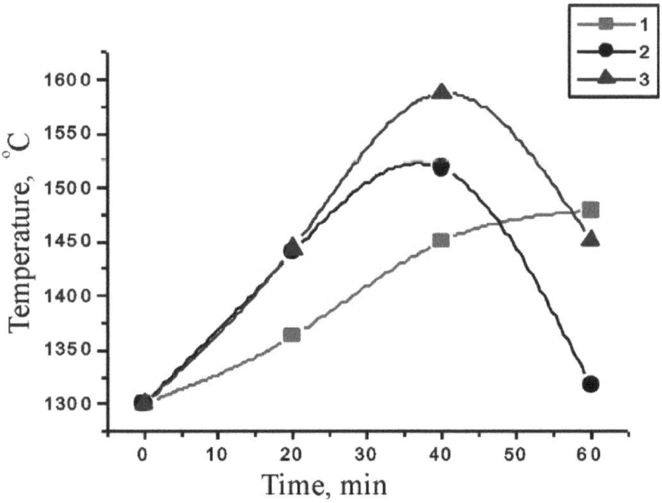

Figure 10.32 Dependence of the maximum combustion temperature of SHS samples with 10% wollastonite after ultrasonic treatment in water on time at a frequency of 40 (1), 100 (2) and simultaneous exposure to 40 and 100 kHz (3).

The figure shows the dependence of the maximum combustion temperature on the conditions of UST of wollastonite. The maximum combustion temperature, almost 1600°C, is shown by samples containing wollastonite after UST for 40 min with simultaneous ultrasonic exposure at 40 and 100 kHz.

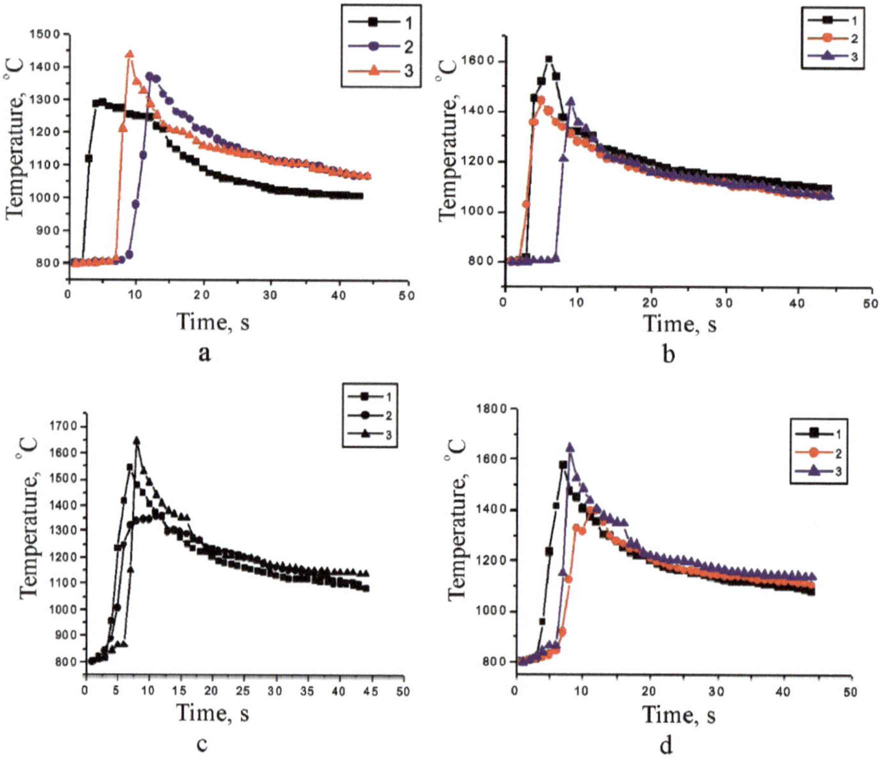

Figure 10.33 Thermograms of the combustion system [(SiO_2 + 37.5% Al] with addition of wollastonite after ultrasound treatment in 10% aqueous solution of $C_3H_5(OH)_3$ (a), $(C_2H_3OH)_n$ (b), $C_9H_8O_4$ (c), $H_2SiO_3 \cdot n(H_2O)$ (d) at a frequency of 40 kHz (1), 100 kHz (2) and simultaneous ultrasonic exposure of 40 and 100 kHz (3).

With introduction of various modifying additives into the medium of ultrasonic treatment of wollastonite, the general tendency to intensify the combustion process is preserved, only the parameters of the SH-synthesis process change somewhat (Fig. 10.33). For samples modified during UST in the presence of polyvinyl alcohol, combustion proceeds more intensively (rate exceeds 120°/s) than in case with wollastonite modified during UST with glycerin, when the combustion rate varies from 50 to 90°/s. The presence of acid-containing modifying agents in the aqueous medium leads to structural changes in the surface of particles, which reduce the combustion rate to 60°/s [90, 91]. However, the combustion temperature at the same time reaches 1600°C. Combustion develops most actively on samples with wollastonite treated with ultrasound under duplicated exposure at frequencies of 40 and 100 kHz.

One of the indicators of the quality of synthesized samples is the tensile strength. If for a synthesized sample of [SiO_2 + 37.5% Al] system with quartz activated for 10 min in a planetary centrifugal mill, the strength is 33 MPa, its values for SHS samples obtained with wollastonite modified at different conditions of UST, vary significantly depending on the conditions of ultrasound modification. The measurement results are presented in Table 10.12.

Table 10.12 Strength of SHS-samples [SiO_2 + 37.5% Al] with 10% wollastonite modified under different conditions of UST depending on the frequency and time in different liquid media

Conditions of US-treatment		The strength of SHS-samples, MPa				
		The medium of US-treatment of wollastonite				
Frequency *f*, kHz	Time, min	H_2O	H_2O+5% $C_3H_5(OH)_3$	H_2O+5% $H_2SiO_3 \cdot nH_2O$	H_2O+5% $(C_2H_3OH)_n$	H_2O+5% $C_9H_8O_4$
40	20	41.8	50.16	25.08	20.9	62.70
	40	71.06	39.71	21.62	22.99	62.70
	60	62.70	41.80	18.81	35.53	52.25
100	20	33.44	58.52	10.45	14.63	73.15
	40	31.35	12.54	16.75	33.44	62.70
	60	37.62	14.63	45.98	31.35	52.25
100 + 40	20	41.80	18.81	52.25	39.71	73.15
	40	73.15	73.15	31.35	41.8	62.70
	60	29.26	20.90	27.17	20.9	62.70

The highest strength was shown by samples containing wollastonite after ultrasonic treatment in water and in an aqueous solution of glycerin (73.15 MPa). Samples with wollastonite modified with silicic acid had the lowest strength. The treatment of wollastonite in salicylic acid solution contributes to a stable increase in the strength of SHS samples (from 52 to 73 MPa), regardless of the ultrasonic treatment conditions, which clearly follows from the data presented in Table 10.12.

In all cases, introduction of wollastonite into the charge after UST contributes to a more complete conversion of the initial components during the combustion of the samples [86]. The greatest amount of corundum and silicon is formed, when using wollastonite treated in water, an aqueous solution of polyvinyl alcohol and salicylic acid. Such samples are characterized by consistently high strength values (up to 73 MPa) and electrical insulating properties.

Thus, the investigation results on the effect of wollastonite modified during ultrasonic treatment on the combustion process showed that its addition significantly activates the combustion process. This is a clear indicator of changes in the structure and state of the material under the influence of UST. The degree of activity of the material is determined by the conditions of the ultrasonic treatment, namely, depends on the frequency and time of exposure. For different media, the optimal conditions for frequency and processing time are different. The most effective, judging by the thermo-kinetic parameters of combustion, is the time of 40 min with duplicated effects of UST at 40 and 100 kHz.

The results of the studies clearly show that preliminary mechanochemical treatment both in dynamic mills and under ultrasonic action in liquid media is an effective way to influence the structure and the state of the material being processed, increases its reactivity, which is manifested in the changes in the thermokinetic characteristics of the combustion process (SHS mode), conversion of the reagents used, the phase composition of synthesis products and the properties of the material obtained.

10.3.4 SH-Synthesis of Composite Systems with Participation of Aluminum Modified during MCT

Activation and modification of aluminum powder during MCT is effectively manifested in the combustion process in the SHS mode of a mixture of aluminum powder as a fuel with silicon dioxide, used as an oxidizer. Silicon dioxide in this case is used in the unactivated state. Mechanochemical treatment of aluminum with graphite and introduction of the obtained powder into the mixture with quartz contribute to a significant reduction in the induction period of ignition, an increase in the rate and temperature at all stages of the combustion process as compared to a non-activated fuel (Fig. 10.34a).

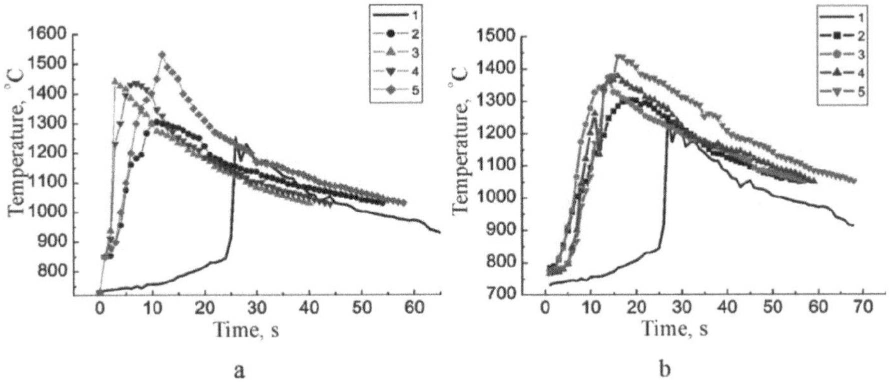

Figure 10.34 Thermograms of combustion systems (SiO_2 + Al) with Al in the initial state and after 20 min of MCT with different amounts of graphite and polyvinyl alcohol. (a) SiO_2 + (Al/C), (b) SiO_2 + Al/$(C_2H_3OH)_n$; (1) Al initial; (2) 3%; (3) 5%; (4) 10%; (5) 2%.

The development of the combustion process vividly indicates the efficiency of using an activated mixture of aluminum and graphite in the charge during the SH-synthesis of systems. The maximum combustion temperature is 1532°C and is fixed when the content of 20% graphite in activated aluminum.

After MCT of aluminum with polyvinyl alcohol, a decrease in the induction period of ignition, an increase in the rate and temperature at all stages of the combustion process are observed (Fig. 10.34b), but somewhat less than for systems containing activated Al/C mixture. A distinctive feature of the combustion of such systems is a more stable development of the process, especially at the post-process stage.

Figure 10.35 presents combustion thermograms of the [SiO_2 + $(Al/C_{17}H_{35}COOH)_{MCT}$] system. The carbon component of stearic acid is actively involved in the reduction of the oxide component of the charge. After MCT of composite ($Al/C_{17}H_{35}COOH$), the induction period of ignition decreases, the rate and temperature of combustion increase. The MCT of aluminum in the presence of a carbon modifier greatly increases the energy intensity of the system and intensifies the combustion process. Preliminary MCT of aluminum with stearic acid provides accelerated ignition of the system, and also provides not only the high temperature of the process, but also a more long-lasting and stage-wise burning of the system. The combustion temperature rises from 1202 to 1366°C with a modifier content of 20% and 3%, respectively. In most cases, the chemical reactions occurring at the same time are of a general nature and represent a radical-chain oxidation process. The generality of the mechanism of oxidation of these substances is primarily due to the presence in their composition of methylene (or methine) units involved in radical-chain oxidative processes.

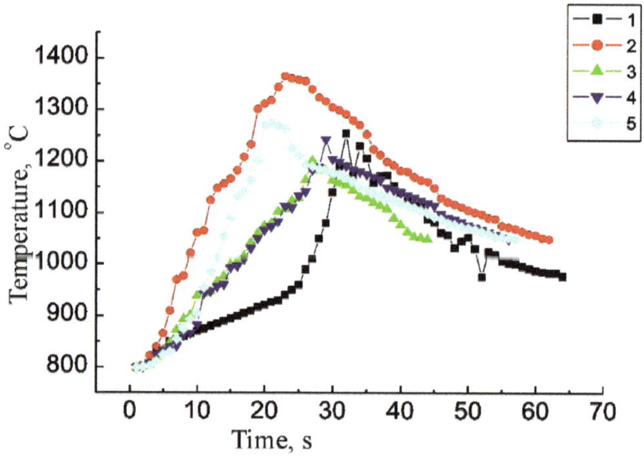

Figure 10.35 Combustion thermograms of the system (SiO_2 + $Al_{initial}$) and [SiO_2 + ($Al/C_{17}H_{35}COOH$)] after MCT. (1) Al initial; (2) Al + 3% $C_{17}H_{35}COOH$; (3) Al + 5% $C_{17}H_{35}COOH$; (4) Al + 10% $C_{17}H_{35}COOH$; (5) Al + 20% $C_{17}H_{35}COOH$.

The change in the reactivity of the system as a result of MCT affects not only the thermokinetic characteristics of the combustion process, but also the composition of the final synthesis product. Table 10.13 presents the data on the phase composition of the SH-synthesis products of (SiO_2 + 37.5% Al) system depending on the state of the charge components after MCT of aluminum with modifying additives. The results of X-ray phase analysis of the synthesized samples based on mechanically activated aluminum showed that the phase formation process is largely determined by the pre-activation of the charge components and the presence of organic additives. Mechanical treatment contributes to a more complete realization of the initial components in the synthesis process. The use of organic additives has shown the effectiveness of their impact on the formation of various forms of silicon carbide in the process of synthesis. Modification of the system contributes to the formation of aluminum nitride, iron and silicon disilicide, and the complex compound $FeAl_3Si_2$. The phase composition of synthesis products determines the properties of the material obtained.

Table 10.13 Results of X-ray phase analysis of SHS samples depending on the conditions of mechanochemical treatment of aluminum with modifiers

	Phase content, %						
	Samples						
Phases	**1**	**2**	**3**	**4**	**5**	**6**	**7**
Al_2O_3	43.6	67.9	77.2	65.8	61.5	65.3	51
γ-Al_2O_3	–	–	–	3.1	10.5	–	–
Si	14.9	15.8	9.8	25.8	25.3	17.5	21.9
Al	10.2	2.0		1.5		5.0	11.7
SiO_2-quartz	29.9	4.1	1.2	2.8	2.7	8.0	4.2
$FeAl_3Si_2$	1.2	–	–	–	–	–	–
AlN	–	1.9	–	1.0	–	1.6	6.0
$FeSi_2$	–	0.6	1.2	–	–	0.7	0.5
SiC- Moissanite 3C	–	2.1	7.8	–	–	–	–
SiC-Moissanite 2H	–	–	2.8	–	–	–	–
$Al_{4.59}Si_{1.41}O_{9.7}$	–	3.4	–	–	–	1.8	4.6

(1) (SiO_2 +Al); (2) [SiO_2 + (Al + 5%C)$_{MCT}$]; (3) [SiO_2 + (Al + 20% C)$_{MCT}$]; (4) [SiO_2 + (Al +3% $(C_2H_3OH)_n$)$_{MCT}$]; (5) [SiO_2 + (Al + 20% $(C_2H_3OH)_n$)$_{MCT}$]; (6) [SiO_2 + (Al + 3%$C_{17}H_{35}COOH$)$_{MCT}$]; (7) [SiO_2 + (Al + 10% $C_{17}H_{35}COOH$)$_{MCT}$].

The X-Ray analysis results of the samples showed the presence of up to 67.9% corundum and 4.1% silica, as well as silicon, iron disilicide and aluminum nitride in the composition of the combustion products of mixtures containing [SiO_2 + (Al + 5% C)$_{MCT}$]. With an increase in the amount of carbon up to 20% in the [SiO_2 + (Al + 20% C)$_{MCT}$] system, aluminum is fully realized in combustion reactions. When modifying the mixture with polyvinyl alcohol, a significant change in the phase composition is observed on samples of [SiO_2 + (Al + 3% $(C_2H_3OH)_n)_{MCT}$] and [SiO_2 + (Al + 20% $(C_2H_3OH)_n)_{MCT}$]. They contain corundum (main phase), a sufficiently large amount of silicon and γ-Al_2O_3.

Stabilization of combustion by modification of the surface of an activated aluminum particle results in the change in the kinetic characteristics of the process, ensuring the completeness of the reaction. Activation of aluminum contributes to formation of aluminum nitride in the combustion process. The presence of carbon [SiO_2 + 37.5% (Al + 20% C)$_{MCT}$] in the modified surface layer of the aluminum particle during MCT leads to formation of silicon carbide in two modifications during the subsequent SH-synthesis. The iron rubbed from the walls of the grinding vessels and the surface of the balls during the grinding of aluminum also interacts with quartz to form iron disilicides.

The strength of the synthesized samples varies from 6.27 to 45.98 MPa (Table 10.14). The high strength of the material is due to both the phase composition and the density of the samples. Modification of aluminum during MCT with stearic acid leads to formation of a finely porous structure due to the burning out of its degradation products during MCT and in the process of SHS.

From Table 10.14 it follows that composition [37.5% (Al + 20% C)$_{MCT}$ + SiO_2] has the maximum combustion temperature but its strength decreases as compared with a sample without carbon. With introduction of carbon into the composition and an increase in its content in the composition of the mixture, the strength of the synthesized sample decreases. This is due to the fact that during the combustion of the mixture [(Al + 3% C)$_{MCT}$ + SiO_2], gaseous products are released, and with an increase in the carbon content of the mixture, the amount of gaseous products increases [92]. The maximum burning rate during SH-synthesis was observed for the [(Al + 5% C)$_{MCT}$ + SiO_2] system and amounts to 118.2°/s. This is probably due to the optimal ratio of the particle size of the components, respectively, with an increase in the packing density, which ensures the density of contact between the oxidizer and the fuel.

Table 10.14 Indicators of maximum temperature, burning rate of SiO_2 mixtures with modified aluminum and strength characteristics of the synthesized samples

The composition of the modified fuel based on aluminum	T_{max}, °C	Burning rate, °C/sec	σ, MPa
Al initial + (SiO_2 20 min ¼)	1319	19.16	37.6
Al + 3% C	1305	41.4	45.98
Al + 5% C	1441	118.2	8.36
Al + 10% C	1436	83.7	12.54
Al + 20% C	1532	56.8	2.11
Al + 3% $(C_2H_3OH)_n$	1326	27.4	22.9
Al + 5% $(C_2H_3OH)_n$	1345	43.8	12.54
Al + 10% $(C_2H_3OH)_n$	1378	38.2	6.89
Al + 20% $(C_2H_3OH)_n$	1439	41.8	1.32
Al + 3% $C_{17}H_{35}COOH$	1366	24.6	17.7
Al + 5% $C_{17}H_{35}COOH$	1202	14.9	10.45
Al + 10% $C_{17}H_{35}COOH$	1267	13.5	6.27
Al + 20% $C_{17}H_{35}COOH$	1275	22.6	18.81

In case of aluminum oxide with polyvinyl alcohol, similar changes occur. With an increase in the amount of the modifying additive, a decrease in the strength of the synthesized samples from 22.9 to 6.89 MPa is observed. After introduction of stearic acid, a decrease in strength is also observed. The decrease in strength is due to the porosity of the samples.

Thus, from the presented data, it follows that by modifying the components of the mixture, one can purposefully influence the combustion process and the formation of a dense or porous structure with ultradispersed phases, which strengthen the material (of the sample being synthesized). In the synthesized samples of the [SiO_2 + 37.5% (Al + 3% C)$_{MCT}$] system, the maximum strength value is equal to 45.98 MPa. The presence of the modifier leads to a decrease in the induction period of the system ignition and an increase in the burning rate. At the same time, in the composition of the charge, the amount of aluminum was lower than a stoichiometric amount (37.5%), since the composites also had an organic modifier.

To increase the degree of aluminum dispersion during MCT with carbon modifiers, silicon dioxide (quartz) in the amount of 5–20% was additionally introduced into the processed mixture, which acted as an abrasive material with respect to aluminum. In addition, it was an activated constituent of the oxide component of the mixture. During the preparation of the samples, the amount of aluminum was recalculated (in a stoichiometric ratio of 37.5%), taking into account SiO_2 present in the activated mixture of aluminum.

Figure 10.36 shows the combustion thermograms of the [SiO_2 + (Al/C + SiO_2)$_{MCT}$] composites with a quartz content of 5–20% and graphite 5–20%. Samples of [SiO_2 + (Al + 20% C + 10% SiO_2)$_{MCT}$] showed the maximum rate and temperature of combustion.

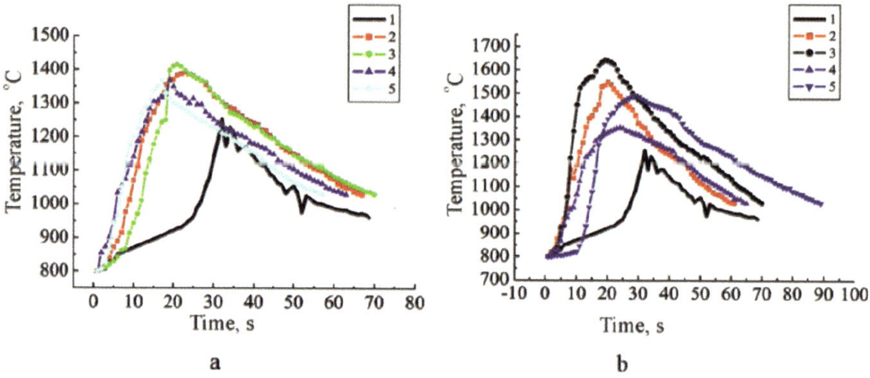

Figure 10.36 Thermograms of combustion systems [SiO_2 + 37.5% Al] with aluminum in the initial state and after MCT with graphite and quartz. (a) 5% of graphite; (б) 20% of graphite; (1) [SiO_2 +37.5% $Al_{initial}$]; (2) [SiO_2 + (Al/C + 5% SiO_2)]; (3) [SiO_2 + (Al/C + 10% SiO_2)]; (4) [SiO_2 + (Al/C + 15% SiO_2)]; (5) [SiO_2 + (Al/C + 20% SiO_2)].

From the obtained thermograms, it follows that after MCT of the mixture the induction period of ignition decreases and the temperature of combustion of thermite mixtures increases. During mechanochemical treatment of metal particles, the concentration of dislocations increases. As a result, the thermo-kinetic characteristics of the combustion process are intensified. The results of the study showed that the sample [SiO_2 + (Al + 20% C + 10% $SiO_2)_{MCT}$] has the maximum burning temperature (~1641°C).

Figure 10.37 shows the combustion thermograms of a composite containing aluminum after MCT with 20% polyvinyl alcohol and from 5% to 20% silicon dioxide. In the SH-system [SiO_2 + (Al + 20% PVA + 5% $SiO_2)_{MCT}$], the ignition time (i.e., the ignition induction period) decreases with formation of γ-Al_2O_3 up to 11.5% in combustion products.

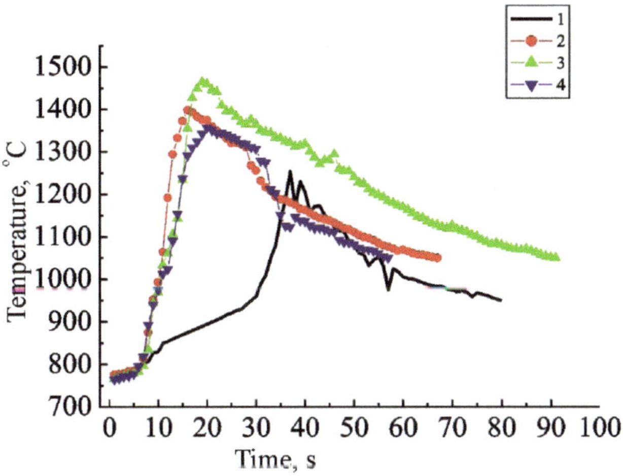

Figure 10.37 Thermograms of combustion systems (SiO_2 + $Al_{initial}$) and [SiO_2 + (Al/20% $(C_2H_3OH)_n$ + $SiO_2)_{MCT}$] after MCT with different SiO_2 content. (1) [SiO_2 +37.5% $Al_{initial}$]; (2) (Al/20% $(C_2H_3OH)_n$ + 5% SiO_2); (3) (Al/20% $(C_2H_3OH)_n$ + 10% SiO_2); (4) (Al/20% $(C_2H_3OH)_n$ + 20% SiO_2).

Figure 10.38 presents thermograms of combustion of composites [SiO_2 + (Al/3% $C_{17}H_{35}COOH$ + $SiO_2)_{MCT}$] with silica content from 5% to 20% in a mixture (Al/3% $C_{17}H_{35}COOH$ + SiO_2). With an increase in the content of silicon dioxide during MCT of aluminum, a decrease in the induction period of ignition and an increase in the combustion temperature are observed.

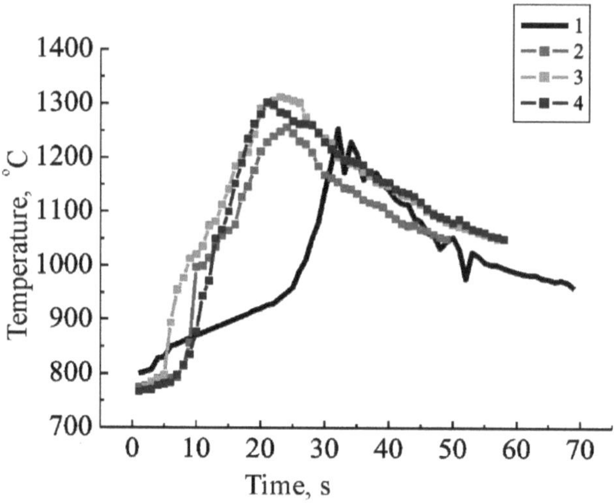

Figure 10.38 Thermograms of combustion systems (SiO_2 + $Al_{initial}$) and [SiO_2 + (Al/3% $C_{17}H_{35}COOH$ + SiO_2)$_{MCT}$] after MCT with different content of SiO_2. (1) [SiO_2 + 37.5% $Al_{initial}$]; (2) (Al/3% $C_{17}H_{35}COOH$ + 5% SiO_2); (3) (Al/3% $C_{17}H_{35}COOH$ + 10% SiO_2); (4) (Al/3% $C_{17}H_{35}COOH$ + 20% SiO_2).

Thus, the use of activated and modified aluminum after its treatment with silicon dioxide leads to an increase in the thermokinetic characteristics of the combustion process. Comparative characteristics depending on the modifying additives used (graphite and polyvinyl alcohol) are presented in Fig. 10.39. Mechanochemical treatment of aluminum with polyvinyl alcohol in the presence of SiO_2 results in a decrease in the induction period of ignition, an increase in combustion rate and temperature at all stages of the process, but somewhat less than for systems containing a mixture of Al/C. A distinctive feature of the combustion of such systems is a more stable development of the process, especially at the post-process stage.

The optimum content of SiO_2 in the processed mixture of aluminum with a carbon modifier is 10%. The maximum burning temperature was recorded in the MCT of aluminum with graphite, and the most stable and long-lasting development of the combustion process is observed, when polyvinyl alcohol and 10% SiO_2 in the MCT of aluminum.

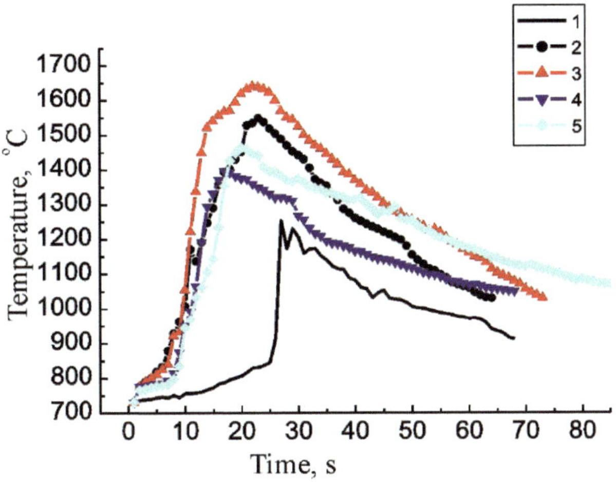

Figure 10.39 Thermograms of combustion systems (SiO_2 + $Al_{initial}$) and [SiO_2 + (Al + modifier + SiO_2)$_{MCT}$]. (1) [SiO_2 + 37.5% $Al_{initial}$]; (2) [SiO_2 + (Al + 20% C + 5% SiO_2)]; (3) [SiO_2 + (Al + 20% C + 10% SiO_2)]; (4) [SiO_2 + (Al + 20% $(C_2H_3OH)_n$ + 5% SiO_2)]; (5) [SiO_2 + (Al + 20% $(C_2H_3OH)_n$ + 10% SiO_2)].

Samples after technological combustion of the mixture based on composites [SiO_2 + (Al/$(C_2H_3OH)_n$ + SiO_2)$_{MCT}$] have a low indicator of strength characteristics due to their high porosity. This is due to the fact that gaseous synthesis products are formed in large quantities in composites. When polyvinyl alcohol is used for modification of the charge, the strength of the samples decreases most sharply. The corresponding level of strength of the samples is due to their density, which can already be clearly seen by the nature of the fracture after testing the strength of the samples (Fig. 10.40). The samples modified with polyvinyl alcohol have the loosest structure.

A finely porous structure with dense partitions in the samples obtained with fuel in the form of aluminum modified with carbon is observed. Porosity is a consequence of formation of a gas phase in the synthesis products. This fact indicates the prospects for the use of such materials for production of heat-insulating systems, which is confirmed by the results of measuring the properties of the synthesized samples (Table 10.15). Thermal conductivity decreases with the increase in the amount of modifying additives used. During combustion of the systems [SiO_2 + ($Al/3\%$ $C_{17}H_{35}COOH$ + $SiO_2)_{MCT}$], the strength of the samples increases to 39.7 MPa. High strength is probably due to the fact that unreacted aluminum is a binder between the products of combustion.

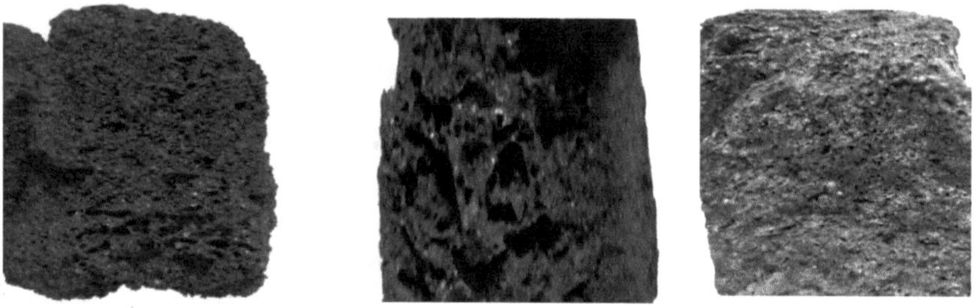

a b c

Figure 10.40 Appearance and fracture after compression testing of SHS samples obtained with modified aluminum. (a) Al/C, (b) $Al/(C_2H_3OH)_n$, (c) $Al/C_{17}H_{35}COOH$.

Table 10.15 Dependence of the thermo-kinetic characteristics of the combustion system (SiO_2 + 37.5% [Al/modifier + SiO_2]) and strength indicators (σ) and thermal conductivity (λ) of the synthesized SHS samples on the conditions of MCT of aluminum

The composition of the charge	T_{max}, °C	Burning rate, degree/sec	σ, MPa	λ, W/mK
Al+5% C+5% SiO_2	1392	25.7	8.36	0.947
Al+5% C+10% SiO_2	1416	29.3	27.17	0.198
Al+5% C+20% SiO_2	1369	33.5	27.99	0.189
Al+20% C+5% SiO_2	1549	37.5	2.09	0.140
Al+20% C+10% SiO_2	1641	44.3	4.18	0.160
Al+20% C+20% SiO_2	1492	23.9	3.22	0.309
Al+20% $(C_2H_3OH)_n$+5% SiO_2	1398	38.9	2.14	0.275
Al+20% $(C_2H_3OH)_n$+10% SiO_2	1464	43.6	3.57	0.200
Al+20% $(C_2H_3OH)_n$+20% SiO_2	1350	29.6	4.98	0.426
Al+3% $C_{17}H_{35}COOH$+5% SiO_2	1330	23.1	14.63	0.163
Al+3% $C_{17}H_{35}COOH$+10% SiO_2	1313	23.7	39.7	0.192
Al+3% $C_{17}H_{35}COOH$+20% SiO_2	1302	25.5	27.17	0.133

X-ray results showed the presence of gamma-alumina, as well as silica, alumina, silicon, and iron silicide. In the synthesized samples, unreacted aluminum remains, which suggests that aluminum does not completely react with silicon oxide. This fact is an advantage for the obtained cermet samples, since the presence of aluminum in the composition of the product increases the strength of the material (Table 10.16).

Table 10.16 Results of X-ray analysis of SHS samples depending on the conditions of mechanochemical treatment of aluminum with modifiers and silicon dioxide

	Phase content, %								
	Samples								
Phases	1	2	3	4	5	6	7	8	9
Al_2O_3	63.7	74.2	71.7	74.7	75.9	62.2	68.3	60.8	60.2
γ-Al_2O_3	–	–	–	–	4.9	11.5	–	–	4.7
Si	17.4	18.0	21.5	11.1	13.7	21.8	21.4	16.9	21.0
Al	0.8	–	–	–	–	–	–	7.3	3.5
SiO_2-quartz	10.6	3.8	1.5	1.9	1.9	2.8	9.6	8.9	6.6
$AlFeO_3$	–	–	–	3.3	–	–	–	–	–
AlN	2.4	–	–	–	–	–	–	6.0	4.0
$FeSi_2$	–	1.0	1.3	1.1	–	–	–	–	–
SiC-Moissanite 3C	5.1	3.0	4.0	6.9	8.5	1.7	0.6	–	–
SiC-Moissanite 2H	–	–	–	–	–	–	–	–	–
$Al_2(SiO_4)O$	–	–	–	1	–	–	–	–	–

(1) SiO_2 + (Al + 5% C + 5% SiO_2); (2) SiO_2 + (Al + 5% C + 15% SiO_2); (3) SiO_2 + (Al + 10% C + 5% SiO_2); (4) SiO_2 + (Al + 20% C + 5% SiO_2); (5) SiO_2 + (Al + 20% C + 10% SiO_2); (6) SiO_2 + (Al+20% $(C_2H_3OH)_n$ + 5% SiO_2); (7) SiO_2 +(Al+20% $(C_2H_3OH)_n$ + 20% SiO_2); (8) SiO_2 + (Al + 3% $C_{17}H_{35}COOH$ + 5% SiO_2); (9) SiO_2 + (Al+3% $C_{17}H_{35}COOH$ + 20%SiO_2).

The presence of carbon in the surface layer of a quartz particle modified with MCT leads to formation of silicon carbide in the course of subsequent SH-synthesis. The iron rubbed from the walls of the grinding vessels and the surface of the balls during the grinding of aluminum also interacts with quartz to form iron disilicides. In addition, activation of aluminum contributes to formation of aluminum nitride in the combustion process.

At the same time, a definite role is played by a carbon modifier, the oxidation (burning) of which in the synthesis process creates an inert atmosphere resulting in the interaction of aluminum with air nitrogen. The reactions of both carbide and nitride formations are exothermic, this providing an increase in the combustion temperature.

Questions

1. What are the main characteristics of the SHS process and their dependence on the calorific value of the mixture, thermal conductivity of the combustion products, concentration of the initial substances and reaction products, the activation energy of the reaction, and other indicators?
2. What effects can control the combustion process, ensuring its stationarity?
3. What are the possibility of using ultrasonic treatment and mechanochemical treatment of charge mixtures to change the energy state of the material and the development of the combustion process?
4. Describe the modification of the surface of particles as a way to stabilize the state of activated systems and conservation of stored energy.
5. How does the encapsulation of particles in the polymer shell in the process of MCT and UST for realization of the energy accumulated during mechanical processing of the substance in the SHS process takes place?
6. Describe the change in the maximum temperature of combustion, the induction period of ignition and the rate of temperature change at the stage of postprocesses depending on the conditions of the MCT of quartz?
7. What are the patterns of the effect of nitrogen-containing compounds on the surface of modified quartz on the development of the combustion process?
8. What are the features of the modification of quartz compounds containing a large amount of bound water, and the realization of these structures in the process and combustion products?
9. Describe the patterns of technological combustion of complex systems (a mixture of quartz and calcite) after the MCT of the mixture components.

10. What is the role of wollastonite in the charge mixture of quartz + calcite depending on the content and preliminary MCT of the preparation in the development of the combustion process?
11. Describe the modification of charge mixture components as a way to enhance the role of gas-phase reactions in the combustion process of an activated system.
12. What are the regularities of changes in the composition of the combustion products and the properties of the synthesized samples depending on the conditions of preliminary MCT?
13. How does the effect of preliminary ultrasonic treatment of quartz in aqueous solutions of various organic compounds (butanol, glycerin, urea) on the change in the thermokinetic characteristics of the combustion process?
14. Describe the regularities of the effect of ultrasonic treatment of a mixture of quartz + calcite in aqueous solutions of various organic compounds (butanol, glycerin, urea) on the change in the thermokinetic characteristics of the combustion process.
15. What variations of the UST conditions for quartz and a mixture of quartz-calcite lead to change in the characteristics of the combustion process, phase formationt and the formation of a porous structure of the synthesized composite material?
16. Describe the regularities of the influence of the UST conditions for wollastonite on the combustion process of the quartz + calcite system, the composition and properties of the final product.

10.4 The Use of Energy-Intensive Powders Based on Aluminum, Obtained by Mechanochemical Treatment, in the Composition of Solid Rocket Fuels

Aluminum is widely used in energy-intensive systems for various purposes, both in hydrogen energy and in technological combustion processes in the synthesis of various composite materials [93]. In the rocket-engine area, aluminum is used mainly to improve the gravimetric and volumetric specific impulse of solid-fuel rocket motors or to increase the regression rate of hybrid fuels [94]. In both cases, the reactivity of the aluminum powder plays a key role, which is related with the processes occurring in the combustion chamber. Micrometric aluminum is the most commonly used in space power plants with a high metal content (for example, 97.7% by weight of aluminum of grade PA4, dispersion of 50 microns) and low toxicity. However, its low reactivity is due to the presence of an oxide film on the surface of the particles, which reduces the specific impulse and burning rate in solid-fuel motors [95]. To increase the activity, different methods are used: chemical activation, mechanical and mechanochemical treatment (MCT) [96]. It is known that addition of metal powders to solid rocket fuel leads to an increase in system performance due to high energy per unit mass.

The use of aluminum powder increases the energy characteristics of the engines, but in the case of compact engines, the metal does not completely burn out. Mixed solid rocket fuel (MSRF) is not a homogeneous material by nature; it consists of chemically and physically different mechanically mixed substances and therefore is characterized by a certain microscopic heterogeneity of structure [97]. Figure 10.41 shows the typical structure of solid propellant based on ammonium perchlorate oxidant (AP), the polymer binder and fuel (hydroxyl-terminated polybutadiene (HTPB)) and metal fuel (Al). The reagents are fairly evenly distributed in the volume of MSRF.

Figure 10.41 Microstructure of a heterogeneous solid rocket fuel.

It should be noted that metal powders in the composition of the fuel burn dispersed. Their burning zone extends much farther than the thickness of the gas-phase flame, and does not affect the heat flow to the burning surface. Therefore, various methods are used to influence the structural characteristics of aluminum powders, which contribute to a change in their burning rate [98].

Studies have been conducted on the effect of mechanochemical treatment (MCT) of aluminum powder on the mechanism of combustion of solid fuels and the structure of combustion. Measurements were made of the burning rate and the visualization of the burning process using a high-speed video camera. To study the effect of MCT on the ballistic properties, we used the standard formulation: AP/Al/HTPB: 68%/18%/14%. The following reagents were used in the manufacture of solid fuels: HTPB (hydroxyl-terminated polybutadiene) of the R-45 grade is used as a binder between the fuel and the oxidizing agent, DAO plasticizer (dioctyl adipate), IPDI hardener (isophorone diisocyanate) and tin catalyst. In all fuels, the mass fraction of solid components is 86%. The fuels were distinguished by organic additives during the mechanochemical synthesis of aluminum powders. The MCT of metal aluminum was carried out for 20 min with the ratio of powder to grinding balls 1/4 M_p/M_b. Graphite, stearic acid and polyvinyl alcohol were used as modifiers. During processing, the amount of modifying additives varied from 3% to 20%, as well as the presence of silicon oxide (SiO_2) in an amount of from 5% to 20%.

Visual observations revealed general patterns of the combustion mechanism of metallized fuels based on ammonium perchlorate and a binder. Al agglomeration is formed in the immediate vicinity of the fuel on the burning surface [98]. Appearance of elements in the form of metallized agglomerates of various structures is observed on the burning surface (Fig. 10.42). The formation of aggregates occurs in stages: they are formed on the surface of the fuel, increase in size, move into the gas phase as the combustion surface moves. This occurs as long as the units do not come off as flowing gas. They undergo various forms from solid structures to a spherical drop of a very irregular shape. Liquid aluminum is rapidly oxidized to Al_2O_3 and evaporates from the gas phase. This can occur until all metallic aluminum is completely oxidized.

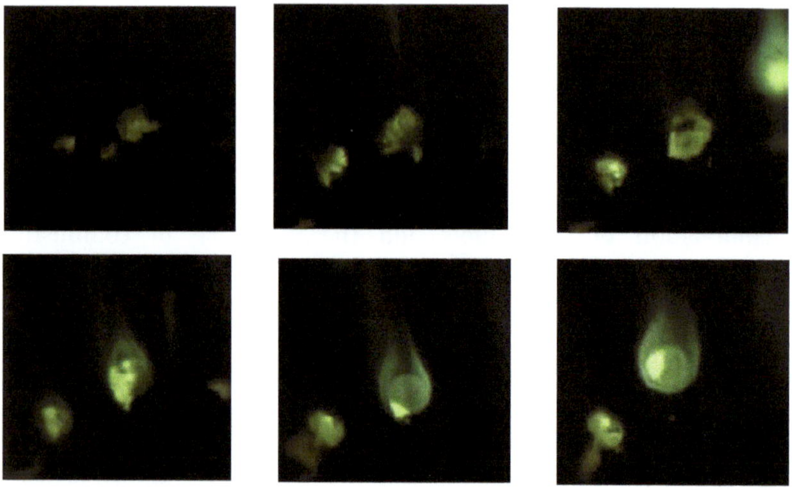

Figure 10.42 The sequence of the agglomerate formation at 5 atm (6000 frames per second).

One of the most important parameters of solid rocket fuel is the burning rate (r_b). A particular interest to this parameter is determined by its influence on the agglomeration processes already studied by many experimenters [99]. The rate of burning of solid fuel depends on various parameters, including the composition, operating conditions, any physical effects, the structure of the MSRF and the production processes of the preparation of the MSRF, and so on. For this reason, its value is usually determined experimentally. The r_b values obtained in various tests at different pressures are interrelated according to Vieille's law:

$$r_b = aP^n \tag{10.4}$$

where, r_b is the burning rate of MSRF, a the proportionality constant, P pressure inside the combustion chamber, and n the pressure exponent.

The sensitivity of the burning rate to the pressure of solid rocket fuel, according to [100], is presented in Fig. 10.43 when using aluminum of different dispersity.

Of particular interest is the change in the rate of combustion depending on the dispersion of the oxidizing agent. A decrease in the particle size of AP leads to an increase in the burning rate (r_b) and the ballistic index n tends to decrease depending on the increase in pressure [101].

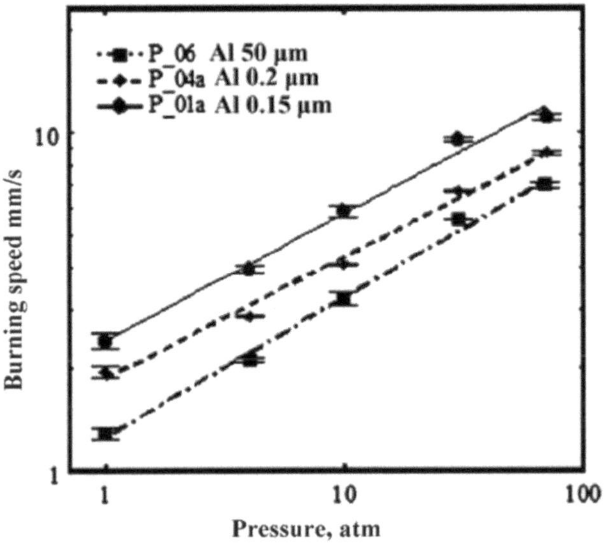

Figure 10.43 The influence of the particle size of aluminum powder on the dependence of the burning rate on pressure in the basic composition PA/HTPB/Al [101].

In this section, solid rocket fuels containing aluminum after MCT with various modifying additives are tested. Table 10.17 shows the data on the characteristics of solid fuels containing aluminum. The theoretical and actual density and porosity of the composition were determined. The actual density of fuel samples compared to theoretical density with the addition of modified aluminum to the composition is noticeably lower. This is due to the fact that the surface of aluminum particles is a shell consisting mainly of an organic film, which reduces the packing density of the particles of the composition. Porosity of the samples depends on the preparation conditions for the metal component of the composition. The value of porosity depends on the type of modifier used in the MCT process. The maximum value of porosity was found for a formulation containing (Al + 5% C + 5% SiO_2) MCT, which is possibly due to the lamellar structure of the metal combustible component and the agglomeration of aluminum particles due to the low content of organic modifier.

Table 10.17 Density (theoretical and actual) and porosity of the used metallized solid rocket fuels

Fuel with formulation AP/HTPB/AlQ	ρ_{teor}, g/cm^3	P_{real}, g/cm^3	Porosity, %
Al (initial)	1.761	1.729	0.018
Al+3% C	1.760	1.658	0.058
Al+20% C	1.753	1.710	0.025
Al+3% $(C_2H_3OH)_n$	1.753	1.690	0.035
Al+20% $(C_2H_3OH)_n$	1.710	1.653	0.033
Al+3% $C_{17}H_{35}COOH$	1.750	1.673	0.044
Al+20% $C_{17}H_{35}COOH$	1.687	1.648	0.023
Al+5% C+5% SiO_2	1.759	1.609	0.074
Al+20% C+10% SiO_2	1.750	1.649	0.057
Al+20% $(C_2H_3OH)_n$+5% SiO_2	1.719	1.682	0.021
Al+20% $(C_2H_3OH)_n$+20% SiO_2	1.718	1.663	0.032
Al+3% $C_{17}H_{35}COOH$+5% SiO_2	1.751	1.706	0.025
Al+3% $C_{17}H_{35}COOH$+20% SiO_2	1.750	1.614	0.071

The measurement of the steady-state burning rate was performed using samples of 4 mm × 4 mm × 30 mm. They were burned in a nitrogen atmosphere in a bomb with windows. As can be seen from in Fig. 10.44, the base fuel containing aluminum in the initial state with a particle size of 50 μm has a low burning rate and a high pressure exponent (n = 0.5572). When replaced with aluminum treated with 3% graphite, an increase in the burning rate by 22% and a decrease in the pressure exponent (n = 0.456) are observed. The obtained data indicate the fact of the combustion process stabilization.

Figure 10.45 presents the results of burning the base solid fuel based on modified and activated aluminum with 5% and 20% polyvinyl alcohol.

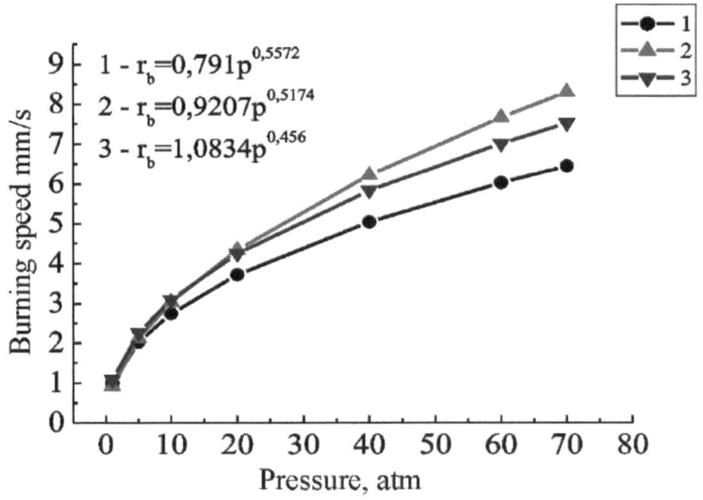

Figure 10.44 Change of the burning rate of AP/HTPB/Al_x fuels with non-activated aluminum and after MCT of composites. (1) Al initial; (2) $[Al+3\% C]_{MCT}$; (3) $[Al+20\% C]_{MCT}$.

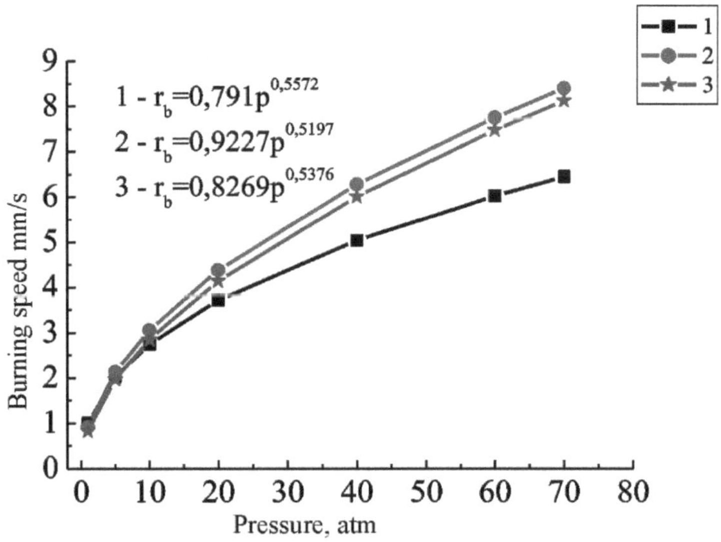

Figure 10.45 Change of burning rate of fuels AP/HTPB/Al_x with non-activated aluminum and after MCT of composites. (1) Al initial; (2) $[Al+3\% (C_2H_3OH)_n]_{MCT}$; (3) $[Al+20\% (C_2H_3OH)_n]_{MCT}$.

From Fig. 10.45 it follows that the burning rate of the PA/HTPB/[Al + 3% $(C_2H_3OH)_n]_{MCT}$ mixture is by 24% higher than in the composition with non-activated aluminum, however, the pressure index is higher than 0.5. This is due to the fact that aluminum particles are coated with polyvinyl alcohol, which is oxidized very actively and accumulates in the combustion zone. In this case, stability of post-combustion burning is disturbed, which can lead to an explosive effect. Perhaps, for solid fuels, polyvinyl alcohol is not an optimal modifier during MCT of aluminum as stability of the fuel system combustion must be maintained in solid fuel engines. Aluminum modified with PVA, can be used for the composition of the gas generators or for explosives.

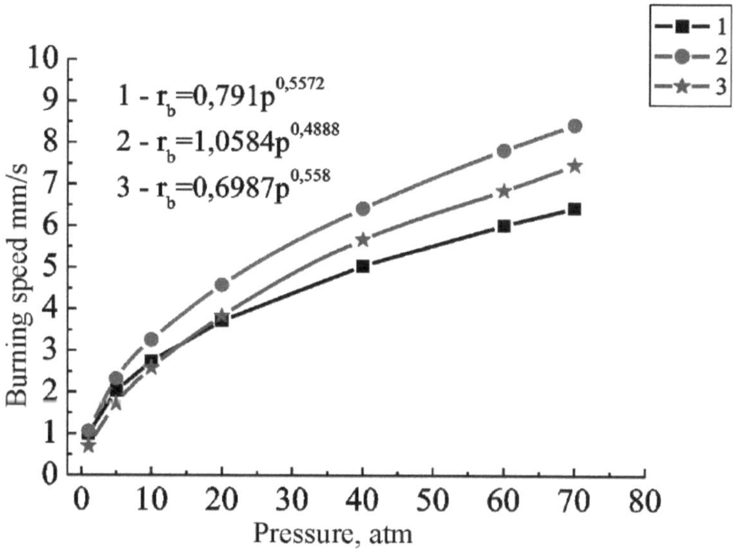

Figure 10.46 Change of burning rate of fuels AP/HTPB/Al_x with non-activated aluminum and after MCT of composites. (1) Al initial; (2) [Al+3% $C_{17}H_{35}COOH]_{MCT}$; (3) [Al+20% $C_{17}H_{35}COOH]_{MCT}$.

Figure 10.46 presents the experimental results on combustion of solid fuels based on metallized combustible on the basis of aluminum with stearic acid. As shown in the figure, the solid fuel of the basic composition has the lowest burning rate. All other compositions containing Al after MCT show significantly higher burning rates. The most stable rate of combustion increases with the addition of Al after MCT to the fuel composition with 3% stearic acid.

An increase in the burning rate and stability of burning solid fuels with a pressure indicator of 0.4721 (Fig. 10.47) are also observed for fuels of the series with aluminum after MCT with graphite and silicon dioxide. Due to the lamellar form of the composite particles, aluminum particles can burn near the surface, as well as in homogeneous and heterogeneous gas-phase modes. With complete combustion of such agglomerates, the flow of gases increases and, accordingly, the rate of combustion of the fuel increases.

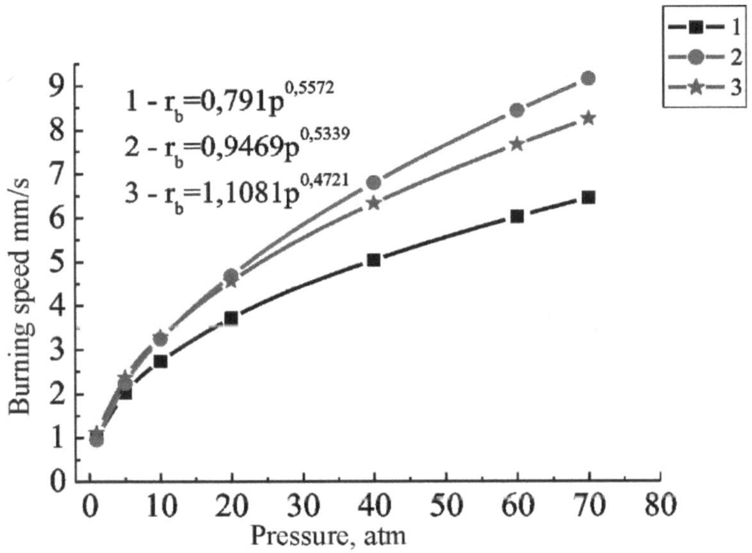

Figure 10.47 Change of burning rate of fuels AP/HTPB/Al_x with non-activated aluminum and after MCT of composites. (1) Al initial; (2) [Al + 5% C + 5% SiO_2]$_{MCT}$; (3) [Al + 20% C + 10% SiO_2]$_{MCT}$.

The ballistic characteristics of solid rocket fuels are presented in Fig. 10.48. The use of activated aluminum contributed to an overall increase in the burning rate of MSRF. It follows from the figure that the burning rate of the composition with [Al + 20% $(C_2H_3OH)_n$ + 20% SiO_2] increases by 25%, and the pressure exponent decreases from 0.5572 to 0.4927.

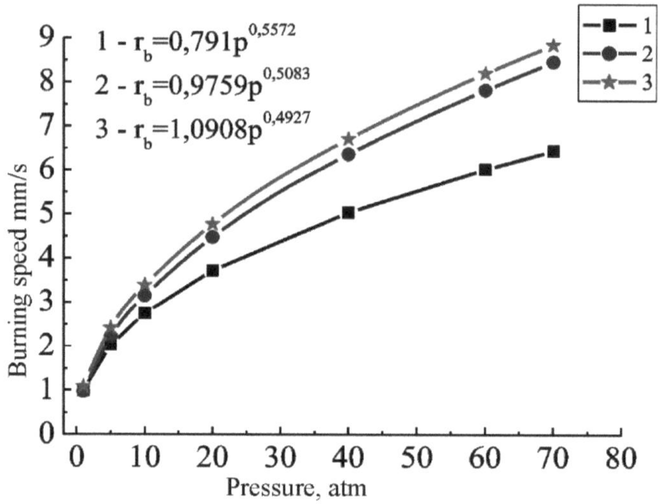

Figure 10.48 Change of burning rate of fuels PA/HTPB/Al_x with non-activated aluminum and after MCT of composites. (1) Al initial; (2) [Al + 20% $(C_2H_3OH)_n$ + 5% $SiO_2]_{MCT}$; (3) [Al + 20% $(C_2H_3OH)_n$ + 10% $SiO_2]_{MCT}$.

Compositions based on PA/HTPB/(Al + $C_{17}H_{35}COOH$ + SiO_2) demonstrated the best result in burning rate and exponential pressure (Fig. 10.49). At present, stearic acid is used for the passivation of aluminum powders on an industrial scale [102]. Firstly, stearic acid is hydrophobic, secondly, with MCT, it fills cracks in the oxide film on the surface of aluminum particles and, thirdly, increases the chemical resistance of aluminum in relation to other fuel components, thereby increasing the shelf life of MSRF. Aluminum particles in solid rocket fuels are initially localized between large oxidizer particles. When the melting point reaches 660°C, the aluminum particles become liquid, but they are still in the bulk of the oxide film, i.e., in an isolated state. The melting point of aluminum oxide is three times higher than the melting point of aluminum. In our case, as in the opinion of Price [103], liquid aluminum can leak due to cracks in the oxide shell formed during the MCT and filled with stearic acid, which can facilitate the ignition of particles. Perhaps the subsequent agglomeration of particles may occur in the heating zone adjacent to the burning surface layer.

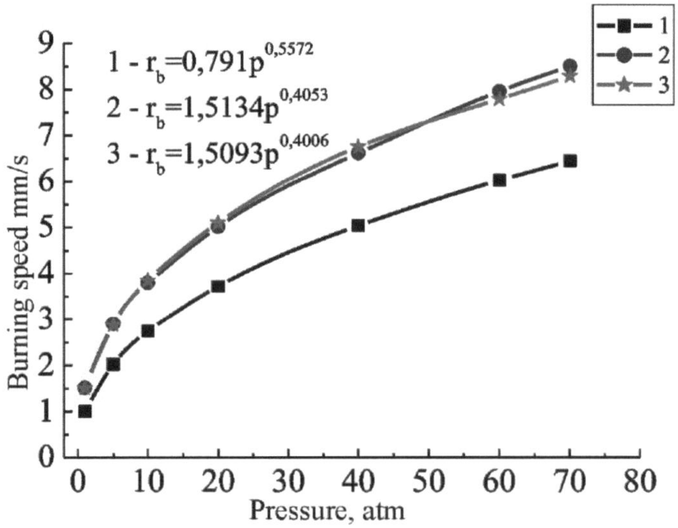

Figure 10.49 Change of burning rate of fuels PA/HTPB/Al_x with non-activated aluminum and after MCT of composites. (1) Al initial; (2) [Al + 3% $C_{17}H_{35}COOH$ + 5% $SiO_2]_{MCT}$; (3) [Al + 3% $C_{17}H_{35}COOH$ + 10% $SiO_2]_{MCT}$.

After introduction of the mechanically activated composite (Al + 3% $C_{17}H_{35}COOH$ + 5% SiO_2) into MSRF, there is a good increase in the burning rate by 13.5% at 5 atm and to 15.9% at 40 atm. The use of powder (Al + 3% $C_{17}H_{35}COOH$ + 20% SiO_2) caused only minor changes in the burning rate as compared to (Al + 3% $C_{17}H_{35}COOH$ + 5% SiO_2) within Δr_b = 17.6% in the considered pressure range. The content of active aluminum, which was determined by the volumetric method, was no more than ~ 85.1% in the system. A possible factor of this result is that the active combustible system is fully composite [Al + 3% $C_{17}H_{35}COOH$ + 20% SiO_2].

The results on combustion of MSRF with aluminum after different conditions of the MCT are systematized in Table 10.18. This table shows the results of testing solid rocket fuels with metallized fuel. When replacing the original aluminum with mechanically activated Al with a modifier, as well as with silicon dioxide, in the MSRF formulation, the effect MCT of aluminum on combustion of MSRF is observed at a pressure of 10 atm. In fuel systems with aluminum after MCT, the metal content is lower than in the initial state, since there are also modifiers and promoter (quartz) present there. When using the composite $[Al + 20\%\ C]_{MCT}$ in MSRF, the burning rate increases by 22%, and in the case of $[Al + 20\%\ (C_2H_3OH)_n + 20\%\ SiO_2]_{MCT}$ by 25% at a pressure of 40 atm. The pressure exponent decreases from 0.5572 to 0.4.

Table 10.18 Burning characteristics of fuels with aluminum powders after MCT

Fuel with formulation AP/ HTPB/Al	**Metal content in the composite**	**r_b, mm/s at different pressures, atm**			***n*, pressure exponent**
		10	**20**	**40**	
Al (initial) KZ (50 μm)	97.70	2.744	3.718	5.039	0.5572
Al+3% C	96.88	3.031	4.338	6.209	0.5174
Al+20% C	88.06	3.095	4.247	5.825	0.456
Al+3% $(C_2H_3OH)_n$	88.29	3.053	4.377	6.275	0.5197
Al+20% $(C_2H_3OH)_n$	70.60	2.851	4.139	6.007	0.5376
Al+3% $C_{17}H_{35}COOH$	87.07	3.262	4.577	6.423	0.4888
Al+20% $C_{17}H_{35}COOH$	51.06	2.584	3.831	5.679	0.568
Al + 5% C + 5% SiO_2	91.7	3.237	4.687	6.786	0.5339
Al + 20% C + 10% SiO_2	75.10	3.286	4.558	6.323	0.4721
Al + 20% $(C_2H_3OH)_n$ + 5% SiO_2	85.6	3.146	4.474	6.364	0.5083
Al + 20% $(C_2H_3OH)_n$ + 20% SiO_2	79.55	3.391	4.773	6.716	0.4927
Al + 3% $C_{17}H_{35}COOH$ + 5% SiO_2	114.4	3.796	5.012	6.615	0.4006
Al + 3% $C_{17}H_{35}COOH$ + 20% SiO_2	85.1	3.848	5.096	6.749	0.4053

Experimental combustion studies were performed on a series of solid fuels to determine the burning rate and particle size of the combustion products. An increase in the burning rate for the PA/HTPB-based system with $(Al + 3\%\ C_{17}H_{35}COOH + 20\%\ SiO_2)$ after the MCT by 25% is observed, and also stabilization of combustion with pressure exponent $n = 0.4$ is provided. The use of aluminum powders after the MCT in the presence of modifiers and silica in MSRF formulations seems to be useful and feasible.

Agglomeration of aluminum particles during combustion of solid fuels was studied using a special installation for collecting condensed products of combustion (CCP). The size of the condensed products of combustion was obtained by dry analysis using MALVERN MASTERSIZER 2000. The collected combustion products were processed and dried in an oven at 60°C for 48 h before testing. For the analysis, both particle size distributions and weighted average mean diameters were considered.

Table 10.19 shows the results of the CCP particle size distribution for the studied solid fuels at 10 atm and 40 atm, respectively. It follows from the table that the combustion products in metallized fuels with aluminum powders after MCT contain a greater amount of highly dispersed particles and to a lesser extent coarse ones. Combustion of fuel based on [Al + 3% $C_{17}H_{35}COOH$ + 20% $SiO_2]_{MCT}$ leads to a significant decrease in the coarse fraction.

Table 10.19 The average particle size of condensed combustion products and the change in their share in the combustion products of solid fuels at 10 and 40 atm

	The average particle size of the CCP, μм		Change in the share of the main fraction	
Solid fuel	***P*** **= 10 atm**	***P*** **= 40 atm**	**%,** ***P*** **= 10 atm**	**%,** ***P*** **= 40 atm**
Al initial	144.0	82.3	—	—
[Al+3% $C_{17}H_{35}COOH]_{MCT}$	95.2	68.7	33.9	16.5
Al + 3% $C_{17}H_{35}COOH$ + 20% $SiO_2]_{MCT}$	48.9	32.6	66.0	60.4

The most significant reduction in particle size was obtained using a powder [Al + 3% $C_{17}H_{35}COOH]_{MCT}$ (namely, 66% at 10 atm and 60.4% at 40 atm). The reason for this may be due to the presence of silicon dioxide and indicates a positive effect of mechanical activation.

Twelve activated aluminum-based powders with a modifier were prepared using a centrifugal mill. After MCT of powders, the morphology of the particles, the metal content in the composite, and its reactivity were studied. The final effect depends on the activation modifier and promoter. Solid additives (silicon dioxide) contributed to finer grinding of aluminum powder and loosening of its surface oxide layer, which led to more effective activation, as well as an increase in the reactivity of the composite. As a result of tests carried out to determine the burning rate, MSRF with the addition of mechanically activated aluminum powders with stearic acid and silicon dioxide showed an increase in the burning rate by 25%, and the particle size of combustion products decreases at 10 atm to 66%. Formulations of powders of aluminum with polyvinyl alcohol after MCT of the burning composition occurs with an explosive effect. Such systems can be used in other energy compositions such as: gas generators, explosives, thermite compositions to obtain porous ceramics.

Questions

1. In what industries is aluminum powder used?
2. What is the definition of solid rocket fuel?
3. Discuss the methods for determining the burning rate of solid rocket fuels.
4. What types of modifiers were used in the mechanochemical treatment of aluminum powder?
5. What was silicon dioxide used for in MCT?
6. What is a pressure exponent?
7. What results were obtained when determining the rate of combustion of solid fuels?

References

1. Nelson-Smith A. *Pollution of the Sea by Oil.* L.: Gidrometeoizdat, 1973. 123 p. (In Russian).
2. Tarasevich Yu. I. *Natural Sorbents in Water Purification Processes.* Naukova Dumka, Kiev, 1981. 208 p. (In Russian).
3. Tarasevich Yu. I., Ovcharenko F. D. The use of natural sorbents for the purification of petroleum products and water. *Proceedings of the Vth All-Union Conference on Adsorbents.* Moscow, 1983. pp. 175–179. (In Russian).
4. Yablokova M. A., Petrov S. I. Complex technology of wastewater treatment from oil and oil products. *Chem. Ind.,* 2003. vol. 80, no. 11. pp. 54–58. (In Russian).

5. Gerasimova V. N. Natural zeolites as adsorbents of petroleum products. *Chem. Interests Sustainable Dev.*, 2003. no. 11. pp. 481–488. (In Russian).
6. Anshits A. G., Nizov V. A. Isolation of magnetic microspheres of constant composition from energy sols and the study of their physicochemical properties. *Chem. Sustainable Dev.*, 1999. no. 7. pp. 105–118. (In Russian).
7. Tarasova Yu. V., Shevchenko T. V. Development of technology for producing porous materials from aluminum production wastes. *Ind. Technol.*, 2002. no. 9. pp. 22–28. (In Russian).
8. Yudakov A. A., Zubets V. N., Ksenik T. V. Preparation and use of hydrophobic sorbents for purification of water contaminated with petroleum products. *Chem. Sustainable Dev.*, 1998. no. 6. pp. 375–379. (In Russian).
9. Tarasevich Yu. I. The relationship between the structure and hydrophilicity of dispersed silica. *Ukrainian Chem. J.*, 1998. vol. 51, no. 2. pp. 133–141. (In Russian).
10. *The Structure and Properties of Adsorbents and Catalysts.*, ed., Linsen B. G. Mir, Moscow, 1973. 653 p. (In Russian).
11. Voronkov M. G., Vlasova N. N., Pozhidaev Yu. N., Belousov L. I., Grigorva O. Yu. Silicone sorbents of noble, non-ferrous, toxic and rare metals. *Sci. Prod.*, 2003. vol. 62, no. 6. pp. 4–9. (In Russian).
12. Efimov K. M., Gembitsky P. A., Nikashina V. A. Organomineral sorbent ZEOPAG in drinking and wastewater treatment. *Water Supply Sanitary Eng.*, 2002. no. 7. pp. 12–15. (In Russian).
13. Alykov N. M., Pashchenko N. M. Quantum-chemical modeling of the reactions of obtaining chemically modified silicas. *Chem. Chem. Technol.*, 2003. vol. 46, no. 6. pp. 35–39. (In Russian).
14. Shapkin N. P., Polyakov V. Yu., Shapkina V. Ya., Sibirtsev Yu. T., Rasskazov V. A. Chemical modification of natural zeolites of the Far East. *Chem. Chem. Techn.*, 2002. vol. 45, no. 2. pp. 101–106. (In Russian).
15. Petrakovskaya E. A., Kukhlevskaya O. P., Pavlov V. F., Kashkina L. D., Zeer E. P. Sorption of inorganic acids by thermal processing products of coal sols. *Chem. Sustainable Dev.*, 2001. no. 9. pp. 679–684. (In Russian).
16. Strelko V. V., Plachenov T. G., Kartel N. T. *Carbon Sorbents.* Science, Moscow, 1983. pp. 172–185. (In Russian).

17. Zubets V. N., Yudakov A. A., Latkin A. S., et al. *Hydrophobization of Dispersed Materials*. Publishing House of the Far-Eastern Branch of the USSR, Vladivostok, 1987. 327 p. (In Russian).
18. Belotserkovsky G. M., Maltseva N. V. The role of mechanochemical activation in obtaining mineral molded adsorbents. In the book: *Adsorbents, Their Preparation, Properties and Application*. L.: Science, 1985. pp. 35–41. (In Russian).
19. Grigoriev T. F., Vorsina I. A., Barinova A. P., Boldyrev V. V. Mechanochemical synthesis of dispersed layered composites based on kaolinite and higher carboxylic acids. *Rep. Acad. Sci.*, 1995. vol. 341, no. 1. pp. 66–68. (In Russian).
20. Kasymbekova D. A., Mofa N. N., Ketegenov T. A., Mansurova R. M. Compositional sorbent on the basis of quartz for purification of oil-containing wastewater. *Vestnik KazNU. Chem. Ser.*, 2003. no. 2 (30). pp. 335–337. (In Russian).
21. Mofa N. N., Shabanova T. A., Mansurov Z. A. Mechanochemical synthesis of nanocomposite powder materials based on quartz. *Vestnik KazNU. Chem. Ser.*, 2005. vol. 39, no. 3. pp. 27–36. (In Russian).
22. Mofa N. N. Mechanochemical processing is a progressive technological process of creating new composite materials. *(Review)/Collection. Chemistry and Chemical Technology. Modern Problems.* Kazak university, Almaty, 2004. pp. 163–198. (In Russian).
23. Mofa N. N., Ketegenov T. A., Kassymbekova D. A., Chervyakova O. V., Mansurova R. M. Mechanochemistry of multifunctional sorbents on the basis of silicon dioxide. *Fouth International Conference on Mechanochemistry and Mechanical Alloying*. Germany, Braunschweing, September 2003. p. 118.
24. Mofa N. N., Chervyakova O. V., Ketegenov T. A., Mansurov Z. A. Magnetic sorbents. Quartz-containing mixtures obtained by mechanochemical treatment. *Chem. Sustainable Dev.*, 2003. no. 11. pp. 755–761. (In Russian).
25. Mofa N. N., Chervyakova O. V., Ketegenov T. A., Mansurov Z. A. Mechanochemical synthesis of magnetic quartz-containing adsorbents modified with carbon compounds. *Sci. News Kazakhstan. Res. Inst. al-Farabi KazNU.*, 2004. no. 2 (81). pp. 59–65. (In Russian).
26. Mansurov Z. A., Mofa N. N., Ketegenov T. A., Chervyakova O. V. Oil spill response on water: Real ways and prospects for solving the problem for the Caspian Sea. *Proceedings of the International Scientific and Practical Conference "Prospects for the Sustainable Development*

of Ecosystems of the Caspian Region", 29–30 June 2004, Almaty, pp. 45–47. (In Russian).

27. Kenzhagaliev A., Serikov F. T., Khasanova A. A. Ecological state of the Northern Caspian Sea and its coastal zone. *Oil Gas*, 1999. no. 1 (5). pp. 101–103. (In Russian).
28. Akhmetova L. A., Kenzhegaliev, A., Pollution of the Northeastern part of the Caspian Sea, *Oil Gas*, 2001. no. 3. pp. 84–87. (In Russian).
29. Izteleulova M. B. Oil spills: Prevention, emergency preparedness, response. *Oil Gas*, 1999. no. 1 (5). pp. 97–100. (In Russian).
30. Mulikov R. R., Aukeshev B. K. Environmental issues in the development of Caspian oil and gas fields. *Oil Gas*, 2000. no. 2. pp. 87–90. (In Russian).
31. Mofa N. N., Ketegenov T. A., Chervyakova O. V. New efficient way to collect oil spills on water. *Oil Gas*, 2001. no. 3. pp. 78–83. (In Russian).
32. Soh D., Ksandopulo G. I., Mofa N. N., Lan B., Ketegenov T. A. Reclaiming property of magnetic adcorbent for oil spell recovery and pollution control. *Proceeging of the KIEEME Annual Autumn Conference*, 2001. vol. 14, no. 1. pp. 296–299.
33. Chervyakova O. V., Ketegenov T. A., Mofa N. N. Oil spills and the use of new technology for cleaning of reservoirs. *Vestnik KazGU, Chem. Ser.*, 2001. no. 2 (22). pp. 344–345. (In Russian).
34. Mansurov Z. A., Mofa N. N., Ketegenov T. A., Chervyakova O. V. Oil spill response in water: Real ways and prospects for solving the problem for the Caspian Sea. *Proceedings of the International Scientific and Practical Conference "Prospects for Sustainable Development of Ecosystems in the Caspian Region"*. Almaty, 2004. pp. 45–47. (In Russian).
35. Kittel H. *Introduction to Solid State Physics*. Science, Moscow, 1978. 345 p. (In Russian).
36. Bugaenko L. T., Kuzmin M. G., Podak L. S., *High Energy Chemistry*. Chemistry, Moscow, 1988. 368 p. (In Russian).
37. Ilyin B. V. *The Nature of the Adsorption Forces*. Science, Moscow, 1952. 124 p. (In Russian).
38. Pre-patent No. 13443 RK, IPC (51) 7 E02B15/04. A device for collecting oil from the surface of the water/Ketegenov T. A., Stakhov O. V., Mofa N. N., Rudenko P. K. Publ. September 15, 2003; Bul. no. 9. (In Russian).
39. *Micelle Formation, Comobilization and Microemulsions.*, ed., Mittel K., trans. from English Goldfeld M.G. Mir, Moscow, 1980. 197 p. (In Russian).

40. Arkhipkin O. P. *Wollastonite, Properties and Applications*. Almaty, 1994. 227 p. (In Russian).
41. Azarov G. M., Mayorova E. V., Oborin M. A., Belyakov A. V. Wollastonite raw materials and areas of its application. *Glass Ceramics*, 1988. no. 1. pp. 19–21. (In Russian).
42. Petrov V. N. *Wollastonite as a Mineral and Mineral Raw Material*. Wollastonite. Science, Moscow, 1982. pp. 5–15. (In Russian).
43. Orlova N. A., Korobschikova T.S. The influence of the optimal ratio of fillers on the physico-mechanical characteristics of epoxy coatings. *Plastics*, 2011. no. 6. pp. 40–43. (In Russian).
44. Korobschikova T. S., Orlova N. A. Study of the effect of wollastonite on the properties of composite materials. *Siberia–Chemistry, Innovations, Technologies: Materials of the Youth Scientific and Technical Forum. Publishing House of the Institute of Catalysis*. I. I. Boreskov SB RAS, Novosibirsk, 2009. pp. 13–14. (In Russian).
45. Korobschikova T. S., Orlova N. A. Study of the granulometric composition of wollastonite of the Sinyukhinskoye field and its influence on the properties of filled polymer compositions. *Paints Varnishes Appl.*, 2010. no. 5. pp. 26–29. (In Russian).
46. Tertykh V. A., Belyakova L. A. *Chemical Reactions Involving the Surface of Silica*. Naukova Dumka, Kiev, 1991. 264 p. (In Russian).
47. Bazhenov C. JI., Berlin A. A., Kulkov A. A., Oshmyan V. G. *Polymer Composite Materials. Durability and Technology*. Intellect, Dolgoprudny, 2010. 352 p. (In Russian).
48. Chernin I. Z., Smekhov F. M., Zherdev Yu. V. *Epoxy Polymers and Compositions*. Chemistry, Moscow, 1982. 232 p. (In Russian).
49. Novikov V. U., Kozlov O. Yu. Fractal approach to the interfacial layer in filled polymers. *Mech. Composite Mater.*, 2000. vol. 35, no. 1. pp. 1–18. (In Russian).
50. Irzhak V. I., Rosenberg B. A., Enikolopov N. S. *Mesh Polymers. Synthesis, Structure, Properties*. Science, Moscow, 1979. 248 p. (In Russian).
51. Rogalev A. V., Khvostov S. A., Ananyeva E. S. Application of the fractal cluster approach for predicting the properties of a polymer filled with nanoscale particles. *Applied Aspects of Chemical Technology of Polymeric Materials and Nanosystems (Polymer 2009): Materials of the III All-Russian Scientific and Practical Conference of Students, Postgraduates and Young Scientists May 29–30, 2009 Alt. State Tech. Univ, BTI*. Publishing house Alt. state tech. University, Biysk, 2009. pp. 88–95. (In Russian).

52. Lipatov, Yu. S. *Physical Chemistry of Filled Polymers*. 2nd ed., Chemistry, Moscow, 1977. 304 p. (In Russian).
53. Cherevatsky A. M. Wollastonite-effective filler for epoxy polymers. *Plastics*, 1982. no. 5. pp. 47–48. (In Russian).
54. Merzhanov A. G., Borovinskaya I. P. Self-propagating high-temperature synthesis. *DAN SSSR*, 1973. vol. 204, no. 2. pp. 366–370. (In Russian).
55. Koizumi M. *Chemistry of Synthesis by Burning*. Mir, Moscow, 1998. 247 p. (In Russian).
56. Ksandopulo G. I., Dubinin V. V. *Chemistry of Gas-Phase Combustion*. Chemistry, Moscow, 1987. 240 p. (In Russian).
57. Lewis B., Elbe G. *Combustion, Flames and Explosions in Gases/Trans. from English*, ed., Kondratiev V. N. Mir, Moscow, 1968. 592 p. (In Russian).
58. Zeldovich Ya. B., Barenblat G. I., Librovich V. B., Makhviladze G. M. *Mathematical Theory of Combustion and Explosion*. Science, Moscow, 1980. 478 p. (In Russian).
59. Volpert V. A., Borzykina R. A. *Theory of Combustion Waves in SH-Systems.* Preprint *The United Institute of Chemical Physics*, USSR Academy of Sciences. ISMAN, Chernogolovka, 1988. 42 p. (In Russian).
60. Levashev E. A., Rogachev A. S., Yukhvid V. I., Borovinskaya I. P. *Physico-Chemical and Technological Bases of Self-Propagating High-Temperature Synthesis.*, ed., BINOM, Moscow, 1999. 176 p. (In Russian).
61. Kaykin B. I., Merzhanov A. G. On the combustion of substances with a solid reaction layer. *DAN USSR*, 1967. vol. 173, no. 6. pp. 1382–1385. (In Russian).
62. Aldushkin A. P., Merzhanov A. G. Gasless combustion with phase transformations. *DAN SSSR*, 1977. vol. 236, no. 5. pp. 1133–1136. (In Russian).
63. Zenin A. A., Merzhanov A. G., Persisyan G. A. Heat wave structure in some SHS processes. *DAN SSSR*, 1980. vol. 250, no. 4. pp. 880–884. (In Russian).
64. Merzhanov A. G. *Self-Propagating High-Temperature Synthesis: Twenty Years of Searching and Discoveries*. Preprint. ISMAN, Chernogolovka, 1989. 50 p. (In Russian).
65. Levashov E. A., Pityulin A. N., Merzhanov A. G., Andreev V. A., Sizov R. A., Khavsky N. N. *Investigation of the CB-synthesis of alloys of the STIM group in an ultrasonic field. Preprint The United*

Institute of Chemical Physics, USSR Academy of Sciences. ISMAN, Chernogolovka, 1987. 47 p. (In Russian).

66. Korchagin M. A., Grigorieva T. F., Bokhonov B. B., Sharafutdinov A. P., Barinova A. P., Lyakhov N. Z. Solid-phase combustion in mechanically activated SHS systems. 1. The effect of the duration of mechanical activation on the characteristics of the process and the composition of the products of combustion. *Phys. Combustion Explosion,* 2003. vol. 39, no. 1. pp. 51–68. (In Russian).
67. Kirichenko O. A., Ushakov V. A., Andryushkova O. A., Ivchenko S. V., Poluboyarov V. A. Phase transformations and mass transfer in mechanically activated low-temperature aluminum oxides. *Inorg. Mater.,* 1999. vol. 35, no. 33. pp. 333–341. (In Russian).
68. Nayborodenko Yu. S., Kasatsky N. G., Lepakova O. K. Features of high-temperature synthesis of intermetallic compounds in mechanically activated systems. *Proceedings of the All-Russian Conference "Combustion and Explosion Processes in Physical Chemistry and Technology of Inorganic Materials",* Moscow, 2002. pp. 287–290. (In Russian).
69. Korchagin M. A., Grigorieva T. F., Bokhonov B. B., Sharafutdinov A. P., Barinova B. B., Lyakhov N. Z. Solid-phase combustion in mechanically activated SHS systems. 2. Influence of mechanical activation modes. *Combustion Explosion Phys.,* 2003. vol. 39, no. 1. pp. 59–68. (In Russian).
70. *The Concept of the Development of Combustion and Explosion as an Area of Scientific and Technological Progress,* ed., Merzhanova A. G. Territory, Chernogolovka, 2001. pp. 78–93. (In Russian).
71. Heinicke G. *Tribochemistry.* Mir, Moscow, 1987. 584 p. (In Russian).
72. Avvakumov E. G. *Mechanical Methods of Activation of Chemical Processes.* Science, Sib. Branch, Novosibirsk, 1986. 300 p. (In Russian).
73. Deryagin B. V., Krotova N. A., Smilga V. P. *Adhesion of Solids.* Science, Moscow, 1973. 280 p. (In Russian).
74. Baramboym N. K. *Mechanochemistry of High-Molecular Compounds.* Chemistry, Moscow, 1978. 384 p. (In Russian).
75. Polukhin L. M., Khrustalev Yu. A. Electrical phenomena during mechanical action on polymer blends. *J. Phys. Chem.,* 1993. vol. 67, no. 4. pp. 795–797. (In Russian).
76. Mofa N. N., Ketegenov T. A., Orynbekov E. S., Mansurov Z. A. SH-synthesis of oxide systems modified by mechanochemical treatment. *XVII Mendeleev Congress.* Kazan, 2003. 189 p. (In Russian).

77. Mofa N. N., Shabanova T. A., Mansurov Z. A. Mechanochemical synthesis of nanocomposite powder materials based on quartz. *Vestnik Kaz.NU. Chem. Ser.*, 2005. no. 3 (39). pp. 27–36. (In Russian).

78. Novitsky B. G. *The Use of Acoustic Vibrations in Chemical-Technological Processes.* Chemistry, Moscow, 1983. 191 p. (In Russian).

79. Flynn G. *Physics of acoustic cavitation in liquids. Physical Acoustics.*, ed., Meson W. Mir, Moscow, 1967. vol. 1, Part. B. 138 p. (In Russian).

80. Ginberg A. M. *Ultrasound in Chemical and Electrochemical Engineering Processes.* Mashgiz, Moscow, 1962. 136 p. (In Russian).

81. Abramov, O. V., Kharbenko, I. G., Shvegla, S. *Ultrasonic Processing of Materials.* Mashinostroenie, Moscow, 1984. 243 p. (In Russian).

82. Margulis M. A. *Sonochemical Reactions and Sonoluminescence.* Chemistry, Moscow, 1986. 288 p. (In Russian).

83. Baranchikov A. E., Ivanov B. K., Tretyakov Yu. D. Sonochemical synthesis of inorganic materials. *Chem. Adv.*, 2007. vol. 76, no. 2. pp. 147–168. (In Russian).

84. Ketegenov, T. A., Urakaev, F. Kh., Tyumentseva, O. A., Mofa, N. N., Mansurov, Z. A. Synthesis of iron silicates on the surface of quartz particles in the process of their mechanical processing. *Rep. Natl. Acad. Sci. Republic Kazakhstan*, 2003. no. 2. pp. 66–72. (In Russian).

85. Mansurov Z. A., Mofa N. N. Mechanical activation and reactivity of SHS-systems on the basis of quartz. *Eurasion Chem. Technol. J.*, 2011. vol. 13, no. 3 4 pp. 125–136.

86. Mansurov Z. A., Mofa, N. N., Shabanova, T. A. Hybride, nano-structurized materials of silicon dioxide. *Adv. Eng. Ceramics Composites*, 2011. vol. 484. pp. 230–240.

87. Mansurov Z. A., Mofa N. N., Sadykov B. S., Shabanova T. A. Activation of the technological combustion systems. *Adv. Ceramic Sci. Eng.*, 2013. vol. 2 , no. 3. pp. 106–112.

88. Mansurov Z. A., Mofa N. N., Sadykov B. S., Antonyuk V. I. Mechanical processing, structural features, properties and reactivity of SHS systems based on natural materials. Part 3: Effect of mechanochemical processing and modification oxide materials for technological combustion. *Eng. Phys. J.*, 2014. vol. 87, no. 5. pp. 1000–1008. (In Russian).

89. Mansurov Z. A., Mofa N. N. Synthesis of nanocomposite ceramics on the basis of a system with modified wollastonite. *Materials of the VII International Symposium "Physics and Chemistry of Carbon Materials/Nanoengineering"*, September 19–21, 2012, Almaty, pp. 161–164. (In Russian).

90. Mansurov Z. A., Mofa N. N., Sadykov B. S., Shoibekova A. B. SH-synthesis of nanocomposition materials based on the system SiO_2+Al+$CaSiO_3$ with wollastonite after ultrasonic treatment. *SHS 2013 XII International Symposium on Self-Propagating High-Temperature Synthesis In Memory of Professor Alexander Merzhanov 21–24 October 2013* (South Padre Island, TX, USA), pp. 342–343.

91. Mansurov Z. A., Mofa N. N., Sadykov B. S., Sabaev Zh. Zh. Ultrasonic processing of wollastonite and obtaining SHS-composite systems. *Proceedings of the III International Scientific Conference "Modern Problems of Condensed Matter Physics, Nanotechnologies and Nanomaterials*, May 15–16, 2014, Almaty, pp. 18–20. (In Russian).

92. Bakkara A. E., Smagulova G. T., Sadykov B. S., Mofa N. N., Lesbaev B. T., Mansurov Z. A., Lyubchik S. B. Studying the effect of the addition of metal nanoparticles on the combustion of condensed systems. *International Conference "Coloids and nanotochnologies in industry"*. Almaty, 2014. 84 p. (In Russian).

93. Mansurov Z. A., Mofa N. N., Sadykov B. S., Sabaev Zh.Zh., Bakkara A. E. Mechanochemical treatment, structural features, properties and reactivity of SHS systems based on natural materials. Part 3: The influence of mechanochemical processing and modification of oxide materials on technological combustion. *Eng. Phys. J.*, 2014. vol. 87, no. 5. pp. 1000–1008. (In Russian).

94. Sutton G. P., Biblarz O. *Rocket Propulsion Elements.*, 8th ed., New Jersey: John Wiley & Sons Inc., 2010. 167 p.

95. DeLuca L. T., Galfetti L., Colombo G., Maggi F., Paravan C., Reina A., Dossi S., Fassina M., Sossi A. *Metal Nanopowders: Production, Characterization, and Energetic Applications.* Weinheim: Wiley-VHC, Germany, 2014. 410 p.

96. Streletsky A. N., Kolbanev I. V., Borunova A. B., Butyagin P. Yu. Mechanical activation of aluminum. 3. The kinetics of the interaction of aluminum with water. *Colloid J.*, 2005. vol. 67, no. 5. pp. 694–701. (In Russian).

97. Paravan C., Manzoni M., Rambaldi G., De Luca, L. T., Analysis of quasi-steady and transient burning of hybrid fuels in a laboratory scale burner by optical technique. *Int. J. Energetic Mater. Chem. Propulsion.*, 2013. vol. 12, no. 5. pp. 385–410.

98. Price E. W., Sigman R. K. Combustion of aluminized solid propellants. *AIAA*, 2000. pp. 663–687.

99. Babuk V. A., Vasiliyev V. A., Malakhov M. S. Condensed combustion products at the burning surface of aluminized solid propellant. *J. Propulsion Power*, 1999. vol. 15, no. 6. pp. 783–793.
100. Cohen N. S. A model for aluminium agglomeration in composite propellants. *AIAA*, 1983. vol. 21. pp. 720–725.
101. DeLuca L. T., Galfetti L., Colombo G., Maggi F., Bandera A., Babuk V. A., Sinditskii V. P. Microstructure effects in aluminized solid rocket propellants. *J. Propulsion Power*, 2010. vol. 24, no. 4. pp. 724–733.
102. Gromov A., Ilyin A., An V., Faubert F., de Izarra C., Espagnacq A., Brunet L. Characterization of aluminum powders. 1: Parameters of reactivity of aluminium powders. *Propellants Explosives Pirotechnics*, 2002. vol. 27, no. 6. pp. 361–364.
103. Price E. W., Sigman R. K., Sambamurthi J. R., Park C. J. Behavior of aluminum in solid propellant combustion. AFOSR-TR-82-0964. *Georgia Inst. Technol.*, 1982. vol. 4. pp. 1344–1351.

Index

abrasion 3, 15, 66, 134, 339–340
abrasive-reactive wear 81, 83, 85, 87, 89–91, 93, 95
abrasive wear 75, 81, 84–85, 90, 92, 94–96
  of steel grinding media 92
acids
  acrylic 134–139, 142–143, 146, 152–156, 164–165, 169, 171–173, 180–181, 183–184, 200, 338, 340–342
  monosilicic 104–105, 195
  polysilicic 175, 195–196, 202
  polystyrene and acrylic 171, 180, 183–184
activation 163, 171
activation energy 46–47, 50–51, 112, 277, 288–289, 376, 414
activation time 87, 124, 126, 128, 158–159, 161–162, 168, 364
additives, amine-containing 367–368, 374
adsorbent particles 352–353
agglomeration 120–121, 124, 130, 218, 236, 241, 301–302, 382, 417, 424
aluminosilicates 289, 391
aluminum 297–302, 304–323, 325–331, 384–385, 388, 393, 396, 402–416, 418–420, 422–426, 428
  activated 403, 405, 420, 423
  activated mixture of 403, 408
  activation of 406, 413
  active 326–329, 425
  grinding of 406, 413
  liquid 320, 324, 417, 424
  mechanical grinding of 298, 330
  mechanochemical treatment of 298, 304, 403, 405, 410, 413
  modified 318, 331, 407, 410, 412, 419
  non-activated 421–425
  oxidation of 320, 396
  unreacted 412–413
aluminum-based powders 313, 315, 317, 319, 321, 323, 325
aluminum carbide 320, 323
aluminum crystallites 308, 310
aluminum melting and complete transformation 320
aluminum nitride 320, 385, 391, 405–406, 413
aluminum oxide 304, 308, 314, 316, 318, 390, 407, 424
aluminum particles 297–299, 301–302, 306–308, 310–311, 313, 326, 328–329, 331, 406, 419, 422–424
  agglomeration of 302, 420, 426
  deformed 311, 314
  dispersible 330
  distribution of 302, 310
  modified 318
  original PA4 301
  in solid rocket fuels 424
  surface area of 302, 309
  treated 298

aluminum powder 297–299, 301, 303, 305, 307, 309, 311, 314, 323, 325–327, 331, 402–403, 415–417, 419, 424, 426–428
  particles 299, 307
  processing 299
  surface of 310, 315
amine groups 370–371
ammonia 229, 232, 238, 241–242, 244, 246, 264–266, 271–273, 366–369, 371, 388, 390, 395
  aqueous solution of 227, 229, 231, 233–234, 237–239, 241, 244, 247, 365, 382
  liquid 238, 272–273, 367
ammonia-modified particles 242
amorphization 6, 20, 42, 92, 94, 107–108, 123, 128, 171, 195, 346

bending angle 363–364, 368–369, 371–372
bending strength index 363
Brinell hardness 361–362, 366–367
bulk density 30, 120–122, 134, 229, 231–232, 237, 241, 261–265, 273
butanol 55, 112, 131–134, 136, 138–139, 141, 146, 150–151, 153–154, 161–163, 169, 171–172, 176, 199, 260, 263–266, 268, 270, 280, 282–284, 286–287, 290–291, 365–369, 374, 392, 396, 415
  aqueous solutions of 271, 394–397
  hydroxyl group of 282, 288
butanol molecules 281, 283, 291

$CaAl_4O_7$ 389–390, 396–397
calcite 2, 235–239, 247–248, 256, 260–264, 356, 380, 385–386, 388, 391–398, 414–415
calcite particles, dispersity of 236
calcium carbonate 235–237, 244, 388–390, 393–397
calcium cations 288–290
calcium ions 280, 283–284
carbon 133–135, 139, 142, 144, 146, 165, 171, 173–174, 177, 179–180, 185–186, 198–199, 201–202, 242, 300, 305, 309, 311, 322, 326, 328, 331, 336–339, 343, 345, 347, 351, 354, 380, 387, 391, 406, 411, 413
  activated 133–134, 177–178, 182
  free 198–200, 202
  solid solution of 171, 173, 179–180, 331
carbon-containing compounds 186, 219, 331
carbon-containing structures 144
carbon modifier 177, 186, 404, 408, 410, 414
cavitation bubbles 252–253, 255–256
centrifugal planetary mill (CPM) 228, 345
chaotic structure 48
chemically modified silica (CMS) 113–114
chemisorption 107, 109–110, 285
  mechanical 10
CMS, *see* chemically modified silica
CO 109–110, 240, 365, 393
CO_2 109–110, 395
combustion 298, 376–377, 379–382, 384–393, 395–396, 398–404, 406–410, 412–415, 417, 423, 426–428

rate of 391, 395, 401, 410, 419, 423
solid-flame 385
technological 380, 415
combustion mechanisms 377, 417
combustion products 376–378, 390, 396–397, 406, 409, 412, 414–415, 426–428
condensed 427
combustion systems 382–383, 386–387, 389, 393, 399–400, 403, 408–412
combustion thermograms 381, 386, 392, 394–395, 403–404, 408–409
composite systems 7, 12, 140, 186, 239, 247–248, 272, 298, 336, 355–357, 359, 361, 363, 365, 367–369, 371–374, 379, 391
composites 193, 235, 239, 300, 302–303, 308–310, 314, 318–319, 321–322, 324–325, 327, 335, 340, 356, 365, 367–372, 374, 392, 402, 408–409, 411, 421–425
composition systems quartz core-polymer shell 229, 231, 233
conventional rotating drum mill (CRDM) 66–67, 70–71
corundum 346, 384–385, 390–391, 402, 406
CPM, *see* centrifugal planetary mill
CRDM, *see* conventional rotating drum mill
crystal lattice 19, 23, 42, 46–49, 51, 55, 124–126, 130, 167, 179, 206–207, 213–214, 268, 278, 291
crystal structure 5, 9, 20, 27, 84, 87, 206, 266, 277–278
crystallization 84–85, 93, 149
crystals, deformed 214, 216

decomposition 3, 8, 10, 139–140, 173, 181, 201, 295, 379, 391, 395
deeply disordered unstable structures 6
deformation energy 20, 46
deformation mobility 34
deformation vibrations 107, 147, 149, 151, 156, 314, 326
deformational transformations 44
density functional theory (DFT) 276
DFT, *see* density functional theory
dielectric constant 208, 210, 258–259, 263, 269–273, 358–361, 363, 366, 372–373, 375
dielectric permeability 259
disperse composition systems 228, 230, 232, 234, 236, 238, 240, 242, 244, 246, 248
dispersion, ultrasonic 253–255, 379, 392
ductility 363, 366, 369, 371–372

elastic deformation 22, 297
elastic waves 27, 29–30
electric charges 55, 103, 206, 214, 221, 234, 309, 378–379
electric field 32, 103, 111, 205, 207, 210–212, 214–216, 220–222, 234, 310, 360
electric field energy 211
electrical resistivity 129, 137, 184

electromagnetic fields 215–216, 219, 221–222, 234
electromagnetic response 215, 221–222, 234
electron beam 233, 235, 243–245, 265, 267, 273
electron emission 31–32, 55, 109, 203–205, 216, 218, 220–221, 378
electron microdiffraction 140–141, 143–144, 165
electron paramagnetic resonance 27, 156, 318
electron paramagnetic resonance (EPR) 27, 156, 160, 186, 318
electrophilic addition 286
electrophysical processes 57
ellipsoids 67–69, 71
energy
  elastic 29, 33, 214
  kinetic 17, 30, 36
  potential 2–3, 288
  stored 19, 21, 28, 48, 52, 379, 414
energy dispersion spectrum 298, 307, 312
energy dissipation 48–49
epoxy compositions 359–360, 363–365, 367, 369–375
  change in dielectric constant of 375
  change in hardness and mechanical properties of 375
epoxy polymers 358–359, 367
epoxy resin 357–359, 361–362, 365–368, 371–372, 374
EPR, *see* electron paramagnetic resonance
esterification 194, 287–288, 290–291
ethanol 131, 161–163, 176, 197, 382
ethylene glycol 132, 151–152, 161–163, 176, 346
exothermic reactions 323, 385, 391, 394

Fe_2O_3 87, 89, 128, 167–169, 179, 181, 254, 343
Fe_3O_4 85, 87, 89, 169, 173, 179, 181, 254, 349
ferric chloride 157–161, 163, 165, 173, 347, 349, 351
ferrite 8, 173, 179–180, 186
ferromagnetic compounds 174, 183, 185–186, 344–345, 355
ferromagnetism 165, 175, 178, 186–187, 213, 353
ferrophase content 162
fettling layer 77, 79–80
free radicals 5–6, 8, 10, 53–54, 56, 204–206, 318

galena 91–96
gaseous medium 50, 52, 204, 219, 221
glass, liquid 257–258, 260, 339
glycerin 176, 264–268, 270–273, 365–369, 371–374, 392, 401, 415
glycerol 132, 142, 260, 264, 270, 273, 368–369, 372–373
graphite 9, 12, 143, 254, 298–302, 304, 308–311, 313–314, 318, 320–323, 327, 329–330, 403, 408, 410, 417, 420, 423
grinding 2–3, 12, 15–17, 19, 21–24, 26–27, 29, 42–45,

49–53, 63, 111, 113, 115, 119, 122, 125, 127, 129, 131–132, 135, 163–164, 195–196, 203, 205, 209–210, 212–213, 215, 229, 231, 239, 241, 243, 253, 296, 298, 309–310, 328, 330, 377, 379
intensive 105–106, 232
grinding apparatus 37–38, 40, 45, 120
grinding balls 42, 156, 165, 209, 219, 222, 233–234, 236–237, 417
grinding bodies 19, 36–37, 40, 64–75, 80–81, 83, 85, 87, 90, 92, 95–96, 229, 379
grinding devices 15, 17, 37, 40
grinding machines 15, 18, 21, 38

helenite 389–391, 394
Hertz theory 38, 45, 64–65, 71
heterogeneity, structural 219, 302
hexane 299, 307, 313–314, 318
hydrogen atoms 280–284
hydrogen bonds 104, 107–108, 111, 149, 281–285, 290–291, 315, 357
hydrometallurgical processes 12

inorganic materials 16, 18, 20, 22, 24, 26, 28, 30, 34, 36, 38, 40, 42–50, 52, 54–56, 81, 101, 134, 140, 164, 175, 239, 298
inorganic particles 43, 379
ionization 8–9, 32, 216, 218, 221
iron 2, 76, 86–87, 89–90, 92–94, 96, 146, 155–159, 161–165, 167, 169, 171–175, 177, 179–181, 185–186, 195–196, 201–202, 220, 228, 342, 349, 385, 405–406, 413
ferric 157–158, 164
metallic 159, 169, 171, 173, 180
trivalent 159
ultrafine particles of 220
iron acrylates 154–156, 174
iron carbide 165, 177, 186
iron-containing structures 157
iron disilicides 406, 413
iron oxides 92, 95, 128, 168, 180–181, 345–347
iron particles, non-interacting 171
iron salts 164, 345, 347
iron silicates 89–90, 201
iron sulfides 93, 95
irreversible processes, thermodynamics of 4, 49

linear thermal expansion coefficient (LTEC) 125, 127–128, 137
LTEC, *see* linear thermal expansion coefficient

macrokinetics 50, 52, 375
macromolecular compounds 9
macromolecules 2, 32
macroradicals 53–55
magnetic adsorbent 348–354
magnetic fields 158, 161–164, 166, 168
magnetic permeability 176–184, 343, 345, 351
magnetization 165, 175–177, 180–181, 184, 186, 343, 355

marble 12, 235–237, 247, 356, 388–389
MCT, *see* mechanochemical treatment
MCT of aluminum 298–299, 302, 304, 307–309, 311, 313–318, 331, 403–405, 409–410, 412, 422
MCT of composites 310, 421–425
MCT of wollastonite 239, 241, 243, 359, 364, 368–369, 372, 375, 387
mechanical action 2–5, 8, 10, 23, 25–30, 33–34, 36–37, 41, 43–44, 50, 53–54, 56–57, 81, 89–90, 107, 110, 120, 125, 127, 158, 177, 194, 202, 211, 214, 216, 232, 234, 291, 299, 316, 329, 378–379
mechanical activation 12, 20, 22–23, 27–28, 34, 43–44, 64, 66–67, 69, 71, 73, 75, 77, 79–80, 86–87, 91–92, 95–96, 106, 108, 111, 128, 148, 158–159, 167, 203, 217, 427
mechanical alloying 8
mechanical deformations 30
mechanical destruction 2, 35, 54, 109, 171, 237
mechanical dispersion 55, 114–115
mechanical energy 2–3, 26, 74, 208
  absorption of 8, 10
  influence of 3, 8
  transformation of 27–28
mechanical energy transformations 27
mechanical grinding 45, 330
  conventional 253, 255
mechanical impact 1, 90, 102, 104, 106, 108, 110, 112–114, 122, 143, 229
mechanical loading 34, 217
mechanical reactors 120, 122–124, 126, 128, 130, 132, 134, 136, 138–140, 142, 144, 146, 148, 150–154, 156, 158, 160, 162–166, 168, 170, 172, 174–176, 178, 180, 182, 184, 186, 193, 204, 206, 209, 213–215, 222, 234, 309, 326, 331, 345–347, 356, 358
mechanical transformations 5
mechanochemical activation 9, 19, 27, 31, 44, 64, 66, 68, 70, 72, 74, 76, 78, 80, 82, 84, 86, 88, 90, 92–96, 175, 186, 289–291, 370, 377, 384
mechanochemical activation of aluminum powders 309, 311
mechanochemical methods 8–11, 343
mechanochemical modification 50, 202, 228, 337, 354–355
mechanochemical processes 5, 7, 22, 26, 30, 40–41, 48–49, 52–53, 55, 75, 79, 81, 83, 85, 87, 89, 92, 101, 204, 215
mechanochemical processing 45, 63, 96, 106, 130, 133, 142, 154–156, 173–175, 177, 179, 185–187, 205, 227, 235, 239, 247, 251, 295, 318, 331, 343, 345, 354
mechanochemical reactions 3, 33, 44
mechanochemical reactors (MR) 38, 64, 72, 74, 81, 91
mechanochemical reduction 96
mechanochemical transformations 5, 8, 52, 101
mechanochemical treatment (MCT) 11–12, 26–27, 45, 47–51, 54, 65, 67, 69, 119, 137–139,

144, 146, 155–156, 164–165, 174–176, 178, 180, 182–183, 185–187, 217–220, 222–223, 227–248, 251, 268, 271, 297–331, 335–340, 342–348, 350, 352–356, 358–362, 364–370, 373–378, 380–386, 388–392, 394, 396, 398, 400, 402–428
mechanochemical treatment of aluminum powders 298–299, 301, 303, 305, 307, 428
mechanochemistry 1–4, 6–12, 22, 52, 106, 193, 195, 197, 199, 201, 203, 205, 251–252, 254, 256, 258, 260, 262, 264, 266, 268, 270, 272, 274, 276, 278, 280, 282, 284, 286, 288, 290
  development of 2–3, 5, 7
mechanochemistry of macromolecular compounds 9
mechanochemistry of organic compounds 53, 55
mechanocracking 32, 52–54, 56, 139–140
metal powders, processing 296
metallic iron nanoparticles 165, 354
metals 1–2, 32, 54, 111, 114, 156, 186, 196, 201, 203, 254, 302, 335, 376, 379, 416, 420
  heavy 335–336
metastable atomic structures 295
metastable structures 48
microencapsulation 114–115
mineral raw materials 11–12, 90, 96, 112, 235, 247–248
  processing of 11, 90
minerals 12, 63, 90–91, 147, 165, 173, 235, 238, 246, 356, 396, 398
mixed solid rocket fuel (MSRF) 318, 416, 418, 423–426, 428
modification of metal systems 295–296, 298, 300, 302, 304, 306, 308, 310, 312, 314, 316, 318, 320, 322, 324, 326, 328, 330
modifiers 130–136, 138–139, 143–144, 146, 151–152, 154–156, 169, 171–172, 174–182, 184, 186–187, 199, 202, 227, 229–230, 232–238, 240–243, 246–248, 258, 260, 268–269, 272, 274, 307–308, 310, 312–313, 326–331, 337–344, 366–371, 374, 384, 386–388, 390–391, 396–397, 405, 411, 413, 417, 419, 425–428
  carbon-containing 142, 387
  destructible 156, 326
  nitrogen-containing 242, 244, 248
  quartz-organic 205
Mössbauer parameters 167–169
Mössbauer spectroscopy 155–156, 165–166, 171, 173, 186, 202
MR, *see* mechanochemical reactors
MSRF, *see* mixed solid rocket fuel
mullite 323, 343, 384–385, 394

NGRS, *see* nuclear gamma resonance spectroscopy
nitrogen-containing compounds 233, 235, 382, 384, 414
nuclear gamma resonance spectroscopy (NGRS) 86–87

organic compounds 7, 32, 53, 55–56, 104, 110–112, 115, 134, 137, 139–140, 144, 165, 186, 196, 204, 206–207, 209, 211, 213, 220, 222, 234, 237, 274, 295, 309, 336–337, 341, 354, 379–380, 392, 415
organic modifiers 136, 139, 171, 186, 201, 241, 299, 309–311, 331, 354, 408, 420

paramagnetic centers (PMCs) 35, 44, 109, 111
parawollastonite 396–397
PCM, *see* planetary-centrifugal mills
PES, *see* potential energy surface
physicochemical processes 7, 9, 11
physicochemical processes of substance transformation 21, 23, 25, 27, 29, 31, 33, 35, 37, 39, 41, 43
physicochemical transformations 7
piezoelectric effect 31, 103, 205–207, 211–213, 215, 217, 234
piezomaterials 203, 207, 213–214, 221, 234
planetary-centrifugal mills (PCM) 18–19, 21, 37, 221
plastic deformation 5, 19–20, 25, 29, 33, 41, 43–44, 48, 65, 214, 330
plasticity 298, 363, 368–371, 374–375
plasticization 11, 366, 368, 370–371, 374
PMCs, *see* paramagnetic centers
polydispersity 121–122, 132, 231, 241
polymer compositions 52, 112, 357–359, 361–362, 364, 366–367, 374–375
polymerization 10, 32, 43, 55–56, 111, 113, 135, 195, 198–200, 203, 205–206, 220, 233, 241, 248, 251, 255–256, 339, 347, 355, 358, 361, 379–380
polymorphic transformations 10, 127, 130
polystyrene 3, 53, 134–139, 143–146, 152–154, 164–165, 169–173, 178–181, 183, 186, 202, 338–340, 342–344, 346–347, 349–351, 355
polyvinyl alcohol 227, 229–231, 233, 235–239, 241–242, 245, 247, 298–302, 305–316, 318, 327–329, 383, 386–388, 390, 400, 402–403, 406–407, 409–411, 417, 420, 422, 428
  destruction of 316, 323
potential energy surface (PES) 285–288, 290
powder systems, ultrasonic treatment of 252, 254, 256, 258, 260, 262, 264, 266, 268, 270, 272, 274, 276, 278, 280, 282, 284, 286, 288, 290

quantum-chemical calculations 275, 277, 279, 281, 283, 285, 287, 289, 291
quantum chemistry 275–277, 291
quartz 125
  activated 89, 107, 110–111, 121, 123–125, 127, 129, 158, 160, 162–163, 168, 389

activated and modified 141, 143, 145, 147, 149, 151, 153, 155, 157, 159, 161, 163, 165, 167, 169, 171, 173
activation and modification of 120, 122, 124, 126, 128, 130, 132, 134, 136, 138, 140, 142, 144, 146, 148, 150, 152, 154, 156, 158, 160, 162, 164, 166, 168, 170, 172, 174, 176, 178, 180, 182, 184, 186
bulk density of 230–231, 262
combustion of 395
crystalline 91, 94, 104, 215, 222
dilatometric curves of 125–126, 138
dispersible 119, 140
EPR spectra of 157, 165
ferromagnetism of 163, 165, 185
fused 31, 91, 93–96, 203
grinding 86, 106, 108, 111, 131, 140, 150–151, 174, 186, 210, 233, 382
magnetization of 164, 176
measurement of magnetic permeability of 178, 180
modification 120, 122, 124, 126, 128, 130, 132, 134, 136, 138, 140, 142, 144, 146, 148, 150, 152, 154, 156, 158, 160, 162, 164, 166, 168, 170, 172, 174, 176, 178, 180, 182, 184, 186, 384
modified 141, 143, 145–147, 149, 151, 153, 155, 157, 159, 161, 163, 165, 167, 169, 171, 173, 182–183, 186–187, 338–339, 354–355, 414
natural 95, 166, 173
non-activated 381, 393
treated 146, 155, 173, 179
ultrasonic treatment of 392, 395–396, 398
unmodified 132, 139
quartz crystals 83, 206, 215
quartz impurities 244, 246
quartz lattice 87–89, 147
quartz particles
activated 87, 90, 175
amorphous 95
crushed 214
crystalline 106
deformable 222
deformed 196, 234
dispersible 165, 384
grinding 202
ground 81, 124
modified 140, 147, 156, 186, 337, 354
quartz powder 131, 133, 135, 137, 139, 230
activated 129
electrical resistivity of 138
ferromagnetism of 175, 177, 179, 181, 183, 185
quartz splitting 203
quartzite 227, 258
quartzite powders 259

reduction 159, 164, 171, 173, 186, 194, 306, 347, 370, 388, 395, 403–404, 427
roll mills 74
rotating drum mill, conventional 66, 70–71

Schrödinger equation 274–275, 290

self-organization 32, 43, 45, 218–219, 221
SH-synthesis 375, 377, 379, 381, 383, 385, 387, 389, 391–393, 395, 397, 399, 401–403, 405–407, 409, 411, 413
silanol groups 104, 110, 195, 197–198, 281–288, 290
silica 11, 35, 89, 101–102, 108–109, 111–113, 115, 146, 159, 182, 186, 196, 201, 246, 280, 282–286, 288–289, 291, 311–312, 322–323, 325, 328, 336, 338, 342, 384, 388–390, 402, 406, 413, 426
  amorphous 23, 104, 106, 111–112, 336
  chemical modification of 112, 115, 338
silica gel 112, 336
silica quartz 163
silicate materials 335–336
silicate structures 104
silicic acid 151, 195–196, 205, 220, 241–242, 246–247, 257–260, 339, 386, 388, 402
  aqueous 257, 259–260, 383
silicon carbide 323, 391, 405–406, 413
silicon dioxide 55, 81, 83, 85–86, 95, 101–106, 108–112, 114, 124, 133–135, 147, 157, 163, 168–169, 173, 178, 184, 198, 227, 239, 243, 258, 277, 280, 310–313, 318–319, 322–325, 329–330, 335–336, 340, 342–343, 381–401, 403–413, 417, 420, 423–428
silicon-oxygen tetrahedra 277–278, 291
  articulated 278, 280
siloxane bonds 105–106, 110, 149, 156, 174, 193–194, 197, 220, 280, 284, 291, 336
sodium chloride 134–135, 138, 142–143, 211, 339, 342, 355
solid fuels 114, 417–420, 422–423, 426–428
solid-phase combustion 318, 378
solid rocket fuels 415, 417, 419, 421, 423, 425, 427
sorbents 7, 114, 184, 335–337, 339–343, 346, 348, 352–354
  mechanochemical modification of 337, 355
stretching vibrations 107, 147, 149, 151, 153, 281, 317
succinic acid 227, 229–231, 233–235, 238, 241, 245, 247, 384, 386–390
supramolecular structure 6

technological combustion 376, 378, 392, 395, 398, 411, 414
thermal expansion 124–126, 130, 137, 212
thermo-kinetic characteristics 396, 409, 412
thermograms of combustion systems 403, 408–411
thermogravimetric analysis of composites 321–322, 324–325
thermokinetic characteristics 380, 384, 388–389, 392, 398–399, 402, 404, 410, 415

ultrasonic processing 251, 253, 255, 259, 268, 398

ultrasonic treatment (UST) 251–274, 276, 278, 280, 282, 284, 286, 288, 290, 297, 356–365, 367–374, 379–380, 392, 394–402, 414–415
ultrasonic vibrations 253, 255, 357, 380
ultrasound 8–9, 31, 254–255, 260, 262, 272, 357, 379, 392, 401
urea 229–231, 233–235, 237–239, 241–242, 245, 247, 260–261, 263–268, 271–273, 280, 282, 284, 289–291, 365–369, 371–373, 392, 395–397, 415
  aqueous solution of 392, 395–396
urea adsorption 284–285
urea-modified powders 241
urea molecules 281, 284–285, 291
urea particles, modification of 371
UST, *see* ultrasonic treatment

water 9, 11, 43, 104–105, 108, 112, 198, 237–238, 257–258, 260, 262–272, 326, 335–337, 339–343, 347–348, 352–354, 360, 362, 364, 367, 370–371, 383, 392–394, 398–399, 401–402
  polluted 340
water molecules 104–105, 107–108, 315, 383
water purification 335–337, 339–343, 345, 348–349, 351, 353–355
water resistance 339–340
water vapor 104, 107–108, 195, 395
wollastonite 12, 235, 239–241, 243, 246–248, 264–265, 268, 270–274, 277–278, 280, 282–289, 291, 356–375, 380, 385–392, 398–402, 415
  crystal structure of 278
  mechanochemical treatment of 247, 375
  modification of 246, 271, 358, 365, 368, 374
  modified 271, 277, 368, 371, 374–375, 388, 390–391
  processed 358
  structure of 278–280, 291
  surface of 281, 283–285, 287
  ultrasonic treatment of 268, 271, 273, 362, 365, 367, 370, 372–373, 375, 400
  ultrasound processing of 367
wollastonite crystal lattice 269–270
wollastonite crystals 239, 277
wollastonite particles
  dispersion of 240
  modification of 274
wollastonite powder particles, bulk density of 265–266

X-ray phase analysis 91–95, 150, 156, 162, 171, 173, 305, 341, 405